MATHEMATICS
of
QUANTUM
COMPUTATION

COMPUTATIONAL MATHEMATICS SERIES

Series Editor Mike J. Atallah

Published Titles

Inside the FFT Black Box: Serial and Parallel Fast Fourier Transform Algorithms
Eleanor Chu and Alan George

Mathematics of Quantum Computation
Ranee K. Brylinski and Goong Chen

Forthcoming Titles

Crytanalysis of Number Theoretic Ciphers
Samuel S. Wagstaff

Fuzzy Automata Theory
John N. Mordeson and Davender S. Malik

COMPUTATIONAL MATHEMATICS SERIES

Series Editor Mike J. Atallah

MATHEMATICS of QUANTUM COMPUTATION

Edited by
Goong Chen
Ranee K. Brylinski

CRC Press
Taylor & Francis Group
Boca Raton London New York

CRC Press is an imprint of the
Taylor & Francis Group, an **informa** business

A CHAPMAN & HALL BOOK

CRC Press
Taylor & Francis Group
6000 Broken Sound Parkway NW, Suite 300
Boca Raton, FL 33487-2742

First issued in paperback 2019

ISBN-13: 978-1-58488-282-4 (hbk)
ISBN-13: 978-0-367-39635-0 (pbk)
Library of Congress Card Number 2001056168

Library of Congress Cataloging-in-Publication Data

Mathematics of quantum computation / Ranee K. Brylinski, Goong Chen, editors.
    p. cm. — (Computational mathematics series)
   Includes bibliographical references and index.
   ISBN 1-58488-282-4 (alk. paper)
    1. Quantum computers. I. Brylinski, Ranee K. II. Chen, Goong, 1950- III. Series.

Q76.889 .M38 2002
004.1—dc21   2001056168

Visit the Taylor & Francis Web site at
http://www.taylorandfrancis.com

and the CRC Press Web site at
http://www.crcpress.com

Preface

One of the most exciting developments in the scientific community these days is the design and construction of the quantum computer (QC). A QC stores and handles data as a collection of two-state *quantum bits*, or *qubits* (e.g., spin 1/2 particles). Quantum computation performs calculations on data densely coded in the entangled states (of qubits) that are the hallmark of quantum mechanics. During the 1980s, D. Deutsch, following the lead of P. Benioff and R. Feynman, made concrete proposals for harnessing some of the peculiar properties of quantum mechanics to obtain unprecedented parallelism in computation. Interest in the field of QC has received a tremendous boost from the following two recent results obtained by P. Shor and L. K. Grover, both researchers at Bell Labs:

1. P. Shor's 1994 discovery of a quantum algorithm for factorization of integers. The Shor algorithm is substantially faster than any known classical algorithm of subexponential complexity and opens the door to decipherment in cryptography.

2. L. K. Grover's discovery in 1996 of a quantum search algorithm that gives a favorable optimal quadratic speed-up in the search of a single object in a large unsorted database, which is suitable for *data mining*.

These stunning improvements in algorithmic complexity, *mathematical in nature*, have been steadily attracting researchers not only from computer science but also from physics, mathematics and chemistry.

The fast miniaturization of electronic circuits and chips, which has proceeded for several decades according to Moore's law, will hit a brick wall in 20 to 25 years. Inevitably, information technology will be entering the realm of quantum mechanics. Quantum information technology

and computing belong to the class of the most innovative and revolution-
ary computing, whose development is now a national project that will
be vital to our future economy and national defense. The rapidly evolving AMO (atomic, molecular and optical) and nanotechnology are now
making the fabrication of QC hardware a distinctly attainable reality.

Mathematics lies at the heart of theoretical quantum computer science. The design of fast and efficient quantum algorithms and the analysis of their complexities belong to the work of a mathematician. Even
the design of quantum circuits and networks will be greatly helped with
mathematicians' insights, whether abstract or quantitative. Mathematics is needed in nearly every aspect of the research and development of
quantum computation and information technology. It actually constitutes the very foundation of this interdisciplinary field.

As rapidly evolving as the contemporary AMO and nanotechnology
are in the race to build the first working QC, there is the consequently
inevitable transient nature of the state-of-the-art quantum computing
devices—they come and go. However, the mathematical foundation of
quantum computation and information, which is device-independent, is
here to stay. It is precisely this *universality* character played by mathematics in the development of quantum computation and information
science that has motivated us to compile and publish this book. We
have invited computer scientists, mathematicians and physicists to write
chapters on the following topics, according to the sequential order of
headings in the Table of Contents:

Quantum Entanglement (Chapters 1, 2 and 3)
Universality of Quantum Gates (Chapter 4)
Quantum Search Algorithms (Chapters 5, 6 and 7)
Quantum Computational Complexity (Chapter 8)
Quantum Error-Correcting Codes (Chapters 9 and 10)
Quantum Computing Algebraic and Geometric Structures (Chapters 11
and 12)
Quantum Teleportation (Chapter 13)
Quantum Secure Communication and Quantum Cryptography (Chapter
14)
Commentary on Quantum Computing (Chapter 15)

Although we have made our best efforts to encompass the topics as
comprehensively as possible, we must concede that a few important ones,
such as Shor's algorithm and quantum information theory, are still not
included. Nevertheless, our list already constitutes a large portion of the
major research interests, as far as mathematics is concerned, since the

active development of QC in the mid-1990s. We hope this book will provide a useful reference source, of appropriate mathematical depth, not only for pure and applied mathematicians doing research on quantum computing, but also for physicists and computer scientists who need to use algebra and mathematical analysis in their work of developing quantum computers and information technology.

Credit for this book publication is to be shared by all the authors who spent months writing and then helping review the chapters. As the book editors, we are deeply indebted to them. Robin Campbell has provided us with her wonderful computer and technical typing service. The work of G. Chen is partially supported by DARPA QuIST grant F49620-01-1-0566. Finally, we express our sincerest thanks to Bob Stern, Senior Editor, and Helena Redshaw, Editorial Supervisor, at CRC Press, for their great enthusiasm and assistance in expediting the publication of the book.

Ranee K. Brylinski
Goong Chen

Contributors

Almut Beige
Max-Planck-Institut für
Quantenoptik
Hans-Kopfermann-Str. 1
85748 Garching
Germany

Jean-Luc Brylinski
Department of Mathematics
Pennsylvania State University
University Park, PA 16802

Ranee Brylinski
Department of Mathematics
Pennsylvania State University
University Park, PA 16802

Goong Chen
Institute for Quantum Studies
and
Department of Mathematics
Texas A&M University
College Station, TX 77843-3368

Jeesen Chen
Department of Mathematical
Sciences
University of Cincinnati
Cincinnati, OH 45221-0025

Berthold-Georg Englert
Atominstitut
Technische Universität Wien
Stadionallee 2
1020 Wien
Austria
and

Departments of Mathematics
and Physics
Texas A&M University
College Station, TX 77743-3368

Stephen A. Fenner
Computer Science and Engineering
Department
University of South Carolina
Columbia, SC 29208

Michael H. Freedman
Microsoft Research
One Microsoft Way
Redmond, WA 98052

Stephen A. Fulling
Institute for Quantum Studies
and
Department of Mathematics
Texas A&M University
College Station, TX 77843-3368

Markus Grassl
Institut für Algorithmen und
Kognitive Systeme
Universität Karlsruhe
Am Fasanengarten 5
D-76128 Karlsruhe
Germany

Lov K. Grover
Bell Laboratories
Lucent Technologies
600-700 Mountain Avenue
Murray Hill, NJ 07974

Kishore T. Kapale
Institute for Quantum Studies
and
Department of Physics
Texas A&M University
College Station, TX 77843-4242

Andreas Klappenecker
Department of Computer Science
Texas A&M University
College Station, TX 77843-3112

Christian Kurtsiefer
Sektion Physik
Universität München
Schellingstr. 4
80799 München
Germany

Feng Luo
Department of Mathematics
Rutgers University
Piscataway, NJ 08903

Nasser Metwally
Sektion Physik
Universität München
Theresienstr. 37
80333 München
Germany

David A. Meyer
Department of Mathematics
University of California-San Diego
9500 Gilman Drive
La Jolla, CA 92093-0112

Martin Rötteler
Institut für Algorithmen
und Kognitive Systeme
Universität Karlsruhe
Am Fasanengarten 5
D-76128 Karlsruhe
Germany

Anirvan M. Sengupta
Bell Laboratories
Lucent Technologies
600-700 Mountain Avenue
Murray Hill, NJ 07974

Shunhua Sun
Department of Mathematics
Sichuan University
Chengdu, Sichuan
China

Nolan Wallach
Department of Mathematics
University of California-San Diego
9500 Gilman Drive
La Jolla, CA 92093-0112

Harald Weinfurter
Max-Planck-Institut für
Quantenoptik
Hans-Kopfermann-Str. 1
85748 Garching
Germany
and
Sektion Physik
Universität München
Schellingstr. 4
80799 München
Germany

M. Suhail Zubairy
Institute for Quantum Studies
and
Department of Physics
Texas A&M University
College Station, TX 77843-4242
and
Department of Electronics
Quaid-i-Azam University
Islamabad
Pakistan

Contents

Quantum Entanglement

Chapter 1

Algebraic measures of entanglement

Jean-Luc Brylinski[*]

Abstract   We study the rank of a general tensor u in a tensor product space $H_1 \otimes \cdots \otimes H_k$. The rank of u is the minimal number p of decomposable states $v_1, \cdots, v_p$ such that u is a linear combination of the v_j's. This rank is an algebraic measure of the degree of entanglement of u. Motivated by quantum computation, we completely describe the rank of an arbitrary tensor in $(\mathbb{C}^2)^{\otimes 3}$ and give normal forms for tensor states up to local unitary transformations. We also obtain partial results for $(\mathbb{C}^2)^{\otimes 4}$; in particular, we show that the maximal rank of a tensor in $(\mathbb{C}^2)^{\otimes 4}$ is equal to 4.

1.1   Introduction

This paper is devoted to the study of the degree of complexity of n-qubit states. An n-qubit state (or simply n-qubit) is a (norm 1) state in the tensor product Hilbert space $(\mathbb{C}^2)^{\otimes n}$ of dimension 2^n. This Hilbert

[*]Research supported in part by NSF Grant No. DMS-9803593.

space has an orthonormal basis consisting of standard n-qubits $|i_1 \cdots i_n\rangle$ where $i_l = 0$ or 1. We think of n-qubit states as tensor product states. We address the question of classifying n-qubit states up to the group $U(2)^n$ of local symmetries. One main ingredient we use is the notion of rank of a tensor (see Definition 1.1).

In Section 1.2 we discuss decomposable vs. entangled tensors in a more general tensor product $H_1 \otimes \cdots \otimes H_k$. The rank of a tensor u is the minimal number p of decomposable states $v_1, \cdots, v_p$ such that u is a linear combination of the v_j's. In Section 1.3 we give the well-known classification of 2-qubit states up to local symmetries. In Section 1.4 we discuss 3-qubit states: here the rank can be 1, 2 or 3. The rank of a general 3-qubit is 2, but there is a complex hypersurface Z where the rank is generically equal to 3. The equation of this hypersurface has degree 4 and is the classical Cayley hyperdeterminant in 8 variables $u_{i_1 i_2 i_3}$. We obtain normal forms for 3-qubit states up to local symmetries. Finally, in Section 1.5 we discuss 4-qubit states. We prove that the maximal rank of a 4-qubit is equal to 4 and we give polynomial equations of the closure of the set of 4-qubit states of rank ≤ 3.

Although the complete classification of n-qubit states up to local symmetries is clearly a highly complex problem for n large, we hope that methods of algebraic geometry will prove fruitful in attacking it.

I thank Ranee Brylinski for her collaboration on related topics and for many useful discussions. I thank Joseph Bernstein for useful discussions. I am grateful to the CPT and IML of the Université de la Méditerranée for their hospitality during part of the time I worked on this paper.

1.2   Rank of a tensor

Let H be a complex Hilbert space; the hermitian scalar product will be denoted by $\langle u|v \rangle$. It is complex linear in v and antilinear in u. A *state* in H is a vector $u \in H$ of norm 1; as usual in quantum mechanics, we consider the states u and $e^{i\alpha}u$ to be equivalent. The reason for that is that the state u can only be observed through the corresponding projection operator $P_u : H \to H$ where $P_u(v) = \langle u|v \rangle$. Clearly u and $e^{i\alpha}u$ yield the same projection operator. We will use the mathematical notation for states (u, v, etc.) as opposed to Dirac kets.

So mathematically speaking, a state in H is a point in the quotient of the unit sphere $S(H)$ by the scaling action of complex numbers $e^{i\alpha}$.

Alternatively, we can view a state as point of the projective space $\mathbb{P}(H)$. By definition, $\mathbb{P}(H)$ is the set as complex lines in H through 0. For H finite-dimensional, $\mathbb{P}(H)$ is a complex algebraic variety. Pick a basis $(e_0, \cdots, e_{d-1})$ of H. Then $\mathbb{P}(H)$ is covered by open sets U_j for $j = 0, 1, \cdots, d - 1$, where $\mathbb{P}(H)$ is the set of complex lines on which the j-th coordinate does not vanish. One identifies U_j with $\mathbb{C}^{d-1}$: to a line $l \subset H$, spanned by a vector $(z_0, \cdots, z_{d-1})$ with $z_j \neq 0$, we attach the point $\left(\frac{z_0}{z_j}, \cdots, \frac{z_{j-1}}{z_j}, \frac{z_{j+1}}{z_j}, \cdots, \frac{z_{d-1}}{z_j}\right)$ of $\mathbb{C}^{d-1}$.

We also observe that P_u is an idempotent hermitian operator of rank 1; in this way we realize $\mathbb{P}(H)$ as the orbit of the unitary group comprised of such operators.

In quantum mechanics, the combination of several quantum systems corresponds to the Hilbert space tensor product $E = H_1 \otimes \cdots \otimes H_k$ of the relevant Hilbert spaces. A state u in E is called *decomposable* if it is a tensor product $\phi_1 \otimes \phi_2 \cdots \phi_k$ of states; otherwise it is called *entangled*. It is easy to characterize decomposable states in terms of homogeneous quadratic equations for the components of the tensor u. If we pick orthonormal bases of each H_j and write $u_{a_1 \cdots a_k}$ for the components of u, we have

PROPOSITION 1.1

The state u in $E = H_1 \otimes \cdots \otimes H_k$ is decomposable iff the following "exchange property" is verified: for any k-tuples $(a_1, \cdots, a_k)$, $(b_1, \cdots, b_k)$, $(c_1, \cdots, c_k)$, $(d_1, \cdots, d_k)$ for which each pair (c_j, d_j) is a permutation of (a_j, b_j), we have

$$u_{a_1, \cdots, a_k} u_{b_1, \cdots, b_k} = u_{c_1, \cdots, c_k} u_{d_1, \cdots, d_k} \tag{1.1}$$

PROOF   Clearly a decomposable tensor satisfies the exchange property. To prove the converse, we proceed by induction over k. Let $(e_0, \cdots, e_{d-1})$ be our orthonormal basis of H_1. Write $u = \sum_i e_i \otimes v_i$ where $v_i \in H_2 \otimes \cdots \otimes H_k$. If $v_i \neq 0$ for some i, the exchange property for the case $a_1 = b_1 = i$ implies that v_i satisfies the exchange property, and so is a decomposable tensor by the inductive hypothesis. Next, if v_i and v_j are non zero, we apply the exchange property to the case where $a_1 = d_1 = i$, $b_1 = c_1 = j$, $a_l = c_l$ and $b_l = d_l$ for $l \geq 2$, and conclude that the tensors v_i and v_j are proportional. It follows that u is a decomposable tensor. ∎

Geometrically, the set of decomposable states is a closed complex algebraic subvariety S of $\mathbb{P}(H)$, isomorphic to the product $\mathbb{P}(H_1) \times \cdots \times \mathbb{P}(H_k)$, which is known to algebraic geometers as the *Segre product*. So its dimension is $d_1 + \cdots + d_k - k$, where $d_j = \dim(H_j)$. Accordingly, the decomposable tensors in $E = H_1 \otimes \cdots \otimes H_k$ form a closed complex algebraic subvariety of dimension $d_1 + \cdots + d_k - k + 1$. Proposition 1.1 gives homogeneous polynomial equations for S.

Entangled states occur naturally both in quantum mechanics and in classical computer science. They are very important in quantum mechanics; cf., e.g., the famous Einstein-Podolsky-Rosen work. Quantum computation lives in the tensor product Hilbert spaces $(\mathbb{C}^2)^{\otimes k}$, and states used in quantum coding and quantum teleportation are typically quite entangled (see, e.g., [6, 13, 14]). Quantum algorithms utilize entanglement in a fundamental way. On the other hand, classical algorithms for matrix multiplication [15, 16] involve studying the entanglement of the tensors which describe matrix multiplication.

So it seems important to study how entangled states can be. The following is a classical notion (see [5]).

DEFINITION 1.1   *We say a state u in $E = H_1 \otimes \cdots \otimes H_k$ has rank $\leq p$ if we can write*

$$u = \sum_{j=1}^{p} \lambda_j v_j \tag{1.2}$$

where each v_j is a decomposable state.

A natural question is to find the highest rank of all states in E; rather little is known about this. We will answer it in some special cases. At least we can give a lower bound:

PROPOSITION 1.2

Let H_j be vector spaces of dimension d_j. Then the highest rank of states in $E = H_1 \otimes \cdots \otimes H_k$ is at least equal to the rational number

$$\frac{d_1 d_2 \cdots d_k}{d_1 + d_2 + \cdots + d_k - k + 1} \tag{1.3}$$

For instance, take $k = 3$, $d_1 = 3$, $d_2 = 4$, $d_3 = 5$; then the highest rank is at least $3 \times 4 \times 5/10 = 6$.

In case $k = 2$, it is easy to describe the rank of any state in classical terms:

PROPOSITION 1.3

The rank of a state u in $E = H_1 \otimes H_2$ is the rank of the matrix u_{ab}.

In particular, for $k = 2$, the rank gives a nice stratification of projective space $\mathbb{P}(E)$. Let S_p denote the set of states of rank $\leq p$. Then S_p is a closed algebraic subvariety of $\mathbb{P}(E)$, defined as the vanishing locus of all order $p + 1$ minors of the matrix u_{ab}. The singular locus of S_p is exactly equal to S_{p-1}. The set $S_p \setminus S_{p-1}$ of states of rank equal to p is then a locally closed complex algebraic subvariety.

There is also a nice analytic characterization of decomposable states ϕ in $H_1 \otimes H_2$, in terms of the projection operator P_ϕ. The partial trace $\rho := Tr_1(P_\phi)$ is a positive hermitian operator on H_2 and we have:

PROPOSITION 1.4

We have $\rho^2 \leq \rho$ with equality iff ϕ is decomposable.

For the proof see Popescu and Rohrlich [13]. There is an interesting relation with the algebraic characterization of decomposable states in Proposition 1.1, which we illustrate for $\mathbb{C}^2 \otimes \mathbb{C}^2$, using the basis (e_0, e_1) of $\mathbb{C}^2$. Here ϕ is given by a matrix (ϕ_{ab}). The $(0, 0)$-component of $\rho - \rho^2$ is equal to $|\phi_{00}\phi_{11} - \phi_{01}\phi_{10}|^2$. Thus the analytic equations characterizing decomposable states are quartic real polynomials which are squares (in general, sums of squares) of absolute values of the quadratic complex polynomial equations.

For $k > 2$ the situation is more complicated. We will still denote by S_p the set of states of rank $\leq p$. Of course we have $S_p \subset S_{p+1}$, but we will see in the next section that the set S_p is not always closed in $\mathbb{P}(E)$.

Note that for $k = 3$ the rank of a tensor is closely connected to the notion of rank of a bilinear map [5, 15, 16]. Indeed, a tensor u in $H_1 \otimes H_2 \otimes H_3$ yields a bilinear form $f : H_1^* \otimes H_2^* \to H_3$, and the rank of u is equal to the rank of f.

The natural symmetry group acting on states in $H_1 \otimes H_2 \otimes \cdots \otimes H_k$ is the product $U(H_1) \times U(H_2) \times \cdots \times U(H_k)$ of the unitary groups. This is the group of local unitary symmetries; we say that two tensor states are *locally equivalent* if they are equivalent under this group.

In this paper we will study n-qubit states for a small n, i.e., states in $(\mathbb{C}^2)^{\otimes n}$.

1.3   Tensors in $(\mathbb{C}^2)^{\otimes 2}$

For a tensor u in $\mathbb{C}^2 \otimes \mathbb{C}^2$, Proposition 1.4 says that u is decomposable if and only if the determinant $u_{00}u_{11} - u_{01}u_{10}$ vanishes.

We have the well-known Schmidt decomposition: u can be written

$$u = \cos\theta f_0 \otimes g_0 + \sin\theta f_1 \otimes g_1 \tag{1.4}$$

where (f_0, f_1) and (g_0, g_1) are orthonormal bases of $\mathbb{C}^2$.

We then have

PROPOSITION 1.5
Any state in $\mathbb{C}^2 \otimes \mathbb{C}^2$ is locally equivalent to a unique state of the form

$$u = \cos\theta e_0 \otimes e_0 + \sin\theta e_1 \otimes e_1 \tag{1.5}$$

where $0 \leq \theta \leq \frac{\pi}{4}$

1.4   Tensors in $(\mathbb{C}^2)^{\otimes 3}$

We study here $E = \mathbb{C}^2 \otimes \mathbb{C}^2 \otimes \mathbb{C}^2$. The structure of the algebra of $SU(2)^3$-invariant polynomial functions on E (viewed as a real vector space) is completely described in [12].

There is in particular a well-known polynomial function D on E which is invariant under $SL(2, \mathbb{C})^3$; this is the hyperdeterminant introduced by Cayley [8, 3]. In the notations of [12], this corresponds to the invariant polynomial f.

D is a homogeneous degree 4 polynomial function on E which is $SL(2, \mathbb{C})^3$-invariant. We will derive the expression for D using an important geometric notion which is prevalent throughout the paper. Given a tensor $u \in (\mathbb{C}^2)^{\otimes 3}$, write $u = e_0 \otimes v_0 + e_1 \otimes v_1$ for $v_j \in \mathbb{C}^2 \otimes \mathbb{C}^2$. Then u gives rise to the *pencil* $xv_0 + yv_1$ of tensors in $\mathbb{C}^2 \otimes \mathbb{C}^2$. Since rescaling the vector (x, y) only has the effect of dilating the tensor $xv_0 + yv_1$, we can think of $xv_0 + yv_1$ as being parameterized by a point in the projective line $\mathbb{CP}^1$, with homogeneous coordinates (x, y).

We expect that for general (x, y) the tensor $xv_0 + yv_1$ has rank 2, and there are 2 points (x, y) of $\mathbb{CP}^1$ for which it has rank 1. However for some choices of (v_0, v_1) it will happen that there is only one point (x, y) of $\mathbb{CP}^1$ for which $xv_0 + yv_1$ has rank 1. To see when this happens, we consider the determinant $\det(xv_0 + yv_1)$, which is a homogeneous degree 2 polynomial in (x, y). We then should look for the discriminant of this quadratic form to vanish. We have the identity for 2 by 2 matrices:

$$\det(xv_0 + yv_1) = x^2 \det(v_0) + xy[Tr(v_0)Tr(v_1) - Tr(v_0v_1)] + y^2 \det(v_1)$$
(1.6)

The hyperdeterminant is then the discriminant of this quadratic form in x and y:

$$D(u) = [Tr(v_0)Tr(v_1) - Tr(v_0v_1)]^2 - 4\det(v_0)\det(v_1)$$
(1.7)

The explicit formula is written neatly using the cube with vertices (i, j, k), $i, j, k = 0, 1$.

$$D(u) = \sum \text{squares of main diagonals}$$

$$-2\sum \text{pairs of opposite edges}$$
(1.8)

$$+4\sum \text{inscribed regular tetrahedra}$$

Figure 1.1 is the cube showing some typical terms of $D(u)$:

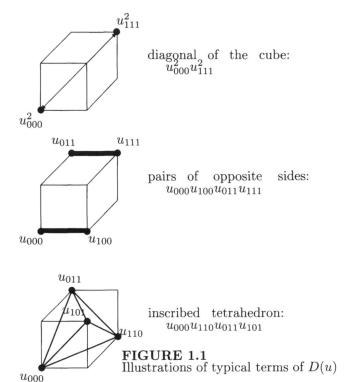

diagonal of the cube:
$u_{000}^2 u_{111}^2$

pairs of opposite sides:
$u_{000} u_{100} u_{011} u_{111}$

inscribed tetrahedron:
$u_{000} u_{110} u_{011} u_{101}$

FIGURE 1.1
Illustrations of typical terms of $D(u)$

This gives:

$$
\begin{aligned}
D(u) = &\ u_{000}^2 u_{111}^2 + u_{001}^2 u_{110}^2 + u_{010}^2 u_{101}^2 + u_{011}^2 u_{100}^2 \\
&-2(u_{000}u_{001}u_{110}u_{111} + u_{000}u_{010}u_{101}u_{111} + u_{000}u_{011}u_{100}u_{111} \\
&+u_{001}u_{010}u_{101}u_{110} + u_{001}u_{011}u_{110}u_{100} + u_{010}u_{011}u_{101}u_{100}) \\
&+4(u_{000}u_{011}u_{101}u_{110} + u_{001}u_{010}u_{100}u_{111})
\end{aligned}
$$

$$(1.9)$$

There is another geometric interpretation of D: for a tensor $u \in (\mathbb{C}^2)^{\otimes 3}$, we have $D(u) = 0$ iff the hyperplane defined by u is tangent to the Segre product S at some point p. This means that $\langle u|v \rangle = 0$ for any tangent vector v to S at p. In other words D is the homogeneous equation of the dual variety Z of S, which is a quartic hypersurface. This description has the advantage of making it clear that D is invariant under permutations of the 3 qubits.

The general theory of hyperdeterminants is developed in [3]. They are defined as the homogeneous equation of the dual variety of S. The review [7] provides interesting comments on the book [3].

For a tensor u in $E = (\mathbb{C}^2)^{\otimes 3}$, there are three additional degrees of entanglement $\delta_1, \delta_2, \delta_3$ to consider: δ_1 is the rank of u viewed as an element of $\mathbb{C}^2 \otimes \mathbb{C}^4$, when we group the second and third factors $\mathbb{C}^2$. δ_2 and δ_3 are defined similarly.

We denote by Y_j the closed algebraic subvariety of $\mathbb{P}(E)$ comprised of states u such that $\delta_j = 1$, i.e., u belongs to Y_1 iff it is decomposable as a tensor in $\mathbb{C}^2 \otimes \mathbb{C}^4$. Note each Y_j has dimension 4 and is contained in the hypersurface Z of equation $D = 0$.

The following result is proved (at least implicitly) in [3]. We include a proof since it uses methods which we will later use for $(\mathbb{C}^2)^{\otimes 4}$.

PROPOSITION 1.6

Let u be a state in $E = (\mathbb{C}^2)^{\otimes 3}$. Then u satisfies exactly one of the following possibilities:

(1) u is a decomposable state.

(2) u is not decomposable but belongs to Y_j for a (unique) $j = 1, 2, 3$.

(3) u is entangled, and $D(u) \neq 0$; in that case u has rank 2, so it is the sum of two decomposable tensors.

(4) $D(u) = 0$, but u belongs to none of the Y_j; then u has rank 3.

PROOF   We will use again the pencil method. So we write $u = e_0 \otimes v_0 + e_1 \otimes v_1$ and associate to u the linear map $T : \mathbb{C}^2 \to \mathbb{C}^2 \otimes \mathbb{C}^2$ such that $T(x, y) = xv_0 + yv_1$. If T has rank 1 then $u \in Y_1$ and we are in case (1). So we may assume T has rank 2. We will consider $T(xe_0 + ye_1)$ as a 2 by 2 matrix. Consider as before the homogeneous degree 2 polynomial $P(x, y) = \det(T(xe_0 + ye_1)) = \det(xv_0 + yv_1)$. There are three cases to consider:

(I) There are exactly two points in $\mathbb{CP}^1$ where P vanishes. Let (x_1, y_1), (x_2, y_2) be homogeneous coordinates for these two points. Then we can make a change of basis in the first $\mathbb{C}^2$ so that these 2 points are $(0, 1)$ and $(1, 0)$. Then both $T(e_0)$ and $T(e_1)$ have rank ≤ 1; they must both be non zero, since otherwise all $T(xe_0 + ye_1)$ would have rank ≤ 1. After a change of basis in the second and third copies of $\mathbb{C}^2$, we may assume $T(e_0) = e_0 \otimes e_0$ and $T(e_1) = e_i \otimes e_j$ for suitable i, j, not both equal to 0. In this case, the tensor u is equal to $e_0 \otimes e_0 \otimes e_0 + e_1 \otimes e_i \otimes e_j$, so it has rank equal to 2. By direct computation, we see that $D(u) \neq 0$ if $i = j = 1$ (case 4), or u belongs to Y_2 (resp. Y_3) if $i = 0$ (resp. $j = 0$), which belongs to case (2).

(II) There is only one point (x, y) of $\mathbb{CP}^1$ at which $P(x, y)$ vanishes. We may assume this point is $(1, 0)$. We can think of T as giving a parameterization of a curve in $\mathbb{CP}^3$ which is tangent to the quadric surface Q consisting of rank 1 matrices. We can change bases in all three copies of $\mathbb{C}^2$ so that $T(e_0) = e_0 \otimes e_0$. As the tangent plane to Q at $e_0 \otimes e_0$ is spanned by $e_0 \otimes e_1$ and $e_1 \otimes e_0$, we can change the basis vector e_1 in the first $\mathbb{C}^2$ so that $T(xe_0 + ye_1) = xe_0 \otimes e_0 + y(\lambda e_0 \otimes e_1 + \mu e_1 \otimes e_0)$. Next, as λ and μ must both be non zero, we can change bases in the other copies of $\mathbb{C}^2$ to arrange that $\lambda = \mu = 1$. Then our tensor u is $u = e_0 \otimes e_0 \otimes e_0 + e_1 \otimes e_1 \otimes e_0 + e_1 \otimes e_0 \otimes e_1$, and has rank exactly 3. Indeed it has the property that for any non zero $v \in \mathbb{C}^2$, the tensor in $\mathbb{C}^2 \otimes \mathbb{C}^2$ obtained by contracting u with w has rank equal to 2; thus u cannot be a sum of two decomposable tensors. We verify easily that $D(u) = 0$. Or we can see geometrically that the corresponding point in $\mathbb{P}(E)$ belongs to the dual variety to the Segre product $S = S_1$, which means that the hyperplane defined by u is tangent to the Segre variety at some point. The relevant point of S is $v = e_0 \otimes e_1 \otimes e_1$: notice that the tangent space to S at v is spanned by tensors of the type $\psi \otimes e_1 \otimes e_1, e_0 \otimes \psi \otimes e_1, e_0 \otimes e_1 \otimes \psi$ where $\psi \in \mathbb{C}^2$. Since u is orthogonal to all these tangent vectors, it is orthogonal to the tangent space of S at v.

(III) The polynomial $P(x, y)$ vanishes identically; this means that the tensor $T(xe_0 + ye_1)$ always has rank ≤ 1. This can happen in either of two ways:

a. There is a vector $\psi \in \mathbb{C}^2$ and a linear map $f : \mathbb{C}^2 \to \mathbb{C}^2$ such that $T(w) = \psi \otimes f(w)$.

b. There is a vector $\psi \in \mathbb{C}^2$ and a linear map $f : \mathbb{C}^2 \to \mathbb{C}^2$ such that $T(w) = f(w) \otimes \psi$.

We need only consider case a. Then we have $u = e_0 \otimes \psi \otimes f(e_0) + e_1 \otimes \psi \otimes f(e_1)$. If $f(e_0)$ and $f(e_1)$ are linearly dependent, the tensor u is decomposable and we are in case 1 of the proposition. Otherwise, u has rank 2 and after changes of bases in the second and third copies of $\mathbb{C}^2$ it takes the form $u = e_0 \otimes e_0 \otimes e_0 + e_1 \otimes e_0 \otimes e_1$. Then u belongs to Y_2. ∎

REMARK 1.1   The Cayley hyperdeterminant D also occurs in the paper [2] in connection with the residual tangle of 3-qubit states. ∎

This also leads to normal forms for tensor states in $(\mathbb{C}^2)^{\otimes 3}$ up to the action of $GL(2,\mathbb{C})^3$. Here $GL(2,\mathbb{C})$ is the group of invertible 2 by 2 matrices. These normal forms are given in [3].

PROPOSITION 1.7
There are exactly 6 orbits of $GL(2,\mathbb{C})^3$ acting on 3-qubit states:

1. A big open orbit U where the rank is 2; this is the orbit of $e_0 \otimes e_0 \otimes e_0 + e_1 \otimes e_1 \otimes e_1$.

[The complement of U is the complex hypersurface Z of equation $D(u) = 0$.]

2. A dense open subset Z^0 of Z on which we have 3-qubit states of rank 3; this is the orbit of $e_0 \otimes e_0 \otimes e_0 + e_1 \otimes [e_0 \otimes e_1 + e_1 \otimes e_0]$.
3. The orbit of $e_0 \otimes [e_0 \otimes e_1 + e_1 \otimes e_0]$: this consists of 3 qubit states of rank 2 which have rank 1 when viewed as states in $\mathbb{C}^2 \otimes \mathbb{C}^4$.
4. and 5. The orbits obtained by permuting the 3 qubits in $e_0 \otimes [e_0 \otimes e_1 + e_1 \otimes e_0]$.
6. The orbit of decomposable states.

For quantum mechanics one needs to consider the group $U(2)^3$ of local symmetries. One obtains normal expressions up to local equivalence:

PROPOSITION 1.8
A state in $(\mathbb{C}^2)^{\otimes 3}$ is locally equivalent to one of the following:
1. A decomposable state is locally equivalent to $e_0 \otimes e_0 \otimes e_0$.
2. A state in Y_1 that is not decomposable is locally equivalent to

$$e_0 \otimes [\cos\theta \, (e_0 \otimes e_0) + \sin\theta \, (e_1 \otimes e_1)] \tag{1.10}$$

States in Y_2 or Y_3 are described similarly.
3. A state of rank 2 that is not in either of the Y_j is locally equivalent to

$$\lambda \, e_0 \otimes e_0 \otimes e_0 + z(\cos\theta_1 e_0 + \sin\theta_1 e_1) \otimes (\cos\theta_2 e_0 + \sin\theta_2 e_1)$$
$$\otimes (\cos\theta_3 e_0 + \sin\theta_3 e_1) \tag{1.11}$$

where $\lambda, \theta_j \in \mathbb{R}$, $z \in \mathbb{C}$ satisfy the relation

$$\lambda^2 + |z|^2 + 2\lambda\Re(z)\cos\theta_1\cos\theta_2\cos\theta_3 = 1$$

(so that the tensor has norm 1). We can assume $\theta_j \in (0, \frac{\pi}{2})$.

4. A state of rank 3 that is locally equivalent to

$$\cos\theta_1 e_0 \otimes e_0 \otimes e_0 + \sin\theta_1 e_1 \otimes [\cos\theta_2 e_0 \otimes (\cos\theta_3 e_0 + \sin\theta_3 e_1) + \sin\theta_2 e_1 \otimes e_0]$$
$$(1.12)$$

In each case there are only finitely many values of the parameters corresponding to a given tensor state.

PROOF   The four cases of the statement correspond to the four cases of Proposition 1.6. Case 1 is obvious. Case 2 follows as $u \in Y_1$ is locally equivalent to $e_0 \otimes v$ for some $v \in \mathbb{C}^2 \otimes \mathbb{C}^2$; by Schmidt's theorem v is locally equivalent to $\cos\alpha\,(e_0 \otimes e_0) + \sin\alpha\,(e_1 \otimes e_1)$.

In case 3, we have $u = v_1 \otimes v_2 \otimes v_3 + w_1 \otimes w_2 \otimes w_3$ where v_i and w_i are linearly independent for each i. There is no harm in assuming that the vectors v_1, v_2, w_1, w_2 have norm 1. By rescaling u by a phase and applying a local transformation we can assume $v_1 = v_2 = e_0$ and $v_3 = \lambda e_0$ for $\lambda > 0$. We can also arrange that $w_i = \cos\theta_i e_0 + \sin\theta_i e_1$ for $i = 1, 2$ and $w_3 = z[\cos\theta_3 e_0 + \sin\theta_3 e_1]$ for some $z \in \mathbb{C}$. This gives the normal form; note the reduction to $\theta_j \in (0, \frac{\pi}{2})$ is easy to achieve by changing the signs of e_0 and e_1 in the j-th factor $\mathbb{C}^2$.

In case 4, the tensor u has the form $u = v_1 \otimes v_2 \otimes v_3 + w_1 \otimes (v_2 \otimes w_3 + w_2 \otimes v_3)$ where v_i and w_i are linearly independent for each i. There are two types of degrees of freedom in the expression of u in this form. First we have the transformation $u = (v_1 + \alpha w_1) \otimes v_2 \otimes v_3 + w_1 \otimes (v_2 \otimes [w_3 - \alpha v_3] + w_2 \otimes v_3)$. With its help we can arrange that $w_1 \perp v_1$. Secondly we have $u = v_1 \otimes v_2 \otimes v_3 + w_1 \otimes (v_2 \otimes [w_3 + \beta v_3] + [w_2 - \beta v_2] \otimes v_3)$. This is used to arrange that $w_2 \perp v_2$. By rescaling the v_i, we may assume that v_1 and v_2 have norm 1. After applying a local unitary transformation, we obtain $v_1 = v_2 = e_0$ and $v_3 = \lambda e_0$ for $\lambda \in \mathbb{C}^*$. Then we have $w_1 = \alpha e_1$ and $w_2 = \beta e_1$ for suitable $\alpha, \beta \in \mathbb{C}^*$. Write $w_3 = \mu(\cos\theta e_0 + \sin\theta e_1)$ for $\mu \in \mathbb{C}^*$. Thus we have $u = \lambda e_0^{\otimes 3} + \alpha e_1 \otimes (\mu e_0 \otimes [\cos\theta e_0 + \sin\theta e_1] + \nu e_1 \otimes e_0)$ for some $\nu \in \mathbb{C}$. Clearly a phase change for u will make λ real; so we can assume $\lambda \in \mathbb{R}$. We can of course assume $\alpha = 1$ by changing μ and ν appropriately.

In the rest of the proof we use the notation $e_i^{(j)}$ to denote the vector e_i in the j-th copy of $\mathbb{C}^2$. We will next do a simultaneous phase change

$$e_0^{(1)} \mapsto e^{i\phi} e_0^{(1)}, e_0^{(3)} \mapsto e^{-i\phi} e_0^{(3)}, e_1^{(3)} \mapsto e^{-i\phi} e_1^{(3)} \qquad (1.13)$$

This operation does not change λ but rescales μ as well as ν; so we can assume μ is real. Finally a phase rescaling of $e_1^{(2)}$ will make ν real without changing λ or μ. This way we easily get the normal form. ∎

There are other results on normal forms for 3-qubits: see [1, 2, 17].

It is interesting to discuss why S_2 is not closed in the case of $(\mathbb{C}^2)^{\otimes 3}$. There is an easy geometric description of the rank, which is well known to algebraic geometers. We start with the Segre product $S = S_1$, which is a closed algebraic subvariety of $\mathbb{P}(E)$. For a $(p-1)$-plane $\Pi \subset \mathbb{P}(E)$, we say that Π is a p-secant plane if Π is spanned by p points $y_1, \cdots, y_p$ of $\Pi \cap S$. For instance, a line is 2-secant if it is a secant line; a 2-plane is 3-secant if it spanned by 3 points of $\Pi \cap S$. Then we have clearly

LEMMA 1.1

A point of $\mathbb{P}(E)$ has rank $\leq p$ iff it belongs to some $(p-1)$-plane $\Pi \subset \mathbb{P}(E)$ which is p-secant to the Segre product S. In other words S_p is the union of all $(p-1)$-planes Π which are p-secant.

The point then is that S_2 need not be closed, because the limit of a sequence of secant lines to S need not be a secant line, but could be a tangent line. For instance, consider the polynomial one-paremeter family of tensors

$$u(t) = \frac{1}{t}[e_1 \otimes e_0 \otimes e_0 - (e_1 + te_0) \otimes (e_0 + te_1) \otimes (e_0 + te_1)] \quad (1.14)$$

For $t \neq 0$, the tensor $u(t)$ has rank 2, and belongs to the secant line through the points $e_1 \otimes e_0 \otimes e_0$ and $(e_1 + te_0) \otimes (e_0 + te_1) \otimes (e_0 + te_1)$ of S. For $t = 0$, we have $u(0) = e_0 \otimes e_0 \otimes e_0 + e_1 \otimes e_1 \otimes e_1 + e_1 \otimes e_0 \otimes e_1$; we have already seen that this state has rank 3. It belongs to the tangent line to S at the point $e_1 \otimes e_0 \otimes e_0$.

The phenomenon of S_p not being closed in general is the geometric reason why the border rank of a bilinear map can be lower than its rank [3, 16]. Indeed a tensor u has border rank $\leq p$ if and only if it belongs to the closure of S_p.

From algebraic geometry we can get some general information about S_p. We will use the Zariski topology of $\mathbb{P}(E)$ for which the closed subsets are the subsets defined by homogeneous complex polynomial equations. The *constructible* sets are then those obtained from the Zariski closed subsets by finite Boolean operations (finite unions, finite intersections, and complementation). A constructible subset F is called *irreducible* if whenever $F = G \cup H$ for G, H closed in F, we have $G = F$ or $H = F$. Note that F is irreducible if and only if its closure is irreducible.

We have the general fact:

PROPOSITION 1.9

The set S_p is an irreducible Zariski constructible subset of $\mathbb{P}(E)$.

PROOF   Let $X \subset S^p \times \mathbb{P}(E)$ be the locally closed algebraic subvariety comprised of $p+1$-tuples $(x_1, \cdots, x_{p+1})$ where $x_1, \cdots, x_p \in S$ are distinct, $x_{p+1} \in \mathbb{P}(E)$ and $(x_1, \cdots, x_{p+1})$ belong to some $(p-1)$-plane. Introduce the algebraic mapping $\Phi : X \to \mathbb{P}(E)$ given by $\Phi(x_1, \cdots, x_{p+1}) = x_{p+1}$. Let T_p be the image of Φ. It is easy to see that X is irreducible; thus standard results in algebraic geometry say that T_p is constructible and its closure is irreducible. We have easily $S_p = \cup_{q \leq p} T_q$ so that S_p is constructible. It is clear that $\bar{T}_q \subseteq \bar{T}_p$ for $q \leq p$, so that $\bar{S}_p$ is irreducible, and so is S_p.
∎

We note that for $(\mathbb{C}^2)^{\otimes 3}$, the set S_2 is not locally closed in $\mathbb{P}((\mathbb{C}^2)^{\otimes 3})$, only constructible.

We think that algebraic geometry methods should prove very useful in studying entanglement. For instance, there are many classical results concerning secant varieties, and some concerning multi-secant varieties.

Another interesting phenomenon is that a real tensor in $(\mathbb{R}^2)^{\otimes 3}$ may have different rank from the same tensor viewed as an element of $(\mathbb{C}^2)^{\otimes 3}$. An example is the tensor $e_0 \otimes (e_0 \otimes e_0 - e_1 \otimes e_1) + e_1(e_0 \otimes e_1 + e_1 \otimes e_0)$, which corresponds to the product law $\mathbb{R}^2 \otimes \mathbb{R}^2 \to \mathbb{R}^2$ on $\mathbb{R}^2 = \mathbb{C}$; this has rank 3 as a real tensor and rank 2 as a complex tensor. We note that this tensor corresponds to the bilinear mapping $\mathbb{R}^2 \otimes \mathbb{R}^2 \to \mathbb{R}^2$ given by multiplication of complex numbers. The fact that it has rank 3 as a real tensor means that multiplying two complex numbers requires 3 multiplications of real numbers.

The method of proof of Proposition 1.6 leads naturally to the following notion:

DEFINITION 1.2   *Let F be a subspace of $E = H_1 \otimes \cdots \otimes H_k$. The rank of F is the smallest integer p such that there exist p decomposable tensors $u_1, \cdots, u_p$ such that F is contained in the span of $u_1, \cdots, u_p$.*

We then have the following easy but useful result:

LEMMA 1.2

[5, Prop. 14.44] Let $u \in E = H_1 \otimes H_2 \otimes \cdots \otimes H_k$, *and let* T *be the corresponding linear map* $T : H_1^* \rightarrow H_2 \otimes \cdots \otimes H_k$. *Then the rank of the tensor* u *is equal to the rank of the range of* T *as a subspace of* $H_2 \otimes \cdots \otimes H_k$.

PROOF   Let p be the rank of u and q the rank of the range of T. Thus u is a linear combination of decomposable tensors $v_1, \cdots, v_p$. Write $v_j = w_j \otimes z_j$ where $w_1 \in H_1$ and $z_j \in H_2 \otimes \cdots \otimes H_k$. Then we have $T(l) = \sum_j \langle l|w_j \rangle z_j$, so that $T(l)$ is a linear combination of the decomposable tensors z_j and $q \leq p$. In the other direction, assume that the range of T is contained in the linear span of the decomposable tensors $\beta_j, 1 \leq l \leq s$. Then there are linear forms v_j on H_1^* (so $v_j \in H_1$) such that $T(l) = \sum_{j=1}^{s} \langle l|v_j \rangle \beta_j$. This means that $u = \sum_{j=1}^{s} v_j \otimes \beta_j$ and $r \leq s$.   ∎

There is a classical example for the rank of a subspace of $M_2(\mathbb{C})^{\otimes 2}$. We identify $M_2(\mathbb{C})$ with its dual, so that $M_2(\mathbb{C})^{\otimes 2}$ identifies with the space of bilinear functionals $(A, B) \mapsto f(A, B)$ of two matrices A, B of size 2. The coefficients of the product AB yield four such bilinear functionals, which span a four-dimensional subspace E of $M_2(\mathbb{C})^{\otimes 2}$. It is a well-known result of Strassen [15, 16] that the rank of this subspace is equal to 7 (instead of the value 8 one might naively expect). This is the basis for fast matrix multiplication. From Lemma 1.2 it ensues that the corresponding tensor in $M_2(\mathbb{C}) \otimes M_2(\mathbb{C}) \otimes M_2(\mathbb{C}) = M_2(\mathbb{C})^{\otimes 3}$ has rank equal to 7. It would be nice to have a geometric interpretation of this fact.

We also note an easy consequence of Lemma 1.2.

LEMMA 1.3

Let $(e_1, \cdots e_{d_1})$ *be a basis of* H_1, *and consider a tensor* $u = \sum_j e_j \otimes v_j \in H_1 \otimes H_2 \otimes \cdots \otimes H_k$. *Then the rank of* u *is at most the sum of the ranks of the* v_j *'s.*

1.5 Tensors in $(\mathbb{C}^2)^{\otimes 4}$

Our results for $E = (\mathbb{C}^2)^{\otimes 4}$ are less complete than for $(\mathbb{C}^2)^{\otimes 3}$. For $E = (\mathbb{C}^2)^{\otimes 4}$, an important invariant is the following: for any permutation (i, j, k, l) of $(1, 2, 3, 4)$, a tensor $u \in (\mathbb{C}^2)^{\otimes 4}$ yields a linear map $\phi_{ijkl} : \mathbb{C}^2 \otimes \mathbb{C}^2 \to \mathbb{C}^2 \otimes \mathbb{C}^2$. For instance $\phi(1234)$ maps $e_i \otimes e_j$ to $\sum_{k,l} \phi_{ijkl} e_k \otimes e_l$. If σ is the permutation coresponding to $(ijkl)$, σ acts on $(\mathbb{C}^2)^{\otimes 4}$ and then $\phi_{ijkl} = (\sigma\phi)_{1234}$, where $\sigma\phi$ is the transform of ϕ under σ.

We consider the determinant $\Delta(ijkl) = \det(\phi_{ijkl})$. We have the following symmetries: $\Delta(ijkl) = -\Delta(jikl) = -\Delta(ijlk) = \Delta(klij)$; so up to sign we have essentially three determinants. Now in fact we have

LEMMA 1.4

$$\Delta(1234) - \Delta(1324) + \Delta(1423) = 0 \qquad (1.15)$$

R. Brylinski and the author give an invariant-theoretic proof of this in [4]. The space of homogeneous degree 4 G-invariant polynomial functions on $(\mathbb{C}^2)^{\otimes 4}$ carries an action of the symmetric group $\mathfrak{S}_4$, and this representation is irreducible of dimension 2.

PROPOSITION 1.10
Let $E = (C^2)^{\otimes 4}$. Then the closure of S_3 in $\mathbb{P}(E)$ is the dimension 13 algebraic variety defined by the equations $\Delta(ijkl) = 0$.

PROOF   The Lie group $G = GL(2, \mathbb{C})^4$ acts naturally on E and preserves each S_p. Let F be the subspace spanned by $e_0^{\otimes 4}$, $e_1^{\otimes 4}$ and $(e_0 + e_1)^{\otimes 4}$. As usual $\mathbb{P}(F)$ denotes the 2-dimensional projective space of complex lines in F.

Clearly the closure of S_3 is the closure of the G-saturation $G \cdot \mathbb{P}(F)$. This is because a general triple of decomposable tensors is G-equivalent to the above triple. We can compute the dimension of $G \cdot \mathbb{P}(F)$ as follows. We consider the infinitesimal equation of the Lie algebra $\mathfrak{g} = \mathfrak{gl}(2, \mathbb{C})^4$ on E and on $\mathbb{P}(E)$. For $u \in F$, we denote by $\mathfrak{h}_u$ the space of $\gamma \in \mathfrak{g}$ such that $\gamma \cdot u \in F$. Then we have

$$\dim(G \cdot \mathbb{P}(F)) = \dim(G) + \dim(F) - min_{u \in F} \dim(\mathfrak{h}_u) - 1$$
$$= 18 - min_{u \in F} \dim(\mathfrak{h}_u) \qquad (1.16)$$

Indeed, let $M : G \times \mathbb{P}(F) \to \mathbb{P}(E)$ be the mapping $M(g, u) = g \cdot u$. Then by Sard's theorem, the dimension of $M(G \times \mathbb{P}(F)) = G \cdot \mathbb{P}(F) \subseteq \mathbb{P}(E)$ is the maximal rank of the differential dM at points (g, u) of $G \times \mathbb{P}(F)$. Now M is G-equivariant where G acts by left translation on the factor G of $G \times \mathbb{P}(F)$. By the G-equivariance of M, dM reaches its maximum rank at points of the form $(1, u)$. Then the kernel of $dM_{(1,u)}$ is the set of vectors $(\xi, -\xi \cdot u)$ for $\xi \in \mathfrak{h}_u$. Thus the rank of dM at $(1, u)$ is equal to $\dim(G) + \dim(\mathbb{P}(F)) - \dim(\mathfrak{h}_u) = \dim(G) + \dim(F) - \dim(\mathfrak{h}_u) - 1$.

Now for any $\delta, \epsilon \in \mathbb{C}^*$, the tensor $u = u_{\delta,\epsilon} = e_0^{\otimes 4} + \delta e_1^{\otimes 4} + \epsilon(e_0 + e_1)^{\otimes 4}$ belongs to F. Let $\mathfrak{k}_u$ be the space comprised of the $\gamma \in \mathfrak{g}$ such that $\gamma \cdot u$ is a linear combination of $e_0^{\otimes 4}$ and $e_1^{\otimes 4}$. Since $\mathfrak{h}_u$ is the direct sum of $\mathfrak{k}_u$ and of the line spanned by $(Id, 0, 0, 0)$, we have $\dim(\mathfrak{h}_u) = \dim(\mathfrak{k}_u) + 1$. So it suffices to compute $\dim(\mathfrak{k}_u)$.

Now for $\gamma = (\gamma_j) \in \mathfrak{g} = \mathfrak{gl}(2, \mathbb{C})^4$ with $\gamma_j = \begin{pmatrix} a_j & b_j \\ c_j & d_j \end{pmatrix}$ we compute:

$$\gamma \cdot u = (a_1 + a_2 + a_3 + a_4 + \epsilon(a_1 + a_2 + a_3 + a_4 + b_1 + b_2 + b_3 + b_4))e_0^{\otimes 4}$$
$$+ (c_4 + \epsilon(a_1 + b_1 + a_2 + b_2 + a_3 + b_3 + c_4 + d_4))e_0^{\otimes 3} \otimes e_1$$
$$+ \text{permutations}$$
$$+ (\delta d_1 + \epsilon(a_1 + b_1 + c_2 + d_2 + c_3 + d_3 + c_4 + d_4))e_0 \otimes e_1^{\otimes 3}$$
$$+ \text{permutations}$$
$$+ \epsilon(a_1 + b_1 + c_2 + d_2 + c_3 + d_3 + c_4 + d_4)e_0^{\otimes 2} \otimes e_1^{\otimes 2}$$
$$+ \text{permutations}$$

$$(1.17)$$

So γ belongs to $\mathfrak{k}_u$ iff the coefficients of $e_0^{\otimes 3} \otimes e_1$, $e_0^{\otimes 2} \otimes e_1^{\otimes 2}$, $e_0 \otimes e_1^{\otimes 3}$, and the other tensors obtained from these by permutations, all vanish. At first sight this is just a system of linear equations in 16 unknowns, but one can essentially separate them according to the 4 groups of 4 variables (a_j, b_j, c_j, d_j) by introducing the sums $\lambda = \sum_j a_j, \mu = \sum_j b_j, \nu = \sum_j c_j, \rho = \sum_j d_j$. One gets the equations:

1. For each j, $\epsilon(a_j + b_j - d_j) - (1 + \epsilon)c_j = \epsilon(\lambda + \mu)$.
2. For each j, $\epsilon(a_j - c_j - d_j) + (\delta + \epsilon)b_j = \epsilon(\nu + \rho)$.
3. For each permutation $(ijkl)$ of (1234) we have $a_i + b_i + a_j + b_j + c_k + d_k + c_l + d_l = 0$.

From (3) we easily see that $a_j + b_j - c_j - d_j$ is independent of j.

By summing each of the three types of equations over all choices of indices (or of permutations for the third), we get consistency requirements for $(\lambda, \mu, \nu, \rho)$; these are easily solved to yield:

$$\mu = \frac{2\epsilon}{\delta - 2\epsilon}\lambda, \nu = \frac{-2\delta\epsilon}{\delta - 2\epsilon}\lambda, \rho = \frac{\lambda\delta(2\epsilon - 1)}{\delta - 2\epsilon} \qquad (1.18)$$

Here λ is a free parameter; once it is chosen we can solve for $(a_j, b_j,$
$c_j, d_j)$ and obtain $\gamma_j = \omega_j Id + \phi_j \xi + \eta$, where $\xi = \begin{pmatrix} \delta + \epsilon + \delta\epsilon & -\epsilon \\ \delta\epsilon & 0 \end{pmatrix}$, $\eta =$
$\begin{pmatrix} \frac{\delta\lambda}{\delta - 2\epsilon} & 0 \\ 0 & 0 \end{pmatrix}$ are matrices independent of j, and ω_j, ϕ_j are some scalars.
The fact that $a_j + b_j - c_j - d_j$ is independent of j then implies that ϕ_j
is too; call this scalar ϕ. Then we need the a_j to sum up to λ, etc. This
gives the value $\frac{-\mu}{4\epsilon}$ for ϕ and the requirement $\sum \omega_j = \rho$.

Counting the free parameters we obtain $\dim(\mathfrak{k}_u) = 4$. It follows that
S_3 has dimension 13. It is clearly contained in the codimension 2 subvariety defined by the vanishing of the $\Delta(ijkl)$; the latter variety is seen
to be irreducible, thus it must equal the closure of S_3.   ∎

THEOREM 1.1

The highest rank of a tensor in $(\mathbb{C}^2)^{\otimes 4}$ is equal to 4.

PROOF   We associate to $u \in (\mathbb{C}^2)^{\otimes 4}$ as before a linear map T :
$\mathbb{C}^2 \to (\mathbb{C}^2)^{\otimes 3}$. If T has rank 1 (as a linear map) it is clear that the
tensor u has rank ≤ 3, so we can assume T is injective. We can think
of T as parameterizing a line l in $\mathbb{P}((\mathbb{C}^2)^{\otimes 3}) = \mathbb{CP}^7$. If this line is not
contained in the hypersurface Z, then 2 of its points have rank ≤ 2, and
it follows using Lemma 1.3 that u has rank $\leq 2 + 2 = 4$. Thus we need
to focus on the case where l is contained in Z.

It could happen that l is contained in Y_j for some j. In that case it
is easy to see that u is of rank ≤ 4. So we can assume that l contains
a point u which belongs to none of the Y_j; so u is $GL(2, \mathbb{C})^3$-conjugate
to $e_0 \otimes e_0 \otimes e_0 + e_1 \otimes e_0 \otimes e_1 + e_1 \otimes e_1 \otimes e_0$. For this choice of vector u
we can write down the equations on a tensor w so that the line through
u and w is contained in Z, i.e., $D(xu + yw)$ vanishes identically. It is
natural to consider w as a vector modulo scaling in the quotient space
$(\mathbb{C}^2)^{\otimes 3}/\mathbb{C}u = \mathbb{C}^7$, i.e., as an element of projective space $\mathbb{CP}^6$.

Look at the equation giving the vanishing of the coefficient of $x^i y^{4-i}$,
as i ranges from 3 to 0. The first equation is $w_{011} = 0$. The second is
$2(w_{111}^2 + w_{001}^2 + w_{010}^2) - (w_{001} + w_{010} + w_{111})^2 = 0$; this is a non-degenerate
quadratic form in three variables. The third equation involves the new
variables $w_{000}, w_{101}, w_{110}$ and is linear as a function of these three variables. The fourth equation involves also the last variable w_{100}, and is
linear in w_{100}. The set A of w such that $D(xu + yw) \equiv 0$ is a closed
algebraic subvariety of $\mathbb{CP}^6$. A has a dense open set which is obtained

by successive fibrations with fibers irreducible algebraic varieties; thus A itself is irreducible and its dimension is equal to 2. Recall Z^0 is the open $SL(2, \mathbb{C})^3$-orbit inside Z, which is the complement of $Y_1 \cup Y_2 \cup Y_3$. Now consider the algebraic variety X comprised of pairs (p, L) where $p \in Z^0$ and L is a line through p which lies entirely inside Z. This is a locally closed subvariety of the product of Z^0 with the Grassmann manifold of lines in $\mathbb{CP}^7$. Then the projection map $X \to Z^0$ is a fibration, because it is $SL(2, \mathbb{C})^3$-equivariant and Z^0 is a single orbit. The fiber of this fibration is A. The dimension of X is therefore $\dim(Z^0) + \dim(A) = 6 + 2 = 8$. Since A and Z are irreducible, so is X.

What we are really after however is the variety V of lines contained in Z and meeting Z^0. This is a locally closed algebraic subvariety of the Grassmannian of lines in $\mathbb{CP}^7$: the set of lines meeting Z is closed, and we delete from it the three closed subvarieties of lines contained in some Y_j. There is an obvious algebraic map $X \to V$ which is a smooth mapping with one-dimensional fibers. Therefore V has dimension $8 - 1 = 7$ and is irreducible.

Now we claim that any line contained in Z and not contained in any Y_j must meet each Y_j. For this purpose consider the set W of lines l such that

1. l gives a point of V, i.e., l is contained in Z and meets Z^0.
2. l meets Y_1.

Then W is a closed subvariety of V. Now we claim that $\dim(W) = 7$. Pick some tensor in Y_1, say $u = e_0^3 + e_0 \otimes e_1^2$, and consider again the set of w such that $D(xu + yw) \equiv 0$. One checks that this forms a subvariety of $\mathbb{CP}^6$ of dimension 3. It follows that the subset W^0 of W comprised of the lines contained in Z and meeting Y_1 in finitely many points has a finite ramified covering which maps to Y_1 with three-dimensional fibers; therefore it has dimension $4 + 3 = 7$. So $\dim(W) \geq 7$. Since V is irreducible, it follows that $W = V$. This means that any line contained in Z must meet Y_1. Similarly it must meet each Y_j.

Thus we can change the basis of the first $\mathbb{C}^2$ so that $T(e_0) \in Y_1$ and $T(e_1) \in Y_2$. Then both these tensors have rank ≤ 2, and by Lemma 1.3 u itself has rank $\leq 2 + 2 = 4$. ∎

It is easy to see that S_2 has dimension 9 and satisfies a number of algebraic equations, namely the 2 by 2 minors of the linear maps $(\mathbb{C}^2)^{\otimes 2} \to (\mathbb{C}^2)^{\otimes 2}$ obtained from the tensor (there are essentially three such linear maps). It is likely the case that these equations precisely describe the closure of S_2.

We note that normal forms for 4-qubit states have been obtained in [18].

References

[1] A. Acin, A. Andrianov, L. Costa, E. Jane, J.I. Latorre, R. Tarrach, Generalized Schmidt decomposition and classification of three-quantum-bit states, *Phys. Rev. Lett.*, 85 (2000) 1560–1563.

[2] A. Acin, A. Andrianov, E. Jane, R. Tarrach, Three-qubit pure-state canonical forms, *J. Phys. A: Math. Gen.*, 34 (2001), 6725; quant-ph/0009107.

[3] D. Bini, M. Capovani, G. Lotti and F. Romani, $O(n^{2.7799})$ complexity for matrix multiplication, *Inf. Proc. Lett.*, 8 (1979), 234–235.

[4] J-L. Brylinski and R. Brylinski, Polynomial invariants for qubits, In Chapter 11 of this book.

[5] P. Bürgisser, M. Clausen and M. A. Shokrollahi, Algebraic Complexity Theory, *Grundl.* vol. 315, Springer–Verlag, New York-Heidelberg-Berlin (1997).

[6] A.R. Calderbank, E. M Rains, P.W. Shor and N.J.A. Sloane, Quantum Error Correction via Codes over GF(4), *IEEE Trans. Inform. Theory*, 44 (1998), 1369–1387; quant-ph/9608006.

[7] F. Catanese, Review of the book [C-R-S-S], *Bull. Amer. Soc.* (2000).

[8] A. Cayley, On the theory of elimination, *Collected Papers*, vol. 1, no. 59, Cambridge University Press, Cambridge, UK (1889), 370–374.

[9] V. Coffman, J. Kundu and W. K. Wootters, Distributed Entanglement, quant-ph/9907047.

[10] I.M. Gelfand, M. Kapranov and A. Zelevinsky, *Discriminants, Resultants and Multidimensional Determinants*, Birkhäuser, Boston (1991).

[11] D. Gottesman, An Introduction to Quantum Error Correction, talk given at AMS Short Course on Quantum Computation in Jan. 2000, `quant-ph/0004072`.

[12] D. Meyer and N. Wallach Invariants for multiple qubits: the case of 3 qubits. In Chapter 3 of this book.

[13] S. Popescu and R. Rohrlich, *The Joy of Entanglement, Introduction to Quantum Computation and Information*, H-K. Lo, S. Popescu and T. Spiller eds, World Scientific, Singapore (1998), 29–48.

[14] A. Steane, Simple Quantum Error Correcting Codes, *Phys. Rev. A*, 54 (1996), 4741.

[15] V. Strassen, Gaussian elimination is not optimal, *Numer. Math.*, 13 (1969), 354–356.

[16] V. Strassen, Rank and optimal computation of general tensors, *Linear Alg. Appl.*, 52/53 (1983), 645–685.

[17] A. Sudbery, On local invariants of pure three-qubit states, *J. Phys. A: Math. Gen.*, 34 (2001), 643–652.

[18] F. Verstraete, J. Dehaene, B. De Moor, H. Verscheld, Four qubits can be entangled in nine different ways, `quant-ph/0109033`.

Chapter 2

Kinematics of qubit pairs

Berthold-Georg Englert and Nasser Metwally

Abstract   We review some of the important properties of the states of qubit pairs. They are specified by 15 numerical parameters that are naturally regarded as the components of two 3-vectors and a 3 × 3-dyadic. There are six classes of families of locally equivalent states in a straightforward scheme for classifying all 2-qubit states; four of the classes consist of two subclasses each. Easy-to-use criteria enable one to check whether a given pair of 3-vectors plus a 3 × 3-dyadic specify a 2-qubit state; and if they do, whether the state is entangled; and if it is, whether it is a separable state. We remark on the Hill–Wootters concurrence, discuss the properties of the fundamental Lewenstein–Sanpera decompositions of 2-qubit states, and report a number of examples for which the optimal decomposition is known explicitly.

2.1   Introduction

A qubit is, in general terms, a binary quantum alternative, for which there are many different physical realizations. Familiar examples include the binary alternatives of a Stern–Gerlach experiment ("spin up" or "spin down"); of a photon's helicity ("left handed" or "right handed");

1-58488-282/4/02/$0.00+$1.50

of two-level atoms ("in the upper state" or "in the lower one"); of Young's double-slit setup ("through this slit" or "through that slit"); of Mach–Zehnder interferometers ("reflected at the entry beam splitter" or "transmitted at it"); and of Ramsey interferometers ("transition in the first zone" or "in the second zone").

The actual physical nature of the qubits in question is irrelevant, however, for the issues dealt with here. We are remarking on entangled states of two qubits, and as far as the somewhat abstract mathematical properties are concerned, all qubits are equal. In particular, the two qubits under consideration could be of quite different kinds, one the spin-$\frac{1}{2}$ degree of freedom of a silver atom, say, the other a photon's helicity. It is even possible, and of experimental relevance [1, 2, 3, 4, 5, 6, 7], that both qubits are carried by the same physical object: the which-way alternative of an atom (photon, neutron, . . .) passing through an interferometer could represent one qubit, for instance, while its polarization (or another internal degree of freedom) is the other.

Entangled qubits are exploited in most schemes proposed for quantum communication purposes, for quantum information processing, or for the secure key distribution procedures known as quantum cryptography. The basic units are entangled qubit pairs. Obviously then, a thorough understanding of the properties of 2-qubit states is desirable. Although there has been considerable progress in this matter recently, the situation is still not fully satisfactory.

Whereas the possible states of a single qubit are easily classified with the aid of a 3-vector (the Bloch vector in one physical context, the Poincaré vector in another, and analogs of both in general — we shall speak of Pauli vectors), the classification of the states of entangled qubit pairs has not been fully achieved as yet. The obvious reason is the richness of the state space, which is parameterized by two 3-vectors, one for each qubit, and a 3×3-dyadic that represents expectation values of joint observables, so that 15 real numbers are necessary to specify an arbitrary 2-qubit state.

The characterization of the 2-qubit states produced by some source requires the experimental determination of these 15 real parameters. Ideally, this is done by measuring a suitably chosen set of five observables [8] that constitute "a complete set of five pairs of complementary propositions" [9]. In an optical model [6], which makes use of single-photon 2-qubit states, these measurements can be realized, and other experimental studies of 2-qubit states can be performed as well.

Then, based on the knowledge of the 15 state-specifying parameters, one can classify the 2-qubit state. We distinguish six classes of families of locally equivalent states. Roughly speaking, local equivalence means that the difference is of a geometrical, not a physical nature. In a certain sense, the 15 parameters can be regarded as consisting of 6 geometrical ones and 9 physical ones.

This classification is straightforward but by far not sufficient. One also needs to know if the 2-qubit state in question is useful for quantum communication purposes. A first important division is the one into entangled states and disentangled ones; a second distinguishes entangled states that are separable from the nonseparable ones. The latter ones differ from each other by various properties. For instance, their so-called Hill–Wootters concurrence [10, 11] is nonzero.

As a rule, calculating the concurrence is a somewhat tedious task, however. But, fortunately, there is the easy-to-use Peres–Horodeccy criterion [12, 13] for deciding whether a given 2-qubit state is separable or not.

For a finer distinction between nonseparable states that shall enable us to tell the states that are more useful for quantum communication from the less useful ones, we ask for their degree of separability as a numerical measure for this usefulness. The degree of separability is part and parcel of the so-called optimal Lewenstein–Sanpera decomposition [14] of a 2-qubit state. The general properties of this important concept are quite well understood, and the optimal decompositions are known for a number of relevant types of states. But a general analytical method for determining the degree of separability of any arbitrary 2-qubit state is still not at hand, despite the progress achieved as the result of considerable effort by various investigators.

At present, it is unclear how the concurrence and the degree of separability are related to each other. One knows that they have to obey an inequality that follows from the Lewenstein–Sanpera decomposition, but the conditions under which it is an equality have not been established. In addition, there may be other intimate relations. In this context, we formulate a conjecture which, if it could be demonstrated, would reveal an intriguing connection.

The qubit pairs that we are concerning ourselves with are the simplest intertwined quantum systems. As such they are prototypical for more complicated systems, much like the hydrogen atom is the prototype of all atoms and molecules. Of course, constituting the simplest system conceivable, qubit pairs do not exhibit all the features that can be found in larger systems. In fact, the simplicity of a 2-qubit system is two-fold:

it consists of two parts only, and each part is a qubit. Both restrictions can be lifted separately or jointly, and the richer-in-structure systems thus obtained are being studied diligently. There is, indeed, a rapidly growing literature, and brief remarks about it are unavoidably superficial and cannot do justice to all investigators involved in this enterprise. Therefore, we shall be content with mentioning two aspects that seem particularly intriguing and important: the phenomenon of "bound entanglement" [15] and the concept of "entanglement witnesses" [16, 17]. These catchwords are, of course, just meant as appetizers that are hoped to trigger the readers' interest in the papers cited. In full awareness of the injustice done to all the good work we do not refer to, we point to a few other recent papers [18, 19, 20, 21, 22, 23] with some relationship to the properties of 2-qubit states discussed here.

2.2   Preliminaries

Analogs of Pauli's spin operators are, as usual, used for the description of the individual qubits: the hermitian set σ_x, σ_y, σ_z for the first qubit, and τ_x, τ_y, τ_z for the second. Upon introducing corresponding sets of three-dimensional unit vectors — $\vec{e}_x$, $\vec{e}_y$, $\vec{e}_z$ and $\vec{n}_x$, $\vec{n}_y$, $\vec{n}_z$, respectively, each set orthonormal and right-handed — we form the vector operators

$$\vec{\sigma} = \sum_{\alpha=x,y,z} \sigma_\alpha \vec{e}_\alpha = (\sigma_x, \sigma_y, \sigma_z) \begin{pmatrix} \vec{e}_x \\ \vec{e}_y \\ \vec{e}_z \end{pmatrix},$$

$$\vec{\tau} = \sum_{\beta=x,y,z} \tau_\beta \vec{n}_\beta = (\tau_x, \tau_y, \tau_z) \begin{pmatrix} \vec{n}_x \\ \vec{n}_y \\ \vec{n}_z \end{pmatrix}. \qquad (2.1)$$

We emphasize that the two three-dimensional vector spaces thus introduced are unrelated and they may have nothing to do with the physical space. Even if the qubits should consist of the spin-$\frac{1}{2}$ degrees of freedom of two electrons, say, so that an identification with the physical space would be natural, we could still define the x, y, and z directions independently for both qubits.

As in [24, 25], we employ a self-explaining notation that distinguishes row vectors from column vectors, related to each other by transposition, as illustrated by

$$\sigma^{\downarrow} = \vec{\sigma}^{\mathrm{T}}, \qquad \vec{\tau} = \tau^{\downarrow\mathrm{T}}, \tag{2.2}$$

for example. Scalar and vector products — denoted by a dot $\cdot$ and a cross $\times$, respectively — such as the ones appearing in the basic algebraic relations

$$\vec{a}_1 \cdot \sigma^{\downarrow} \, \vec{a}_2 \cdot \sigma^{\downarrow} = \vec{a}_1 \cdot a_2^{\downarrow} + \mathrm{i} \left(\vec{a}_1 \times \vec{a}_2 \right) \cdot \sigma^{\downarrow},$$

$$\vec{b}_1 \cdot \tau^{\downarrow} \, \vec{b}_2 \cdot \tau^{\downarrow} = \vec{b}_1 \cdot b_2^{\downarrow} + \mathrm{i} \left(\vec{b}_1 \times \vec{b}_2 \right) \cdot \tau^{\downarrow},$$

$$\vec{a} \cdot \sigma^{\downarrow} \, \vec{b} \cdot \tau^{\downarrow} = \vec{b} \cdot \tau^{\downarrow} \, \vec{a} \cdot \sigma^{\downarrow}, \tag{2.3}$$

where $\vec{a}_1, \vec{a}_2, \vec{a}$ and $\vec{b}_1, \vec{b}_2, \vec{b}$ are arbitrary numerical vectors, involve rows and columns of the same type, that is: two of e-type or two of n-type. Numerical summands in operator equations, such as $\vec{a}_1 \cdot a_2^{\downarrow}$ in the first statement of (2.3), are to be read as multiples of the identity operator.

Products of the "column times row" kind are dyadics, for which $\sigma^{\downarrow}\vec{\tau}$ is an important example; it is a column of e-type combined with a row of n-type. The transpose of such a en-dyadic is a ne-dyadic; there are also ee-dyadics and nn-dyadics. Suppose that $\overset{\ulcorner\urcorner}{A}$ and $\overset{\ulcorner\urcorner}{B}$ are two en-dyadics, so that their transposes $\overset{\ulcorner\urcorner}{A}^{\mathrm{T}}$, $\overset{\ulcorner\urcorner}{B}^{\mathrm{T}}$ are ne-dyadics. Then $\overset{\ulcorner\urcorner}{A}^{\mathrm{T}} \cdot \overset{\ulcorner\urcorner}{B}$, for example, is a nn-dyadic and $\overset{\ulcorner\urcorner}{B} \cdot \overset{\ulcorner\urcorner}{A}^{\mathrm{T}}$ is of ee-type. Yet another product of $\overset{\ulcorner\urcorner}{A}$ and $\overset{\ulcorner\urcorner}{B}$ is the symmetric two-fold vector product $\left\{ \overset{\ulcorner\urcorner}{A}, \overset{\ulcorner\urcorner}{B} \right\} = \left\{ \overset{\ulcorner\urcorner}{B}, \overset{\ulcorner\urcorner}{A} \right\}$, which is the en-dyadic defined by*

$$\frac{1}{2} \left(\vec{\sigma} \cdot \overset{\ulcorner\urcorner}{A} \cdot \tau^{\downarrow} \vec{\sigma} \cdot \overset{\ulcorner\urcorner}{B} \cdot \tau^{\downarrow} + \vec{\sigma} \cdot \overset{\ulcorner\urcorner}{B} \cdot \tau^{\downarrow} \vec{\sigma} \cdot \overset{\ulcorner\urcorner}{A} \cdot \tau^{\downarrow} \right)$$

$$= \mathrm{Sp}\left\{ \overset{\ulcorner\urcorner}{A}^{\mathrm{T}} \cdot \overset{\ulcorner\urcorner}{B} \right\} + \vec{\sigma} \cdot \left\{ \overset{\ulcorner\urcorner}{A}, \overset{\ulcorner\urcorner}{B} \right\} \cdot \tau^{\downarrow}. \tag{2.4}$$

All properties of $\left\{ \overset{\ulcorner\urcorner}{A}, \overset{\ulcorner\urcorner}{B} \right\}$ follow from its linearity in both $\overset{\ulcorner\urcorner}{A}$ and $\overset{\ulcorner\urcorner}{B}$ in conjunction with

$$\left\{ a_1^{\downarrow}\vec{b}_1, a_2^{\downarrow}\vec{b}_2 \right\} = a_1^{\downarrow} \times a_2^{\downarrow} \, \vec{b}_2 \times \vec{b}_1, \tag{2.5}$$

---

*We write Sp $\{\ \}$ for the trace of a dyadic and tr $\{\ \}$ for the quantum mechanical operator trace.

where $a_1^\downarrow, a_2^\downarrow$ are any two columns of e-type and $\vec{b}_1, \vec{b}_2$ are any two rows of n-type. In particular, we have

$$\left\{ \overleftrightarrow{A}, \overleftrightarrow{A} \right\} = -2\, \overleftrightarrow{A}_{\text{sub}} \tag{2.6}$$

and

$$\overleftrightarrow{A}^{\text{T}} \cdot \left\{ \overleftrightarrow{A}, \overleftrightarrow{A} \right\} = -2 \det\left\{ \overleftrightarrow{A} \right\} \overleftrightarrow{1}_{nn} \,, \tag{2.7}$$

where the *en*-dyadic $\overleftrightarrow{A}_{\text{sub}}$ consists of the signed subdeterminants, the cofactors, of $\overleftrightarrow{A}$, and $\overleftrightarrow{1}_{nn}$ is the unit dyadic of nn-type. The implied identities

$$\left(\overleftrightarrow{A}_{\text{sub}} \right)_{\text{sub}} = \overleftrightarrow{A} \det\left\{ \overleftrightarrow{A} \right\} \tag{2.8}$$

and

$$2\operatorname{Sp}\left\{ \overleftrightarrow{A}^{\text{T}}_{\text{sub}} \cdot \overleftrightarrow{A}_{\text{sub}} \right\} = \left(\operatorname{Sp}\left\{ \overleftrightarrow{A}^{\text{T}} \cdot \overleftrightarrow{A} \right\} \right)^2 - \operatorname{Sp}\left\{ \left(\overleftrightarrow{A}^{\text{T}} \cdot \overleftrightarrow{A} \right)^2 \right\} \tag{2.9}$$

are worth remembering.

As an immediate consequence of (2.3) all functions of $\vec{\sigma}$ and $\vec{\tau}$ are linear in these Pauli vector operators. An arbitrary 2-qubit state is therefore specified by a statistical operator of the form

$$\rho = \frac{1}{4}\left(1 + \vec{\sigma} \cdot s^\downarrow + \vec{t} \cdot \tau^\downarrow + \vec{\sigma} \cdot \overleftrightarrow{C} \cdot \tau^\downarrow \right) \,; \tag{2.10}$$

it determines the expectation values $\langle f(\vec{\sigma}, \vec{\tau}) \rangle$ of all operator functions $f(\vec{\sigma}, \vec{\tau})$ in accordance with $\langle f(\vec{\sigma}, \vec{\tau}) \rangle = \operatorname{tr}\{ f(\vec{\sigma}, \vec{\tau}) \rho \}$. Rather than distinguishing pedantically between a 2-qubit state and its statistical operator ρ, we'll simply speak of "the state ρ." It involves the real *cross dyadic* $\overleftrightarrow{C}$,

$$\overleftrightarrow{C} = \left\langle \sigma^\downarrow \vec{\tau} \right\rangle = \left(e_x^\downarrow, e_y^\downarrow, e_z^\downarrow \right) \begin{pmatrix} C_{xx} & C_{xy} & C_{xz} \\ C_{yx} & C_{yy} & C_{yz} \\ C_{zx} & C_{zy} & C_{zz} \end{pmatrix} \begin{pmatrix} \vec{n}_x \\ \vec{n}_y \\ \vec{n}_z \end{pmatrix} \,, \tag{2.11}$$

in addition to the two real *Pauli vectors* $s^\downarrow$ and $\vec{t}$,

$$s^\downarrow = \left\langle \sigma^\downarrow \right\rangle = \left(e_x^\downarrow, e_y^\downarrow, e_z^\downarrow \right) \begin{pmatrix} s_x \\ s_y \\ s_z \end{pmatrix} \,, \qquad \vec{t} = \left\langle \vec{\tau} \right\rangle = (t_x, t_y, t_z) \begin{pmatrix} \vec{n}_x \\ \vec{n}_y \\ \vec{n}_z \end{pmatrix} \,. \tag{2.12}$$

The 2-qubit observable	
which identifies the joint eigenstates of	determines the three expectation values
σ_x and τ_x	$\langle\sigma_x\rangle$, $\langle\tau_x\rangle$, $\langle\sigma_x\tau_x\rangle$
σ_y and τ_y	$\langle\sigma_y\rangle$, $\langle\tau_y\rangle$, $\langle\sigma_y\tau_y\rangle$
σ_z and τ_z	$\langle\sigma_z\rangle$, $\langle\tau_z\rangle$, $\langle\sigma_z\tau_z\rangle$
$\sigma_x\tau_y$ and $\sigma_y\tau_z$	$\langle\sigma_x\tau_y\rangle$, $\langle\sigma_y\tau_z\rangle$, $\langle\sigma_z\tau_x\rangle$
$\sigma_y\tau_x$ and $\sigma_z\tau_y$	$\langle\sigma_y\tau_x\rangle$, $\langle\sigma_z\tau_y\rangle$, $\langle\sigma_x\tau_z\rangle$

Table 2.1 A set of five 2-qubit observables whose measurement supplies all fifteen parameters that characterize the state ρ of (2.10). (From [24], with permission from Taylor & Francis, http://www.tandf.co.uk.)

Note that ρ is properly normalized to unit trace by construction, but restrictions apply to $\vec{s}$, $t^{\downarrow}$, and $\overleftrightarrow{C}$ to ensure its positivity, $\rho \geq 0$; see Section 2.5 for particulars.

The 15 expectation values that constitute $\vec{s}$, $t^{\downarrow}$, and $\overleftrightarrow{C}$ can be obtained by measuring 5 well-chosen 2-qubit observables, such as the ones specified in Table 2.1. These five observables are pairwise complementary and thus represent an optimal set in the sense of Wootters and Fields [8]. Or, as Brukner and Zeilinger would put it, the left column of Table 2.1 lists "a complete set of five pairs of complementary propositions" [9].

In addition to the pre-chosen xyz coordinate systems, we also need to consider 123 coordinate systems that are adapted to the 2-qubit state of interest. Then

$$\vec{\sigma} = (\sigma_1, \sigma_2, \sigma_3) \begin{pmatrix} \vec{e}_1 \\ \vec{e}_2 \\ \vec{e}_3 \end{pmatrix}, \quad \tau^{\downarrow} = \left(n_1^{\downarrow}, n_2^{\downarrow}, n_3^{\downarrow}\right) \begin{pmatrix} \tau_1 \\ \tau_2 \\ \tau_3 \end{pmatrix} \tag{2.13}$$

are the respective parameterizations of $\vec{\sigma}$ and $\tau^{\downarrow}$, and $\overleftrightarrow{C}$ is represented by the 9 numbers $\vec{e}_j \cdot \overleftrightarrow{C} \cdot n_k^{\downarrow}$ with $j, k = 1, 2, 3$. A 4×4-matrix representation of the operators, one that is often particularly convenient (it is essentially identical with what Hill and Wootters call the "magic basis"

[10, 11]), has imaginary antisymmetric matrices for the Pauli operators,

$$\vec{\sigma}\cdot s^{\downarrow} + \vec{t}\cdot\tau^{\downarrow} \stackrel{\wedge}{=} \begin{bmatrix} 0 & -\mathrm{i}(s_1+t_1) & \mathrm{i}(s_2+t_2) & -\mathrm{i}(s_3-t_3) \\ \mathrm{i}(s_1+t_1) & 0 & \mathrm{i}(s_3+t_3) & \mathrm{i}(s_2-t_2) \\ -\mathrm{i}(s_2+t_2) & -\mathrm{i}(s_3+t_3) & 0 & \mathrm{i}(s_1-t_1) \\ \mathrm{i}(s_3-t_3) & -\mathrm{i}(s_2-t_2) & -\mathrm{i}(s_1-t_1) & 0 \end{bmatrix}, \quad (2.14)$$

and real symmetric matrices for their products,

$$\vec{\sigma}\cdot\overleftrightarrow{C}\cdot\tau^{\downarrow}$$

$$\stackrel{\wedge}{=} \begin{bmatrix} C_{11}+C_{22}-C_{33} & C_{23}+C_{32} & C_{13}+C_{31} & -C_{12}+C_{21} \\ C_{23}+C_{32} & C_{11}-C_{22}+C_{33} & -C_{12}-C_{21} & -C_{13}+C_{31} \\ C_{13}+C_{31} & -C_{12}-C_{21} & -C_{11}+C_{22}+C_{33} & C_{23}-C_{32} \\ -C_{12}+C_{21} & -C_{13}+C_{31} & C_{23}-C_{32} & -C_{11}-C_{22}-C_{33} \end{bmatrix}.$$

$$(2.15)$$

Now, if the state ρ is represented by a 4×4 matrix of this kind, then the transposed matrix represents the related state

$$\rho^{\mathsf{T}} = \frac{1}{4}\left(1 - \vec{\sigma}\cdot s^{\downarrow} - \vec{t}\cdot\tau^{\downarrow} + \vec{\sigma}\cdot\overleftrightarrow{C}\cdot\tau^{\downarrow}\right), \quad (2.16)$$

the so-called *(total) transpose* of ρ. Although the transformation

$$\vec{\sigma} \to -\vec{\sigma}, \qquad \tau^{\downarrow} \to -\tau^{\downarrow} \quad (2.17)$$

is not unitary, the two states ρ and ρ^{T} have the same eigenvalues and are, therefore, unitarily equivalent, but the unitary operator U that effects $\rho \to U^{-1}\rho U = \rho^{\mathsf{T}}$ is not universal. It depends on the particular ρ under consideration. Equivalently, we can think of $\rho \to \rho^{\mathsf{T}}$ as resulting from

$$s^{\downarrow} \to -s^{\downarrow}, \qquad \vec{t} \to -\vec{t}, \qquad \overleftrightarrow{C} \to \overleftrightarrow{C}. \quad (2.18)$$

Consistent with its definition in terms of transposed matrices, but also as implied by the two-fold Pauli algebra specified by (2.3), the total transpose of a product $\rho_1\rho_2$ is given by $\rho_2^{\mathsf{T}}\rho_1^{\mathsf{T}}$. This natural property of a transposition is not possessed by the two kinds of *partial transposition* introduced by

$$\rho \to \rho^{\mathsf{T}_1} = \frac{1}{4}\left(1 - \vec{\sigma}\cdot s^{\downarrow} + \vec{t}\cdot\tau^{\downarrow} - \vec{\sigma}\cdot\overleftrightarrow{C}\cdot\tau^{\downarrow}\right),$$

that is: $\vec{\sigma} \to -\vec{\sigma}, \quad \tau^{\downarrow} \to \tau^{\downarrow}$

or, equivalently, $s^{\downarrow} \to -s^{\downarrow}, \quad \vec{t} \to \vec{t}, \quad \overleftrightarrow{C} \to -\overleftrightarrow{C}$

or, compactly, $\rho \to \rho^{\mathsf{T}_1} = \frac{1}{2}\left(\vec{\sigma}\cdot\rho\sigma^{\downarrow} - \rho\right)$ (2.19) for the first qubit and analogously for the second qubit by

$$\rho \rightarrow \rho^{\mathrm{T}_2} = \frac{1}{2}\left(\vec{\tau} \cdot \rho\tau^{\downarrow} - \rho\right)$$

$$\text{or} \quad s^{\downarrow} \rightarrow s^{\downarrow}, \quad \vec{t} \rightarrow -\vec{t}, \quad \overleftarrow{C} \rightarrow -\overleftarrow{C}. \tag{2.20}$$

As the two compact versions emphasize, these partially transposed states are weighted sums of four unitarily equivalent states with three weights of $+\frac{1}{2}$ and one weight of $-\frac{1}{2}$. As a rule, therefore, the common eigenvalues of ρ^{T_1} and ρ^{T_2} are different from the common eigenvalues of ρ and ρ^{T}, and the partial transposes of a given ρ are not assuredly positive. They are not guaranteed to be states themselves.

Although $(\rho_1\rho_2)^{\mathrm{T}_1} = \rho_2^{\mathrm{T}_1}\rho_1^{\mathrm{T}_1}$ does *not* hold in general, traces of products do behave benignly inasmuch as

$$\mathrm{tr}\left\{\rho_1\rho_2^{\mathrm{T}}\right\} = \mathrm{tr}\left\{\rho_1^{\mathrm{T}}\rho_2\right\} = \mathrm{tr}\left\{\rho_1^{\mathrm{T}_1}\rho_2^{\mathrm{T}_2}\right\} = \mathrm{tr}\left\{\rho_1^{\mathrm{T}_2}\rho_2^{\mathrm{T}_1}\right\} \tag{2.21}$$

is true for all 2-qubit states ρ_1, ρ_2. In this context the positivity of ρ_1 and ρ_2 is not essential. With $s_1^{\downarrow}, \vec{t}_1, \overleftarrow{C}_1$ and $s_2^{\downarrow}, \vec{t}_2, \overleftarrow{C}_2$ parameterizing ρ_1 and ρ_2, respectively, we have, for example,

$$\mathrm{tr}\{\rho_1\rho_2\} = \frac{1}{4}\left(1 + \vec{s}_1 \cdot s_2^{\downarrow} + \vec{t}_1 \cdot t_2^{\downarrow} + \mathrm{Sp}\left\{\overleftarrow{C}_1^{\mathrm{T}} \cdot \overleftarrow{C}_2\right\}\right),$$

$$\mathrm{tr}\{\rho_1^{\mathrm{T}_1}\rho_2\} = \frac{1}{4}\left(1 - \vec{s}_1 \cdot s_2^{\downarrow} + \vec{t}_1 \cdot t_2^{\downarrow} - \mathrm{Sp}\left\{\overleftarrow{C}_1^{\mathrm{T}} \cdot \overleftarrow{C}_2\right\}\right),$$

$$\mathrm{tr}\{\rho_1^{\mathrm{T}}\rho_2\} = \frac{1}{4}\left(1 - \vec{s}_1 \cdot s_2^{\downarrow} - \vec{t}_1 \cdot t_2^{\downarrow} + \mathrm{Sp}\left\{\overleftarrow{C}_1^{\mathrm{T}} \cdot \overleftarrow{C}_2\right\}\right), \tag{2.22}$$

as explicit numerical statements about such traces.

Further we note these properties of transpositions, all of which are immediate consequences of (2.16)–(2.20): they are linear, trace-conserving mappings; the adjoint of a transpose is the transpose of the adjoint; partially transposing the first qubit and the second, in either order, amounts to a total transposition; two successive transpositions of the same kind compensate for each other; and, in a sequence of successive transpositions, the order in which they are executed does not matter. We shall make extensive use of these properties, particularly in Section 2.6.2.

Following Hill and Wootters [10, 11], we associate a nonnegative *concurrence* $\mathcal{C}$ with each 2-qubit state ρ. It is given by

$$\mathcal{C}(\rho) = \max\left\{0, 2\max_k\{h_k\} - \sum_k h_k\right\}, \tag{2.23}$$

where h_1, h_2, h_3, h_4 are the four nonnegative eigenvalues of

$$\left| \sqrt{\rho^{\mathsf{T}}} \sqrt{\rho} \right| = \sqrt{\sqrt{\rho}\, \rho^{\mathsf{T}} \sqrt{\rho}} \; . \tag{2.24}$$

Roughly speaking, the concurrence vanishes unless one of these eigenvalues is exceedingly large. The roles of ρ and ρ^{T} can be interchanged in this definition of $\mathcal{C}$; thus the concurrence of ρ^{T} is equal to the concurrence of ρ. It is of practical importance that $h_1^2, \ldots, h_4^2$ are the eigenvalues of the products $\rho^{\mathsf{T}}\rho$ and $\rho\rho^{\mathsf{T}}$ which, as a rule, are not hermitian themselves. Its convexity,

$$\mathcal{C}\big(x\rho_1 + (1-x)\rho_2\big) \leq x\mathcal{C}(\rho_1) + (1-x)\mathcal{C}(\rho_2) \quad \text{for} \quad 0 \leq x \leq 1 \,, \tag{2.25}$$

is a particularly important property of the Hill–Wootters concurrence.

2.3   Basic classification of states

Unitary transformations that affect only one of the qubits or both qubits independently are *local* transformations. Geometrically speaking, local transformations rotate $\vec{\sigma}$ and $\tau^{\downarrow}$,

$$\vec{\sigma} \rightarrow \vec{\sigma} \cdot \overleftrightarrow{O}_{ee} \,, \qquad \tau^{\downarrow} \rightarrow \overleftrightarrow{O}_{nn} \cdot \tau^{\downarrow} \,, \tag{2.26}$$

where $\overleftrightarrow{O}_{ee}$ is a unimodular, orthogonal ee-dyadic, and $\overleftrightarrow{O}_{nn}$ is one of nn-type. Equivalently, we can think of rotating the Pauli vectors and the cross dyadic,

$$s^{\downarrow} \rightarrow \overleftrightarrow{O}_{ee} \cdot s^{\downarrow} \,, \qquad \vec{t} \rightarrow \vec{t} \cdot \overleftrightarrow{O}_{nn} \,, \qquad \overleftrightarrow{C} \rightarrow \overleftrightarrow{O}_{ee} \cdot \overleftrightarrow{C} \cdot \overleftrightarrow{O}_{nn} \,. \tag{2.27}$$

Two states that can be turned into each other by such a local transformation are *locally equivalent*. Physically speaking, they are essentially the same states, distinguished only by the conventional choices for the two xyz coordinate systems.

To decide whether two given 2-qubit states belong to the same family of unitarily equivalent states, one may put them into a generic form that is uniquely fixed by convenient conventions. A set of conventions that we found quite natural is described in [25].

A central feature of this classification are the 123 bases in which the cross dyadic $\overleftrightarrow{C}$ is diagonal,

$$\overleftrightarrow{C} = \pm(e_1^{\downarrow}c_1\vec{n}_1 + e_2^{\downarrow}c_2\vec{n}_2 + e_3^{\downarrow}c_3\vec{n}_3) \quad \text{for} \quad \begin{cases} \det\{\overleftrightarrow{C}\} \geq 0, \\ \det\{\overleftrightarrow{C}\} < 0, \end{cases} \tag{2.28}$$

with its characteristic values ordered in accordance with

$$c_1 \geq c_2 \geq c_3 \geq 0. \tag{2.29}$$

Their squares are the common eigenvalues of $\overleftrightarrow{C}\cdot\overleftrightarrow{C}^{\mathrm{T}}$ and $\overleftrightarrow{C}^{\mathrm{T}}\cdot\overleftrightarrow{C}$; the eigencolumns of $\overleftrightarrow{C}\cdot\overleftrightarrow{C}^{\mathrm{T}}$ constitute the orthonormal right-handed set $e_1^{\downarrow}, e_2^{\downarrow}, e_3^{\downarrow}$, and the corresponding $\vec{n}_1, \vec{n}_2, \vec{n}_3$ are eigenrows of $\overleftrightarrow{C}^{\mathrm{T}}\cdot\overleftrightarrow{C}$. The unimodular orthogonal *en*-dyadic

$$\overleftrightarrow{O}_{en} = e_1^{\downarrow}\vec{n}_1 + e_2^{\downarrow}\vec{n}_2 + e_3^{\downarrow}\vec{n}_3 = (e_1^{\downarrow}, e_2^{\downarrow}, e_3^{\downarrow})\begin{pmatrix} \vec{n}_1 \\ \vec{n}_2 \\ \vec{n}_3 \end{pmatrix} \tag{2.30}$$

turns the two 123 bases into each other. It also appears in

$$\overleftrightarrow{C} = \pm\overleftrightarrow{O}_{en} \cdot \left|\overleftrightarrow{C}\right| = \pm\left|\overleftrightarrow{C}^{\mathrm{T}}\right| \cdot \overleftrightarrow{O}_{en}, \tag{2.31}$$

which relate the cross dyadic $\overleftrightarrow{C}$ to its modulus

$$\left|\overleftrightarrow{C}\right| = \sqrt{\overleftrightarrow{C}^{\mathrm{T}}\cdot\overleftrightarrow{C}} = n_1^{\downarrow}c_1\vec{n}_1 + n_2^{\downarrow}c_2\vec{n}_2 + n_3^{\downarrow}c_3\vec{n}_3 \tag{2.32}$$

(a *nn*-dyadic) and to the modulus of $\overleftrightarrow{C}^{\mathrm{T}}$ (a *ee*-dyadic).

Thus, of the fifteen parameters associated with the Pauli vectors and the cross dyadic in (2.10) or with the expectation values in Table 2.1, six are used up for the geometrical purpose of specifying the two 123 bases relative to the a priori *xyz* bases, so that nine parameters are left for the characterization of the essential physical properties of the state ρ in question. Three of them are the characteristic values of $\overleftrightarrow{C}$; the other six are the coefficients of $s^{\downarrow}$ and $\vec{t}$ in the 123 bases, which appear in the matrix (2.14). Since the 123 bases are not quite unique, one can exploit this freedom of choice to make as many as possible of these coefficients

Class	Characterized by	Count of parameters
A	$c_1 = c_2 = c_3 = 0$	2 (both nonnegative)
B	$c_1 = c_2 = c_3 > 0$	4 (3 nonnegative)
C	$c_1 > c_2 = c_3 = 0$	5 (4 nonnegative)
D	$c_1 > c_2 = c_3 > 0$	7 (5 nonnegative)
E	$c_1 = c_2 > c_3$	7 (5 nonnegative)
F	$c_1 > c_2 > c_3$	9 (5 nonnegative)

Table 2.2 The six classes of families of locally equivalent 2-qubit states. In classes A and C the $+$ sign of (2.28) applies; both signs can occur in classes B, D, E, and F which, therefore, consist of two subclasses each.

vanish and to give definite signs to as many as possible of the remaining ones. This procedure results in the classification of Table 2.2. Additional details are given in [25].

By definition, arbitrary local unitary transformations turn members of a family into other members of the same family, but it is possible that some local transformations have no effect at all, as exemplified by the local unitary operator $\exp(i\varphi\sigma_1 + i\phi\tau_1)$ acting on the class-A state $\rho = \frac{1}{4}(1 + s\sigma_1 + t\tau_1)$ with $0 \leq s, t \leq 1$. Therefore, some families are larger than others, and determining a family's size is a problem of considerable interest. Recent progress on this front is reported by Kuś and Życzkowski [26].

It is clear that the concurrence of a 2-qubit state is not altered by the local transformations (2.26) or (2.27). The concurrence is a family property.

2.4   Projectors and subspaces

2.4.1   Rank 1

An elementary but important example is the pure states ρ_{pure}, the projectors of rank 1. For ρ_{pure} of the general form (2.10), the purity condition $\rho_{\mathrm{pure}}(1 - \rho_{\mathrm{pure}}) = 0$ requires

$$s^{\downarrow} = \overleftrightarrow{C} \cdot t^{\downarrow} \,, \qquad \vec{t} = \vec{s} \cdot \overleftrightarrow{C} \,, \qquad \overleftrightarrow{C} = s^{\downarrow} \vec{t} - \overleftrightarrow{C}_{\mathrm{sub}} \,,$$

$$3 = \vec{s} \cdot s^{\downarrow} + \vec{t} \cdot t^{\downarrow} + \mathrm{Sp}\left\{ \overleftarrow{C}^{\mathsf{T}} \cdot \overrightarrow{C} \right\} , \tag{2.33}$$

which imply the generic form

$$\rho_{\text{pure}} = \tfrac{1}{4}\left(1 + p\sigma_1 - p\tau_1 - \sigma_1\tau_1 - q\sigma_2\tau_2 - q\sigma_3\tau_3\right)$$
$$\text{with} \quad 0 \le p \le 1, \; q = \sqrt{1 - p^2} \ge 0 . \tag{2.34}$$

Thus, any pure 2-qubit state is essentially characterized by the common length p of its Pauli vectors or by its concurrence q, available as $q^2 = \mathrm{tr}\left\{\rho_{\text{pure}}^{\mathsf{T}}\rho_{\text{pure}}\right\}$. For $p = 0$, $q = 1$ we have the family of Bell states (frequently called "maximally entangled states"),

$$\rho_{\text{Bell}} = \tfrac{1}{4}\left(1 - \sigma_1\tau_1 - \sigma_2\tau_2 - \sigma_3\tau_3\right) , \tag{2.35}$$

which is in class B; the $p = 1, q = 0$ family consists of the product states $\tfrac{1}{2}(1 + \sigma_1)\tfrac{1}{2}(1 - \tau_1)$ and is in class C; and the $0 < p < 1$ families belong to class D. These families are of different sizes [26]: three-dimensional, four-dimensional, and five-dimensional, respectively.

Local unitary transformations turn the generic Bell state (2.35) into any other Bell state. Accordingly, the general form of a Bell state is given by

$$\rho_{\text{Bell}} = \frac{1}{4}\left(1 - \vec{\sigma} \cdot \overrightarrow{O}_{en} \cdot \tau^{\downarrow}\right) , \tag{2.36}$$

where $\overleftarrow{O}_{en}$ is any unimodular, orthogonal *en*-dyadic.

Since $\rho_{\text{pure}}^{\mathsf{T}}$ is obtained from ρ_{pure} of (2.34) by $p \to -p$, the local transformation $(\sigma_1, \sigma_2, \sigma_3; \tau_1, \tau_2, \tau_3) \to (-\sigma_1, -\sigma_2, \sigma_3; -\tau_1, -\tau_2, \tau_3)$ turns them into each other. Therefore, ρ_{pure} and $\rho_{\text{pure}}^{\mathsf{T}}$ belong to the same family of pure states. The same unitary transformation relates the two partial transposes. Their spectral decompositions are given by

$$\left.\begin{matrix} \rho_{\text{pure}}^{\mathsf{T}_1} \\ \rho_{\text{pure}}^{\mathsf{T}_2} \end{matrix}\right\} = \frac{1 \pm p}{2}\rho_{\text{pure}}^{(1)} + \frac{1 \mp p}{2}\rho_{\text{pure}}^{(2)} + \frac{q}{2}\rho_{\text{pure}}^{(3)} - \frac{q}{2}\rho_{\text{pure}}^{(4)} , \tag{2.37}$$

where $\rho_{\text{pure}}^{(1)}$, $\rho_{\text{pure}}^{(2)}$ are product states and $\rho_{\text{pure}}^{(3)}$, $\rho_{\text{pure}}^{(4)}$ are Bell states,

$$\left.\begin{matrix} \rho_{\text{pure}}^{(1)} \\ \rho_{\text{pure}}^{(2)} \end{matrix}\right\} = \tfrac{1}{4}(1 \mp \sigma_1)(1 \mp \tau_1) , \qquad \left.\begin{matrix} \rho_{\text{pure}}^{(3)} \\ \rho_{\text{pure}}^{(4)} \end{matrix}\right\} = \tfrac{1}{4}(1 - \sigma_1\tau_1 \pm \sigma_2\tau_2 \pm \sigma_3\tau_3) . \tag{2.38}$$

Matters are particularly simple for Bell states for which we have

$$\rho_{\text{Bell}}^{\mathsf{T}} = \rho_{\text{Bell}} , \qquad \rho_{\text{Bell}}^{\mathsf{T}_1} = \rho_{\text{Bell}}^{\mathsf{T}_2} = \tfrac{1}{2} - \rho_{\text{Bell}} . \tag{2.39}$$

We learn here — what has been noted by Sanpera, Tarrach, and Vidal [27], for instance — that the partial transpose of a pure state with positive concurrence q has three positive and one negative eigenvalue and that, in particular, the eigenstate associated with the negative eigenvalue is a Bell state. More generally, the common eigenstates of $\rho_{\text{pure}}^{\mathsf{T}_1}$ and $\rho_{\text{pure}}^{\mathsf{T}_2}$ to the eigenvalue pair $\pm\frac{1}{2}q$ are Bell states. These are their one negative eigenvalue and the middle one of their three positive eigenvalues.

While we are at it, let us note that

$$\mathrm{tr}\,\{\rho_{\text{pure}}\rho_{\text{Bell}}\} \leq \tfrac{1}{2}(1+q) = \tfrac{1}{2} + \tfrac{1}{2}\big(\mathrm{tr}\,\{\rho_{\text{pure}}^{\mathsf{T}}\rho_{\text{pure}}\}\big)^{\frac{1}{2}} \qquad (2.40)$$

for any pure state (2.34) and all Bell states (2.36). If $q > 0$, the equal sign holds only for the Bell state $\rho_{\text{pure}}^{(4)}$ in (2.37). We exploit this observation about the concurrence of pure states in Section 2.6.2.

Pure states are often represented by the Hilbert state vectors to which they project, and then the Schmidt decomposition plays a role in the discussion. In our formalism, where all is said in terms of statistical operators and Hilbert space vectors do not appear, the Schmidt decomposition is of no particular interest. Therefore, we just offer a brief remark to establish contact with other approaches.

Generally speaking, given any two 2-qubit states ρ_1 and ρ_2, we can form superposition states in accordance with

$$\rho = \big(\sqrt{w_1\rho_1}\,U + \sqrt{w_2\rho_2}\,\big)\big(U^\dagger\sqrt{w_1\rho_1} + \sqrt{w_2\rho_2}\,\big)$$
$$= w_1\rho_1 + w_2\rho_2 + \sqrt{w_1 w_2}\big(\sqrt{\rho_1}\,U\sqrt{\rho_2} + \sqrt{\rho_2}\,U^\dagger\sqrt{\rho_1}\,\big)\,, \qquad (2.41)$$

where w_1, w_2 are nonnegative weights, U can be chosen unitary, and

$$w_1 + w_2 + 2\sqrt{w_1 w_2}\,\mathrm{Re}\Big(\mathrm{tr}\,\{\sqrt{\rho_1}\,U\,\sqrt{\rho_2}\,\}\Big) = 1 \qquad (2.42)$$

ensures the proper normalization of ρ. Now, the Schmidt decomposition of a pure state (2.34) amounts to the superposition specified by

$$\left.\begin{array}{c} w_1 \\ w_2 \end{array}\right\} = \tfrac{1}{2}(1 \pm p)\,, \quad \left.\begin{array}{c} \rho_1 \\ \rho_2 \end{array}\right\} = \tfrac{1}{4}(1 \pm \sigma_1)(1 \mp \tau_1)\,, \quad U = -\sigma_2\tau_2\,, \quad (2.43)$$

so that two orthogonal pure product states are superimposed.

For $p > 0$, this Schmidt decomposition is unique. (It is true that there is a choice between different Us but that does not matter because they are all equivalent; the uniqueness of ρ_1, ρ_2 and w_1, w_2 counts.) But for Bell states (which have $p = 0$) the direction $\vec{e}_1$ is arbitrary and, accordingly, there is a two-parametric set of Schmidt decompositions for each Bell state. This is of some importance in Section 14.2.2 of Chapter 14.

Abouraddy, Saleh, Sergienko, and Teich reported recently [28] that it is also possible to write a given pure state as the superposition of a Bell state ρ_1 and an orthogonal product state ρ_2 with respective weights $w_1 = q$ and $w_2 = 1 - q$. For ρ_{pure} of (2.34) one such decomposition is specified by

$$
\begin{aligned}
\rho_1 &= \tfrac{1}{4}\big[1 - \big(\sigma_1\tau_1 + \sigma_3\tau_3\big)\cos(2\vartheta) - \sigma_2\tau_2 - \big(\sigma_1\tau_3 - \sigma_3\tau_1\big)\sin(2\vartheta)\big] \,, \\
\rho_2 &= \tfrac{1}{4}\big(1 + \sigma_1\sin\vartheta + \sigma_3\cos\vartheta\big)\big(1 - \tau_1\sin\vartheta + \tau_3\cos\vartheta\big) \,, \\
U &= 1 - \rho_1 - \rho_2 + 2^{-\frac{3}{2}}\big[\big(\sigma_1 - \tau_1\big)\cos\vartheta - \big(\sigma_3 + \tau_3\big)\sin\vartheta \\
&\quad - \big(\sigma_1\tau_1 + \sigma_3\tau_3\big)\sin(2\vartheta) + \big(\sigma_1\tau_3 - \sigma_3\tau_1\big)\cos(2\vartheta)\big] \qquad (2.44)
\end{aligned}
$$

with $\cos\vartheta = \sqrt{2q/(1+q)}$, $\sin\vartheta = \sqrt{(1-q)/(1+q)}$. Admittedly, this looks somewhat incomprehensible; it is, however, easily understood as soon as the rank-2 machinery of the next section is at hand, and we shall return to these issues at the end of Section 2.4.2. Right now we just remark that the local transformations generated by $\sigma_1 + \tau_1$ have no effect on the pure state of (2.34) but they change ρ_1 and ρ_2 of (2.44). Thus there is a corresponding one-parametric set of decompositions of this kind, all with the same weights $w_1 = q$ and $w_2 = 1 - q$ but different Bell states ρ_1 and product states ρ_2.

2.4.2 Rank 2

By arguments similar to the ones that led us to (2.34) one finds that the generic form of a projector of rank 2 is given by

$$
\Sigma_0 = \tfrac{1}{2}(1 + u\sigma_3 + v\tau_3 + z_1\sigma_1\tau_1 + z_2\sigma_2\tau_2) \qquad (2.45)
$$

with the nonnegative parameters u, v, z_1, z_2 restricted by $z_1 \geq z_2$ as well as $uv = z_1 z_2$ and $u^2 + v^2 + z_1^2 + z_2^2 = 1$, and

$$
\begin{aligned}
u &= \cos\gamma_1\cos\gamma_2 \,, & v &= \sin\gamma_1\sin\gamma_2 \,, \\
z_1 &= \sin\gamma_1\cos\gamma_2 \,, & z_2 &= \cos\gamma_1\sin\gamma_2 \,, \\
& \text{with} \quad \tfrac{1}{2}\pi \geq \gamma_1 \geq \gamma_2 \geq 0 & & (2.46)
\end{aligned}
$$

is a convenient, unambiguous way of writing them. The projector property $\Sigma_0(1 - \Sigma_0) = 0$ is easily verified, and $\text{tr}\{\Sigma_0\} = 2$ is immediate.

Clearly, $\rho = \frac{1}{2}\Sigma_0$ has the properties of a 2-qubit state; its concurrence is z_2, which cannot exceed $\frac{1}{2}$, the value it acquires for $\gamma_1 = \gamma_2 = \pi/4$.

Since the subspace specified by Σ_0 is kinematically equivalent to the state space of a single qubit, the general state in this subspace must be of the 3-parametric form

$$\rho_{\mathrm{rk2}} = \frac{1}{2}\left(\Sigma_0 + x_1\Sigma_1 + x_2\Sigma_2 + x_3\Sigma_3\right) \quad \text{with} \quad x_1^2 + x_2^2 + x_3^2 \leq 1 , \quad (2.47)$$

where $\Sigma_{1,2,3}$ are analogs of Pauli's spin operators, traceless hermitian operators with the basic algebraic properties

$$\Sigma_0 \Sigma_k = \Sigma_k \qquad \text{for } k = 0, 1, 2, 3 ,$$

$$\Sigma_j \Sigma_k = \delta_{jk}\Sigma_0 + \mathrm{i}\sum_{l=1}^{3} \epsilon_{jkl}\Sigma_l \qquad \text{for } j, k = 1, 2, 3 . \qquad (2.48)$$

The choice

$$\Sigma_1 = \frac{1}{2}\left(\sin\gamma_1\,\sigma_1 + \cos\gamma_2\,\tau_1 + \sin\gamma_2\,\sigma_1\tau_3 + \cos\gamma_1\,\sigma_3\tau_1\right) ,$$

$$\Sigma_2 = \frac{1}{2}\left(\sin\gamma_2\,\sigma_2 + \cos\gamma_1\,\tau_2 + \sin\gamma_1\,\sigma_2\tau_3 + \cos\gamma_2\,\sigma_3\tau_2\right) ,$$

$$\Sigma_3 = \frac{1}{2}\left(v\sigma_3 + u\tau_3 - z_2\sigma_1\tau_1 - z_1\sigma_2\tau_2 + \sigma_3\tau_3\right) \qquad (2.49)$$

is particularly convenient. It is such that the limiting situations of $u = 1$, when $\rho_{\mathrm{rk2}} = \frac{1}{4}(1 + \sigma_3)(1 + x_1\tau_1 + x_2\tau_2 + x_3\tau_3)$, and $v = 1$, when $\rho_{\mathrm{rk2}} = \frac{1}{4}(1 + x_1\sigma_1 + x_2\sigma_2 + x_3\sigma_3)(1 + \tau_3)$, are parameterized most naturally.

As we see, in these limits of $\pi/2 = \gamma_1 = \gamma_2$ and $\gamma_1 = \gamma_2 = 0$, respectively, all states (2.47) are product states. By contrast, if $z_1 = z_2 > 0$, that is $\pi/2 > \gamma_1 = \gamma_2 > 0$, there is but a single product state, the pure state $\frac{1}{2}(\Sigma_0 + \Sigma_3)$. And for $z_1 > z_2 \geq 0$, there are only two product states, the pure states that obtain for $x_1 = \pm\sqrt{1 - (z_2/z_1)^2}$, $x_2 = 0$, $x_3 = z_2/z_1$. Similar observations about the product states in a rank-2 subspace have been made by Hill and Wootters [10], and by Sanpera, Tarrach, and Vidal [27].

The pure state $\frac{1}{2}(\Sigma_0 - \Sigma_3)$ is the one with the shortest Pauli vectors, its family has the smallest p parameter and the largest concurrence q, namely $q = z_1 + z_2 = \sin(\gamma_1 + \gamma_2)$. Thus, the subspace specified by Σ_0 of (2.45) contains Bell states only if $\gamma_1 + \gamma_2 = \pi/2$. For, $\pi/2 > \gamma_1 = \pi/2 - \gamma_2 \geq \pi/4$, there is only one Bell state, the state with $x_1 = x_2 = 0$, $x_3 = -1$. For $\pi/2 = \gamma_1$, $\gamma_2 = 0$, all states with $x_1 = 0$ and $x_2^2 + x_3^2 = 1$ are Bell states.

Let us now briefly return to the question of writing a given pure state as a superposition of two other particular pure states. We get the Schmidt decompositions of (2.43) by considering the case of $z_1 = 1$, $z_2 = u = v = 0$. Then $\rho_1 = \frac{1}{2}(\Sigma_0 + \Sigma_1)$ and $\rho_2 = \frac{1}{2}(\Sigma_0 - \Sigma_1)$ are orthogonal pure product states, and $w_1 = \frac{1}{2}(1 + p)$, $w_2 = \frac{1}{2}(1 - p)$ together with $U = 1 - \Sigma_0 + \Sigma_3$ in (2.41) give the pure superposition state $\rho = \frac{1}{2}(\Sigma_0 + p\Sigma_1 + q\Sigma_3)$ with concurrence q. Local unitary transformations turn it into all other pure states with this concurrence, and so the Schmidt decomposition of any given pure 2-qubit state is available. In particular, you arrive at the generic form (2.34) by performing the transformation $(\tau_1, \tau_2, \tau_3) \to (-\tau_1, \tau_2, -\tau_3)$.

Similarly we get the "Bell state and product state" decompositions of [28] by considering the case $z_1 = z_2 = u = v = \frac{1}{2}$. Then $\rho_1 = \frac{1}{2}(\Sigma_0 - \Sigma_3)$ is a Bell state and $\rho_2 = \frac{1}{2}(\Sigma_0 + \Sigma_3)$ is a product state orthogonal to it. With $w_1 = q$, $w_2 = 1 - q$, $U = 1 - \Sigma_0 + \Sigma_1$ in (2.41) this gives the pure superposition state $\rho = \frac{1}{2}\left[\Sigma_0 + 2\sqrt{q(1 - q)}\Sigma_1 + (1 - 2q)\Sigma_3\right]$, which has concurrence q. Local unitary transformations produce all other members of its family. In particular, you get to (2.44) by combining the transformations $(\sigma_1, \sigma_2, \sigma_3) \to (\sigma_1 \cos \vartheta - \sigma_3 \sin \vartheta, \sigma_2, \sigma_3 \cos \vartheta + \sigma_1 \sin \vartheta)$ and $(\tau_1, \tau_2, \tau_3) \to (-\tau_1 \cos \vartheta - \tau_3 \sin \vartheta, -\tau_2, \tau_3 \cos \vartheta - \tau_1 \sin \vartheta)$.

When determining the respective concurrences of the superpositions referred to in the last two paragraphs we made use of two simple special cases of the general formula given in (2.162) below. The examples needed here, namely $\mathcal{C} = (x_2^2 + x_3^2)^{\frac{1}{2}}$ for $z_1 = 1$, $z_2 = u = v = 0$ and $\mathcal{C} = \frac{1}{2}(1 - x_3)$ for $z_1 = z_2 = u = v = \frac{1}{2}$, are quite easily obtained without first finding the general answer of (2.162).

2.4.3   Rank 3

The 2-qubit state that corresponds to a projector of rank 3 is of the form $\rho = \frac{1}{3}(1 - \rho_{\text{pure}})$ with any pure state ρ_{pure}; this ρ has vanishing concurrence. All states that are orthogonal to the chosen ρ_{pure} are in the subspace thus defined. With ρ_{pure} in its generic form (2.34), all ρ's in this subspace are obtained by adding to $\rho = \frac{1}{3}(1 - \rho_{\text{pure}})$ linear combinations of the 8 hermitian operators

$$\sigma_1 + \tau_1 \, , \ \sigma_2\tau_2 - \sigma_3\tau_3 \, , \ \sigma_2\tau_3 + \sigma_3\tau_2 \, ,$$
$$p(\sigma_2 - \tau_2) + (1 + q)(\sigma_1\tau_2 + \sigma_2\tau_1) \, , \ (1 + q)(\sigma_2 + \tau_2) - p(\sigma_1\tau_2 - \sigma_2\tau_1) \, ,$$
$$p(\sigma_3 - \tau_3) + (1 + q)(\sigma_1\tau_3 + \sigma_3\tau_1) \, , \ (1 + q)(\sigma_3 + \tau_3) - p(\sigma_1\tau_3 - \sigma_3\tau_1) \, ,$$
$$2\sigma_1\tau_1 + p(\sigma_1 - \tau_1) - q(\sigma_2\tau_2 + \sigma_3\tau_3) \, ,$$

$$\tag{2.50}$$

which commute with ρ_{pure}, are traceless, and have traceless products among each other.

Unfortunately, the systematic study of the states of rank 3 has not been completed as yet and our understanding of them is still wanting. We just report that there is no lack of pure product states here. For $p = 1$, the products $\frac{1}{4}(1 + \vec{\sigma} \cdot e^{\downarrow})(1 + \tau_1)$ and $\frac{1}{4}(1 - \sigma_1)(1 + \vec{n} \cdot \tau^{\downarrow})$ are orthogonal to ρ_{pure} for whatever choice is made for the unit vectors $e^{\downarrow}$ and $\vec{n}$, respectively. For $p < 1$, all products $\frac{1}{4}(1 + \vec{\sigma} \cdot e^{\downarrow})(1 + \vec{n} \cdot \tau^{\downarrow})$ will do, provided that the coefficients of the unit vectors $e^{\downarrow}$ and $\vec{n}$ obey

$$(1 + pe_1)(1 - pn_1) = q^2 , \qquad \frac{(e_2, e_3)}{\sqrt{1 + pe_1}} = \frac{(n_2, n_3)}{\sqrt{1 - pn_1}} , \qquad (2.51)$$

so that either $e^{\downarrow}$ or $\vec{n}$ can be chosen freely.

In this context, we should not fail to mention the connection with Hardy's neat observation [29] of quantum correlations that defy the common sense of everyday logic. If the three pairs of unit vectors $(e_a^{\downarrow}, \vec{n}_a)$, $(-e_a^{\downarrow}, \vec{n}_b)$, $(e_b^{\downarrow}, -\vec{n}_a)$ specify three pure product states in the subspace orthogonal to ρ_{pure}, then the product state associated with the pair $(e_b^{\downarrow}, \vec{n}_b)$ cannot be in this subspace as well (unless ρ_{pure} has $p = 0$ or $p = 1$, which we thus exclude here). So, if the ρ_{pure} considered were the state of the qubit pair, then joint measurements on the two qubits would reveal (i) that the values $\vec{\sigma} \cdot e_a^{\downarrow} = 1$ and $\vec{n}_a \cdot \tau^{\downarrow} = 1$ never occur together, (ii) that the values $\vec{\sigma} \cdot e_a^{\downarrow} = -1$ and $\vec{n}_b \cdot \tau^{\downarrow} = 1$ never occur together, and (iii) that the values $\vec{\sigma} \cdot e_b^{\downarrow} = 1$ and $\vec{n}_a \cdot \tau^{\downarrow} = -1$ never occur together. Under these circumstances $\vec{\sigma} \cdot e_b^{\downarrow} = 1$ seems to imply first that (iii) $\vec{n}_a \cdot \tau^{\downarrow} = 1$, then (i) $\vec{\sigma} \cdot e_a^{\downarrow} = -1$, and finally (ii) $\vec{n}_b \cdot \tau^{\downarrow} = -1$. Accordingly, one is apparently forced to conclude that (iv) the values $\vec{\sigma} \cdot e_b^{\downarrow} = 1$ and $\vec{n}_b \cdot \tau^{\downarrow} = 1$ never occur together, but in fact sometimes they do! Somewhat amusingly, the ratio $g = \frac{1}{2}(\sqrt{5} - 1)$ of the golden section determines the concurrence q of the optimal choice of ρ_{pure}, namely $q = 2g^2$, when the probability of finding the "forbidden" values is largest, namely g^5, slightly in excess of 9%.

Concerning Bell states in the subspace orthogonal to ρ_{pure}, we note that, if $p > 0$ in (2.34), there is the one-parametric set whose members

$$\frac{1}{4}\left(1 + \sigma_1\tau_1 + (\sigma_2\tau_2 - \sigma_3\tau_3)\cos\theta + (\sigma_2\tau_3 + \sigma_3\tau_2)\sin\theta\right) \qquad (2.52)$$

are labeled by the value of θ, any (real) value being permitted. If $p = 0$, so that ρ_{pure} is a Bell state itself, then the Bell states orthogonal to ρ_{pure} are of the form

$$\frac{1}{4}\left[1 - \vec{\sigma} \cdot \left(2e^{\downarrow}\vec{n} - \overleftarrow{O}_{en}\right) \cdot \tau^{\downarrow}\right] , \qquad (2.53)$$

where $\overset{\rightharpoonup}{O}_{en}$ is the *en*-dyadic that (2.36) associates with ρ_{pure}, $e^{\downarrow}$ is an arbitrary unit *e*-column, and $\vec{n} = \vec{e} \cdot \overset{\rightharpoonup}{O}_{en}$ is the corresponding unit *n*-row.

2.4.4   Rank 4

Finally, the projector of the full rank 4 is, of course, the identity operator. The corresponding properly normalized statistical operator $\rho_{\text{chaos}} = \frac{1}{4}$ is the so-called "chaotic state" of the qubit pair. It forms a single-state family of its own in class A.

2.5   Positivity and separability

An arbitrary choice for the (real) coefficients in (2.11) and (2.12) or, equivalently, of the nine family-defining parameters plus the 123 coordinate systems specifies a hermitian ρ of unit trace, but its positivity must be ensured by imposing restrictions on the Pauli vectors $s^{\downarrow}$, $\vec{t}$, and the cross dyadic $\overset{\rightharpoonup}{C}$. It is expedient to switch the emphasis from ρ to the traceless operator K introduced by

$$K = 1 - 4\rho = -\vec{\sigma} \cdot s^{\downarrow} - \vec{t} \cdot \tau^{\downarrow} - \vec{\sigma} \cdot \overset{\rightharpoonup}{C} \cdot \tau^{\downarrow}, \qquad (2.54)$$

so that $\rho = \frac{1}{4}(1 - K) \geq 0$ requires

$$K \leq 1. \qquad (2.55)$$

Convex sums of two states correspond to weighted sums of their K's. Admixing ρ_{chaos} to a given ρ amounts to multiplying its K by a factor.

One could, of course, check the positivity criterion (2.55) by calculating the eigenvalues of K, possibly with the aid of the 4×4-matrices (2.14) and (2.15). But precise knowledge of the actual eigenvalues of K is not needed if we only want to verify (2.55).

Since K is traceless, its eigenvalues κ_j ($j = 1, 2, 3, 4$) have a vanishing sum and solve a quartic equation without a cubic term,

$$\kappa^4 - A_2\kappa^2 + A_1\kappa - A_0 = 0, \qquad (2.56)$$

where

$$A_2 = \tfrac{1}{2}\text{tr}\left\{K^2\right\}, \qquad A_1 = -\tfrac{1}{3}\text{tr}\left\{K^3\right\},$$

$$A_0 = \tfrac{1}{4}\mathrm{tr}\left\{K^4\right\} - \tfrac{1}{8}\left[\mathrm{tr}\left\{K^2\right\}\right]^2 . \tag{2.57}$$

These three numbers are invariant under arbitrary (local or not) unitary transformations; they are three independent *global* invariants of the given ρ.

Upon establishing

$$K^2 = \mathrm{Sp}\left\{\overleftrightarrow{C}^{\mathrm{T}}\cdot\overleftrightarrow{C}\right\} + \vec{s}\cdot s^{\downarrow} + \vec{t}\cdot t^{\downarrow}$$

$$+ 2\vec{\sigma}\cdot\overleftrightarrow{C}\cdot t^{\downarrow} + 2\vec{s}\cdot\overleftrightarrow{C}\cdot\tau^{\downarrow} + 2\vec{\sigma}\cdot\left(s^{\downarrow}\vec{t} - \overleftrightarrow{C}_{\mathrm{sub}}\right)\cdot\tau^{\downarrow} \tag{2.58}$$

with the aid of (2.4) and (2.6), it is easy to evaluate the traces in (2.57) and express A_2, A_1, A_0 in terms of $s^{\downarrow}$, $\vec{t}$, and $\overleftrightarrow{C}$. Explicitly they read

$$A_2 = 2\,\mathrm{Sp}\left\{\overleftrightarrow{C}^{\mathrm{T}}\cdot\overleftrightarrow{C}\right\} + 2\vec{s}\cdot s^{\downarrow} + 2\vec{t}\cdot t^{\downarrow},$$

$$A_1 = 8\vec{s}\cdot\overleftrightarrow{C}\cdot t^{\downarrow} - 8\det\left\{\overleftrightarrow{C}\right\},$$

$$A_0 = -\left(\tfrac{1}{2}A_2\right)^2 + 4\,\mathrm{Sp}\left\{\left(s^{\downarrow}\vec{t} - \overleftrightarrow{C}_{\mathrm{sub}}\right)^{\mathrm{T}}\cdot\left(s^{\downarrow}\vec{t} - \overleftrightarrow{C}_{\mathrm{sub}}\right)\right\}$$

$$+ 4\vec{s}\cdot\overleftrightarrow{C}\cdot\overleftrightarrow{C}^{\mathrm{T}}\cdot s^{\downarrow} + 4\vec{t}\cdot\overleftrightarrow{C}^{\mathrm{T}}\cdot\overleftrightarrow{C}\cdot t^{\downarrow}$$

$$= \left(\mathrm{Sp}\left\{\overleftrightarrow{C}^{\mathrm{T}}\cdot\overleftrightarrow{C}\right\}\right)^2 - 2\,\mathrm{Sp}\left\{\left(\overleftrightarrow{C}^{\mathrm{T}}\cdot\overleftrightarrow{C}\right)^2\right\}$$

$$- 2\left(\vec{s}\cdot s^{\downarrow} + \vec{t}\cdot t^{\downarrow}\right)\mathrm{Sp}\left\{\overleftrightarrow{C}^{\mathrm{T}}\cdot\overleftrightarrow{C}\right\} - \left(\vec{s}\cdot s^{\downarrow} - \vec{t}\cdot t^{\downarrow}\right)^2$$

$$+ 4\vec{s}\cdot\overleftrightarrow{C}\cdot\overleftrightarrow{C}^{\mathrm{T}}\cdot s^{\downarrow} + 4\vec{t}\cdot\overleftrightarrow{C}^{\mathrm{T}}\cdot\overleftrightarrow{C}\cdot t^{\downarrow} - 8\vec{s}\cdot\overleftrightarrow{C}_{\mathrm{sub}}\cdot t^{\downarrow}, \tag{2.59}$$

where (2.9) is used in obtaining the second version of A_0.

As we see, the traces of (2.57) involve nine different local polynomial invariants of $s^{\downarrow}$, $\vec{t}$, and $\overleftrightarrow{C}$, and it is clear that their values are determined by the nine family-specifying parameters of classes A, ..., F. Suggestive as it is, the converse is not true, as can be demonstrated by a counter example [25]. According to Makhlin [30], there are 18 polynomial invariants whose values uniquely characterize the family in question (actually, of 9 of them only the sign matters). In addition to the nine invariants in (2.59), which exhaust the polynomials of degree 4 or lower, there are nine invariants of higher degree in Makhlin's set that do not enter the three global invariants A_0, A_1, A_2.

So, if one just wants to check if two given states are locally equivalent, without identifying the family they belong to, one could evaluate

Makhlin's 18 polynomial invariants. We note that Makhlin's set is minimal in this respect; a general method for generating all polynomial invariants of any degree has been given by Grassl, Rötteler, and Beth [31].

All solutions of the quartic equation (2.56) are real by construction— it is, after all, the characteristic polynomial of a hermitian operator. Then, if all solutions are in the range $\kappa \leq 1$, this polynomial and its derivatives must be nonnegative for $\kappa \geq 1$. Consequently, the positivity requirement (2.55) implies

$$A_2 - A_1 + A_0 \leq 1 , \quad 2A_2 - A_1 \leq 4 , \quad A_2 \leq 6 . \tag{2.60}$$

It is reasonably obvious, and can be demonstrated in a rather simple manner [25], that the converse is also true. If these three inequalities are obeyed, the four real solutions of (2.56) are in the range $\kappa \leq 1$, so that $K \leq 1$ and $\rho \geq 0$. In other words, the restrictions on $s^{\downarrow}$, $\vec{t}$, and $\overleftrightarrow{C}$ alluded to at the beginning of this section are just the inequalities (2.60).

If $\overleftrightarrow{C} = s^{\downarrow}\vec{t}$, the state in question is of product form,

$$\rho = \frac{1}{2}\left(1 + \vec{\sigma} \cdot s^{\downarrow}\right) \frac{1}{2}\left(1 + \vec{t} \cdot \tau^{\downarrow}\right) , \tag{2.61}$$

so that results of measurements on the first qubit show no correlations whatsoever with measurement results concerning the second qubit. Under these circumstances the 2-qubit system is *not entangled*. Entangled qubit pairs, $\overleftrightarrow{C} \neq s^{\downarrow}\vec{t}$, may be in a mixed state blended from disentangled ingredients,

$$\rho = \sum_n w_n \frac{1}{2}\left(1 + \vec{\sigma} \cdot s_n^{\downarrow}\right) \frac{1}{2}\left(1 + \vec{t}_n \cdot \tau^{\downarrow}\right)$$

$$\text{with} \quad w_n > 0 , \ \sum_n w_n = 1 ; \tag{2.62}$$

then all correlations found in the measurement data can be understood classically. States of this kind are called *separable*.* The decomposition (2.62) of a separable state into the convex sum of product states is not unique; if one wishes one can always use pure product states for this

---

*Others prefer a different terminology and call the product states (2.61) "uncorrelated," the separable states (2.62) "classically correlated," and the nonseparable states "entangled."

purpose. More generally, as demonstrated by Wootters [11], one can always write a given ρ as a mixture of four or fewer pure states with the same concurrence as ρ.

By construction, the subspace of separable states is convex and compact. As Hill and Wootters have shown [10, 11], the concurrence of a separable state vanishes, and nonseparable states have truly positive concurrences.

Correlations of a genuine quantum character require a nonseparable state ρ. Peres [12] observed that the partial transposes of a separable ρ are nonnegative, that they are 2-qubit states themselves, and his conjecture of the converse, namely that ρ is separable if $\rho^{T_1} \geq 0$, was proven by M., P., and R. Horodecki [13]. This profound insight is the very important **Peres–Horodeccy criterion:**

> A 2-qubit state ρ is separable if its partial transpose ρ^{T_1} is nonnegative, and only then. $\qquad$ (2.63)

As a matter of record we remark that Peres actually considered partial transpositions of the form $\rho \to \sigma_y \rho^{T_1} \sigma_y$ rather than $\rho \to \rho^{T_1}$. Inasmuch as the sandwiching by the two σ_y's is a local unitary transformation, it is irrelevant for the separability criterion (2.63). Equivalently, one could single out the x or z components of $\vec{\sigma}$ or any other one, but $\rho \to \rho^{T_1}$, which treats all components on equal footing, has its obvious advantages. This has been noted by others as well; see [32], for instance.

To decide whether a given 2-qubit state is separable or not, we could calculate its Hill–Wootters concurrence or employ the Peres–Horodeccy criterion. The latter is easier to use in practice because the positivity of ρ^{T_1}, or

$$1 - 4\rho^{T_1} = K^{T_1} = \frac{1}{2}\left(\vec{\sigma} \cdot K\sigma^{\downarrow} - K\right) \leq 1 \,, \qquad (2.64)$$

can be checked analogously to the positivity of ρ. The quartic equation solved by the common eigenvalues of K^{T_1} and K^{T_2} is obtained from (2.56) by the replacements

$$A_1 \to A_1 + 16 \det\left\{\overset{\ulcorner}{\vec{C}}\right\} \,, \qquad A_0 \to A_0 + 16\vec{s} \cdot \overset{\rightharpoonup}{\vec{C}}_{\text{sub}} \cdot t^{\downarrow} \,. \qquad (2.65)$$

With the aid of the identity

$$\det\left\{\overset{\ulcorner}{\vec{A}} - \overset{\ulcorner}{\vec{B}}\right\} = \det\left\{\overset{\ulcorner}{\vec{A}}\right\} - \det\left\{\overset{\ulcorner}{\vec{B}}\right\}$$

$$+ \operatorname{Sp}\left\{\overset{\ulcorner}{\vec{A}}{}^{\mathrm{T}} \cdot \overset{\rightharpoonup}{\vec{B}}_{\text{sub}}\right\} - \operatorname{Sp}\left\{\overset{\ulcorner}{\vec{B}}{}^{\mathrm{T}} \cdot \overset{\rightharpoonup}{\vec{A}}_{\text{sub}}\right\} \,, \qquad (2.66)$$

an implication of (2.6), here used for $\overrightarrow{\ulcorner A} = \overrightarrow{\ulcorner C}$ and $\overrightarrow{\ulcorner B} = s^{\downarrow} \overrightarrow{t}$, we thus find that

$$A_2 - A_1 + A_0 \leq 1 + 16 \det \left\{ \overrightarrow{\ulcorner C} - s^{\downarrow} \overrightarrow{t} \right\}, \quad 2A_2 - A_1 \leq 4 + 16 \det \left\{ \overrightarrow{\ulcorner C} \right\} \tag{2.67}$$

are equivalent to (2.64); the third inequality, $A_2 \leq 6$, is always obeyed by a positive ρ. So, a nonseparable state must violate either the first or the second inequality in (2.67), or both. The equal sign holds in the first inequality, if the partial transposes of the given ρ have a zero eigenvalue; both are equalities, if they have two zero eigenvalues. Accordingly, the partial transposes of a nonseparable ρ can at most have one zero eigenvalue and thus must be of rank 3 or 4. While we are at it, let us also mention that ρ^{T_1}, ρ^{T_2} of any state ρ can have at most a single negative eigenvalue; see Reference [27] and below at (2.76).

Thus the separability of a given ρ is checked as easily as its positivity. Neither test requires actual knowledge of the solutions of (2.56) or the quartic equation resulting from the replacements (2.65). They could, of course, be stated analytically but these explicit expressions are not very transparent unless special relations exist among the coefficients of the quartic equations.

As a simple application of the Peres–Horodeccy criterion, in the form of the inequalities (2.67), we note that a state ρ with $\det \left\{ \overrightarrow{\ulcorner C} \right\} \geq 0$ and $\det \left\{ \overrightarrow{\ulcorner C} - s^{\downarrow} \overrightarrow{t} \right\} \geq 0$ is surely separable. Therefore, for example, all states in classes A and C are separable.

In passing, we note that a violation of Bell's inequality [33], or rather of the more appropriate Clauser–Horne inequality [34], occurs whenever the characteristic values of the cross dyadic are such that $c_1^2 + c_2^2 > 1$, as observed by R., M., and P. Horodecki [35], for example. A violation is indicative of a nonseparable state. It is, however, not a reliable criterion because there are nonseparable states that obey the Clauser–Horne inequality; an example is given at the end of Section 2.7.1.

2.6 Lewenstein–Sanpera decompositions

As an immediate consequence of (2.37), a pure state (2.34) is not entangled and therefore separable if $p = 1, q = 0$, and it is entangled and not separable if $p < 1, q > 0$. Somewhat more generally, states of the form

$$(1 - x)\rho_{\text{chaos}} + x\rho_{\text{pure}} \quad \text{with} \quad -\tfrac{1}{3} \leq x \leq 1 \qquad (2.68)$$

are separable if $(1 + 2q)x \leq 1$.

Turning to the rank-2 states (2.47), we note that all of them are separable if $z_1 = 0$, and if $z_1 > 0$ then the states with $x_1^2 \leq 1 - (z_2/z_1)^2$, $x_2 = 0$, $x_3 = z_2/z_1$ are surely separable because each of them is either a pure product state or the convex sum of two pure product states. We draw two conclusions from this observation: (1) If a separable state is of rank 1 or 2, its partial transposes are of the same rank. And therefore the partial transposes of a separable state of rank 3 or 4 must also be of rank 3 or 4. (2) Any rank-2 state is either separable or the convex sum of a separable state and a pure state, because each point inside the unit sphere defined by the inequality in (2.47) lies on a straight line connecting $(x_1, x_2, x_3) = (0, 0, z_2/z_1)$ with a point on the surface, and since these two endpoints specify a separable state and a pure state, respectively, all points on the line represent states that are convex sums of them.

Now consider a general 2-qubit state. Its spectral decomposition

$$\rho = \sum_{k=1}^{4} r_k \rho_{\text{pure}}^{(k)} \quad \text{with} \quad r_1 \geq r_2 \geq r_3 \geq r_4 \geq 0 \quad \text{and} \quad \sum_{k=1}^{4} r_k = 1 \qquad (2.69)$$

involves four pure states that are mutually orthogonal. Having dealt with $r_2 = 0$ (rank 1) and $r_2 > r_3 = 0$ (rank 2) above, we take $r_3 > 0$ for granted. Then upon writing

$$\rho = (r_1 + r_2 - 2r_3)\left(\frac{r_1 - r_3}{r_1 + r_2 - 2r_3}\rho_{\text{pure}}^{(1)} + \frac{r_2 - r_3}{r_1 + r_2 - 2r_3}\rho_{\text{pure}}^{(2)}\right)$$

$$+ (3r_3 + r_4)\left(\frac{4r_3}{3r_3 + r_4}\rho_{\text{chaos}} + \frac{r_4 - r_3}{3r_3 + r_4}\rho_{\text{pure}}^{(4)}\right) \qquad (2.70)$$

we recognize that ρ is the convex sum of a rank-2 state* and a separable state of the kind (2.68) with $x = -(r_3 - r_4)/(3r_3 + r_4) \leq 0$. In view of what we just observed about these states, we conclude that any state

---

*Strictly speaking, it is a rank-2 state only if $r_2 > r_3$, but what follows applies equally well to the degenerate situations of $r_1 > r_2 = r_3$ (rank 1) or $r_1 = r_3$ (vanishing weight).

ρ can be written as a convex sum of a separable state ρ_{sep} and a pure state ρ_{pure},

$$\rho = \lambda \rho_{\text{sep}} + (1 - \lambda)\rho_{\text{pure}} \quad \text{with} \quad 0 \leq \lambda \leq 1 \,. \tag{2.71}$$

We owe this important insight to Lewenstein and Sanpera [14]; it is actually a special case of a more general theorem about bipartite systems in which each component could be kinematically larger than a qubit.

Rare exceptions aside, the *Lewenstein–Sanpera decomposition* (LSD) of a given (nonseparable) ρ is not unique. There is usually a continuum of LSDs to choose from. Among them is the *optimal LSD*, the one with the largest value of λ,

$$\rho = \mathcal{S} \varrho_{\text{s}} + (1 - \mathcal{S})\varrho \quad \text{with} \quad \mathcal{S} = \max\{\lambda\} \,. \tag{2.72}$$

We denote the separable state and the pure state of the optimal LSD by ϱ_{s} and ϱ, respectively, and call $\mathcal{S}$, the maximal λ value, the *degree of separability* of ρ. Roughly speaking, a state ρ is the more useful for quantum communication purposes, the smaller its degree of separability.* Just like its concurrence, the degree of separability is not an individual property of a 2-qubit state. It is a family property; locally equivalent states have the same value of $\mathcal{S}$.

Here, then, is the challenge:

> Find an analytical method that determines
> the optimal LSD of any given 2-qubit state ρ. (2.73)

By "analytical" we mean that the answer is surely known after a finite number of steps. There is, of course, the option of using iterative methods which produce a series of LSDs that converge toward the optimal LSD. Such methods are not regarded as analytical, unless they assuredly give the answer after a finite number of iterations. Incidentally, the first ρ considered by Lewenstein and Sanpera as an illustrating example [14], one for which the optimal LSD is now known analytically [see (2.130) below], was then decomposed by a numerical iteration of the nonanalytical kind.

---

*In technical terms that we do not wish to elaborate upon, $\mathcal{S}$ is a separability monotone (this follows from a more general observation by Eisert and Briegel [36]) and, as noted by Wellens and Kuś [37], the product $(1 - \mathcal{S})q$, where q is the concurrence of ϱ, possesses the three most crucial ones of the properties that Vedral, Plenio, Rippin, and Knight [38, 39] require from a good entanglement measure.

The complete solution of this problem is not known as yet. But there has been considerable progress, and we know optimal LSDs for a variety of 2-qubit states. Before turning to these matters in detail, let us offer a few general remarks.

The mapping $\rho \rightarrow S$ is concave,

$$S(x\rho_1 + (1-x)\rho_2) \geq xS(\rho_1) + (1-x)S(\rho_2) \quad \text{for} \quad 0 \leq x \leq 1 . \quad (2.74)$$

Unless $x = 0$ or $x = 1$, the equal sign only holds if the same pure state shows up in the optimal LSDs of both ρ_1 and ρ_2. Otherwise the convex sum of the two ϱ's is a rank-2 state that has LSDs of its own and a nonzero degree of separability.

Since the concurrence of a separable state vanishes, applying the convexity property (2.25) to (2.72) implies [37]

$$C \leq (1-S)q \leq 1-S , \quad (2.75)$$

where C and q are the concurrences of ρ and ϱ, respectively. The concurrence of a 2-qubit state thus sets an upper bound on its degree of separability.

The spectral decomposition of the partial transpose of a pure state (2.34) is of the generic form (2.37). Therefore, the partial transpose of any 2-qubit state ρ can be written as

$$\rho^{T_1} = (1+x)\rho' - x\rho_{\text{Bell}}, \quad 0 \leq x \leq \tfrac{1}{2}(1-S) \quad (2.76)$$

with some state ρ' and a Bell state ρ_{Bell}. As a consequence, ρ^{T_1} and ρ^{T_2} can have at most one negative eigenvalue.

Since ρ' is a mixture of four or fewer pure states, (2.76) shows that the partial transpose of a nonseparable state is a pseudo-mixture of up to five pure states with one negative weight only, carried by a Bell state. There is a very similar observation by Sanpera, Tarrach, and Vidal [27] about ρ itself. It can always be presented as a pseudo-mixture of four or five separable pure states; as an immediate consequence its partial transpose is also such a pseudo-mixture.

We end these general remarks about LSDs with a digression. Lewenstein and Sanpera themselves call the separable remainder ϱ_s the "best separable approximation" to ρ. As a rule, however, the separable state nearest to ρ, in any of the popular distance measures, is different from ϱ_s and, therefore, we decided to not adopt this terminology. One could, of course, be interested in such "nearest separable states," and a recent paper by Verstraete, Dehaene, and De Moor deals with some aspects of this issue [40]. What is "near" depends, however, on the distance measure chosen. If one opts for the Hilbert–Schmidt distance, as these authors do, then there is a unique separable state that is nearest to the given nonseparable state. As an extreme example, consider the pure state ρ_{pure} of (2.34). Its nearest separable state is

$$\rho_{\text{HS}-\text{nearest}} = \tfrac{1}{4}\big[1 + s(\sigma_1 - \tau_1) - c\sigma_1\tau_1 - \tfrac{1}{2}(1-c)(\sigma_2\tau_2 + \sigma_3\tau_3)\big] \quad (2.77)$$

with

$$s = p \quad \text{and} \quad c = 1 - \tfrac{2}{3}q \quad \text{if} \quad q \geq \tfrac{2}{3} \qquad (2.78)$$

or

$$s \text{ solves } 3s^3 = 2(1-q)s + p \quad \text{and} \quad c = s^2 \quad \text{if} \quad q \leq \tfrac{2}{3}. \qquad (2.79)$$

The resulting Hilbert–Schmidt distance between ρ_{pure} and $\rho_{\text{HS}-\text{nearest}}$ is given by

$$\sqrt{\text{tr}\big\{|\rho_{\text{pure}} - \rho_{\text{HS}-\text{nearest}}|^2\big\}} = \frac{1}{2}\sqrt{(s-p)^2 + \tfrac{3}{2}\big(c - 1 + \tfrac{2}{3}q\big)^2 + \tfrac{4}{3}q^2} \qquad (2.80)$$

which thus equals $3^{-\frac{1}{2}}q$ for $q \geq \tfrac{2}{3}$ and $\big(\tfrac{5}{14}\big)^{\frac{1}{2}}q$ for $q \ll 1$. But in fact, the Hilbert–Schmidt distance is not the most natural distance measure for statistical operators; one should rather use the trace-class distance. Then a unique nearest separable state is an exception — as a rule one will have quite a few equally distant separable states that are all nearest to the given nonseparable state.

2.6.1 Basic properties of optimal LSDs

The optimal LSD (2.72) has a number of properties that help in decomposing given states in the optimal way. We consider some particularly important ones.

Existence: The degree of separability S is really the maximum of all possible λ values in (2.71), not just their supremum, because the subset of separable states is compact. Therefore, a LSD with $\lambda = S$ does exist.

Uniqueness: If we have two different LSDs with the same nonzero value of λ, their symmetric convex sum also equals the given ρ. It contains the convex sum of the two different ρ_{sep}'s, which is separable, and the convex sum of the two ρ_{pure}'s, which is a rank-2 state that has LSDs of its own. Either one of them contains a separable part, so that we get a new LSD of the ρ in question with a larger λ value. Consequently, the common λ of the original two LSDs is not maximal, and it follows that the optimal LSD is unique. As a formal statement we thus note that

$$\varrho_s + (1/\mathcal{S} - 1)(\varrho - \rho_{\text{pure}})$$
is either nonpositive or nonseparable
for each $\rho_{\text{pure}} \neq \varrho$. $\qquad\qquad$ (2.81)

This does not imply that one can always find another LSD with the same λ value if $\lambda < \mathcal{S}$. There are states with a continuum of LSDs in which each value of λ occurs only once. Examples are the rank-2 states of (2.47) that obey inequality (2.158) below.

As an immediate consequence of the uniqueness, we note that the optimal LSD of ρ^{T} is obtained by transposing the one of ρ. Therefore, ρ and ρ^{T} have the same degree of separability.

ϱ_s **is barely separable:** Consider the optimal LSD of some nonseparable ρ and a parameter ε in the range $0 < \varepsilon \leq 1 - \mathcal{S}$. In

$$\rho = (\mathcal{S} + \varepsilon) \left[\frac{\mathcal{S}}{\mathcal{S} + \varepsilon} \varrho_s + \frac{\varepsilon}{\mathcal{S} + \varepsilon} \varrho \right] + (1 - \mathcal{S} - \varepsilon)\varrho \qquad (2.82)$$

the convex sum in square brackets is surely nonnegative, but cannot be separable. Because, if it were, we would have found a LSD with $\lambda > \mathcal{S}$. Therefore,

the state $\rho_\epsilon = (1 + \epsilon)^{-1}(\varrho_s + \epsilon\varrho)$
is nonseparable for $\epsilon > 0$. $\qquad\qquad$ (2.83)

Thus, $\rho_\epsilon^{\mathsf{T}_1}$ has a negative eigenvalue for $\epsilon > 0$, but none for $\epsilon = 0$. Since the eigenvalues are continuous functions of ϵ, it follows that $\varrho_s^{\mathsf{T}_1}$ and $\varrho_s^{\mathsf{T}_2}$ must have at least one zero eigenvalue. Formally,

$$\varrho_s^{\mathsf{T}_1}, \varrho_s^{\mathsf{T}_2} \geq 0 \quad \text{but not} \quad \varrho_s^{\mathsf{T}_1}, \varrho_s^{\mathsf{T}_2} > 0; \qquad (2.84)$$

for ϱ_s, the equal sign holds in the first inequality of (2.67). A useful terminology calls ϱ_s *barely separable* with respect to ϱ.

When searching for the optimal LSD of a given ρ it is, therefore, sufficient to consider LSDs with ρ_{sep}'s that are barely separable with respect to the ρ_{pure} with which they are paired in (2.71). If the ρ_{sep} of some LSD does not have this property, one adds the appropriate amount of the respective ρ_{pure} to it (in the sense of a convex sum, of course) and gets a barely separable ρ_{sep}.

Pairing property: Since the infinitesimal neighborhoods of ϱ_s and ϱ are critical in (9.3) and (9.4), the actual value of $\mathcal{S}$ is irrelevant and, as a consequence, we note the *pairing property*:

> If $\rho_\lambda = \lambda\rho_{\text{sep}} + (1-\lambda)\rho_{\text{pure}}$ is the optimal LSD
> for one value of λ in the range $0 < \lambda < 1$,
> then it is optimal also for all other λ values.   (2.85)

It is also implied by the concavity (2.74). Obviously, a systematic method for identifying all ρ_{sep}'s that pair with a given ρ_{pure}, or vice versa, would be quite helpful, but we are not aware of one.

Local invariance is passed on: Suppose that the state ρ considered is invariant under some local unitary transformation, $U_{\text{loc}}^\dagger \rho U_{\text{loc}} = \rho$. Then its ϱ_s and ϱ must be invariant under this local transformation as well. Otherwise we could apply it to the optimal LSD and get another LSD with the same λ value, in conflict with the uniqueness of the optimal LSD. This argument builds on the elementary observation that local transformations do not affect the purity and separability of a state.

The limitations resulting from this "inheritance of local invariance" can facilitate the search for the optimal LSD substantially. See Sections 2.7.1 and 2.7.2 for applications.

Swapping invariance is passed on: Similarly one finds that the ϱ_s and ϱ of a ρ that is invariant under a swapping transformation, such as

$$\sigma_k \leftrightarrow \tau_k \quad \text{for} \quad k = 1, 2, 3 \tag{2.86}$$

or, more generally,

$$\sigma^{\downarrow} \rightarrow \overrightarrow{O}_{en} \cdot \tau^{\downarrow}, \qquad \vec{\tau} \rightarrow \vec{\sigma} \cdot \overrightarrow{O}_{en} \tag{2.87}$$

with some unimodular orthogonal *en*-dyadic $\overrightarrow{O}_{en}$, must be invariant themselves because swapping does not affect the separability or the purity of a state. Clearly, this swapping invariance is only possible if the Pauli vectors $s^{\downarrow}$ and $\vec{t}$ are of equal length.

Orthogonality is passed on: If the state ρ in question is orthogonal to a certain other state $\rho_\perp$,

$$\text{tr}\{\rho\rho_\perp\} = 0, \tag{2.88}$$

then the ρ_{sep}'s and ρ_{pure}'s of all LSDs of ρ are also orthogonal to $\rho_\perp$ because both traces in

$$0 = \lambda\text{tr}\{\rho_{\text{sep}}\rho_\perp\} + (1-\lambda)\text{tr}\{\rho_{\text{pure}}\rho_\perp\} \tag{2.89}$$

must be nonnegative, so both must vanish. In particular, the ϱ_s and ϱ of ρ must have this orthogonality property. Thus, the optimal LSD of a rank-2 state (2.47), for example, must involve ϱ_s and ϱ from the same rank-2 subspace.

No locally optimal LSDs: Suppose that $\rho = \lambda \rho_{\text{sep}} + (1 - \lambda)\rho_{\text{pure}}$ is a nonoptimal LSD of ρ. Then, if ρ_{sep} is *not* barely separable with respect to ρ_{pure}, there is a range of positive ϵ values such that $\lambda \rightarrow (1 + \epsilon)\lambda$ and $\rho_{\text{sep}} \rightarrow (1 + \epsilon)^{-1}(\rho_{\text{sep}} + \epsilon \rho_{\text{pure}})$ specify other LSDs with a continuum of larger λ values. If, however, ρ_{sep} *is* barely separable with respect to ρ_{pure}, then $\rho_{\text{pure}} \neq \varrho$ and the convex sums of the optimal LSD and the nonoptimal LSD give further, different LSDs upon decomposing the rank-2 state formed by the convex sum of ϱ and ρ_{pure}. In all of these new LSDs, the separable parts carry weights that are larger than λ and less than $\mathcal{S}$. Therefore, the given nonoptimal LSD is not only globally nonoptimal, it is also locally nonoptimal. There must be neighboring LSDs with larger λ values.

2.6.2 Optimal LSDs of truly positive states

Since any state ρ can be regarded as the $0 < x \rightarrow 0$ limit of the rank-4 states $x\rho_{\text{chaos}} + (1 - x)\rho$, the generic situation is that of $\rho > 0$, rather than $\rho \geq 0$, as it is the case for the rank-2 states in Section 2.4.2 or the rank-3 states in Section 2.4.3. Accordingly, we shall assume throughout this section that $\rho > 0$. The following results are excerpts from recent work by Karnas and Lewenstein [41] and, in particular, by Wellens and Kuś [37].

So, we consider a nonseparable ρ of rank 4. The separable state ϱ_s of its optimal LSD is then either of rank 3 or rank 4, and its partial transposes $\varrho_s^{T_1}$, $\varrho_s^{T_2}$ are of rank 3, as follows from (2.84) and the observation after (2.68). The unique pure states ρ_1 and ρ_2 that are associated with the nondegenerate null eigenvalue of $\varrho_s^{T_1}$ and $\varrho_s^{T_2}$, respectively,

$$\varrho_s^{T_1}\rho_1 = \rho_1\varrho_s^{T_1} = 0 \,, \qquad \varrho_s^{T_2}\rho_2 = \rho_2\varrho_s^{T_2} = 0 \,, \tag{2.90}$$

are related to each other by transposition,

$$\rho_1^{\mathsf{T}} = \rho_2 \,, \qquad \rho_2^{\mathsf{T}} = \rho_1 \,, \qquad \rho_1^{T_1} = \rho_2^{T_2} \,. \tag{2.91}$$

Now, for $\epsilon > 0$, the "barely separable" condition (9.4) requires the existence of a pure state $\rho_{\text{pure}}^{(\epsilon)}$ such that

$$\text{tr}\left\{(\varrho_s + \epsilon\varrho)^{T_1} \rho_{\text{pure}}^{(\epsilon)}\right\} < 0 \,, \tag{2.92}$$

and since the eigenvalues and eigenstates of $(\varrho_s + \epsilon\varrho)^{\mathsf{T}_1}$ depend continuously on ϵ, we have $\rho_{\text{pure}}^{(\epsilon)} \to \rho_1$ as $\epsilon \to 0$ and get

$$\text{tr}\left\{\varrho^{\mathsf{T}_1}\rho_1\right\} \leq 0 \quad \text{or} \quad \text{tr}\left\{\varrho\rho_1^{\mathsf{T}_1}\right\} = \text{tr}\left\{\varrho\rho_2^{\mathsf{T}_2}\right\} \leq 0 \qquad (2.93)$$

in this limit.

Next, we exploit the uniqueness condition (9.3) where we put

$$\Delta_\epsilon\rho = \varrho - \rho_{\text{pure}} = \varrho - e^{-i\epsilon G}\varrho\, e^{i\epsilon G} = i\epsilon[G, \varrho] + O(\epsilon^2) \qquad (2.94)$$

with a hermitian generator G that does not commute with ϱ. Then, for $\epsilon \neq 0$, there exists either a pure state $\rho_{\text{pure}}^{(\epsilon)}$ such that

$$\text{tr}\left\{\left[\varrho_s + (1/\mathcal{S} - 1)\Delta_\epsilon\rho\right]\rho_{\text{pure}}^{(\epsilon)}\right\} < 0 \qquad (2.95)$$

or a pure state $\overline{\rho}_{\text{pure}}^{(\epsilon)}$ such that

$$\text{tr}\left\{\left[\varrho_s + (1/\mathcal{S} - 1)\Delta_\epsilon\rho\right]^{\mathsf{T}_1}\overline{\rho}_{\text{pure}}^{(\epsilon)}\right\} < 0\,. \qquad (2.96)$$

In the limit $\epsilon \to 0$, the positivity violation (2.95) can only occur if ϱ_s is of rank 3.

2.6.2.1 Separable part has full rank

Let us, therefore, first deal with the situation in which ϱ_s has full rank, $\varrho_s > 0$, so that the separability violation (2.96) must be the case. Here, too, the eigenvalues and eigenstates of $[\cdots]^{\mathsf{T}_1}$ depend continuously on ϵ, and $\overline{\rho}_{\text{pure}}^{(\epsilon)} \to \rho_1$ obtains as $\epsilon \to 0$. But, in marked contrast to the reasoning that took us from (2.92) to (2.93), ϵ is not restricted to positive values here (alternatively, if G is a permissible generator, so is $-G$), and so we get

$$\text{tr}\left\{i[G, \varrho]^{\mathsf{T}_1}\rho_1\right\} = 0 \quad \text{or} \quad \text{tr}\left\{[\rho_1^{\mathsf{T}_1}, \varrho]G\right\} = 0\,. \qquad (2.97)$$

This must hold for all G's that do not commute with ϱ, and since it is always true for those that do, it must in fact hold for all hermitian G's. Therefore, ϱ commutes with $\rho_1^{\mathsf{T}_1}$, it is an eigenstate of $\rho_1^{\mathsf{T}_1} = \rho_2^{\mathsf{T}_2}$,

$$\rho_1^{\mathsf{T}_1}\varrho = \varrho\rho_1^{\mathsf{T}_1} = -\frac{1}{2}\overline{q}\varrho \quad \text{with} \quad \overline{q} = \sqrt{\text{tr}\left\{\rho_1^{\mathsf{T}}\rho_1\right\}} = \sqrt{\text{tr}\left\{\rho_1\rho_2\right\}}\,. \qquad (2.98)$$

That the eigenvalue is related in this way to the concurrence $\bar{q}$ of the family of pure states to which ρ_1 and ρ_2 belong follows from (2.37) in conjunction with (2.93).

Taking the trace in (2.90), $\mathrm{tr}\{\varrho_s^{T_1}\rho_1\} = \mathrm{tr}\{\varrho_s\rho_1^{T_1}\} = 0$, tells us that $\rho_1^{T_1} \not\geq 0$, since $\varrho_s > 0$. Accordingly, $\bar{q} > 0$ and ϱ is the Bell state $\rho_{\mathrm{pure}}^{(4)}$ that (2.37) and (2.38) associate with ρ_1, and

$$\mathrm{tr}\{\rho_1\varrho\} = \tfrac{1}{2}(1+\bar{q}) , \qquad \varrho^{T_1} = \varrho^{T_2} = \tfrac{1}{2} - \varrho \qquad (2.99)$$

follow immediately. Then, partially transposing the optimal LSD of ρ establishes

$$\rho^{T_1} = \mathcal{S}\varrho_s^{T_1} - (1-\mathcal{S})\varrho + \tfrac{1}{2}(1-\mathcal{S}) , \qquad (2.100)$$

so that

$$\mathrm{tr}\{\rho^{T_1}\rho_{\mathrm{pure}}\} \geq \tfrac{1}{2}(1-\mathcal{S}) - (1-\mathcal{S})\mathrm{tr}\{\varrho\rho_{\mathrm{pure}}\} \qquad (2.101)$$

for all pure states ρ_{pure}, with the equal sign holding only for $\rho_{\mathrm{pure}} = \rho_1$, and since ϱ is a Bell state, (2.40) implies

$$\mathrm{tr}\{\rho^{T_1}\rho_{\mathrm{pure}}\} \geq -\frac{1}{2}(1-\mathcal{S})\sqrt{\mathrm{tr}\{\rho_{\mathrm{pure}}^T\rho_{\mathrm{pure}}\}}$$

$$\text{with "=" only for } \rho_{\mathrm{pure}} = \rho_1 = \rho_2^T . \qquad (2.102)$$

Accordingly, for $\rho_{\mathrm{pure}} = \mathrm{e}^{\mathrm{i}\epsilon G}\rho_1\,\mathrm{e}^{-\mathrm{i}\epsilon G}$ with any hermitian generator G, the first-order terms must take care of each other, so that

$$\mathrm{tr}\{\rho^{T_1}\mathrm{i}[G,\rho_1]\} = -\frac{1-\mathcal{S}}{2\bar{q}}\mathrm{tr}\{\rho_2\mathrm{i}[G,\rho_1]\} \qquad (2.103)$$

for all G, and therefore ρ_1 is an eigenstate of $\rho^{T_1} + (2\bar{q})^{-1}(1-\mathcal{S})\rho_2$,

$$\left(\rho^{T_1} + \frac{1-\mathcal{S}}{2\bar{q}}\rho_2\right)\rho_1 = 0 . \qquad (2.104)$$

That the eigenvalue is 0 follows from the vanishing trace of the left-hand side as required by the "=" case of (2.102). Equally well, we could have employed partial transpositions of the second qubit in this line of reasoning, so that

$$\left(\rho^{T_2} + \frac{1-\mathcal{S}}{2\bar{q}}\rho_1\right)\rho_2 = 0 \qquad (2.105)$$

must hold as well. Indeed, total transposition turns one of the equations into the other. Jointly they state

$$\rho^{T_2}\rho^{T_1}\rho_1 = \tfrac{1}{4}(1-\mathcal{S})^2\rho_1 \quad \text{or} \quad \rho^{T_1}\rho^{T_2}\rho_2 = \tfrac{1}{4}(1-\mathcal{S})^2\rho_2 , \qquad (2.106)$$

where $\rho_1\rho_2\rho_1 = \overline{q}^2\rho_1$ and $\rho_2\rho_1\rho_2 = \overline{q}^2\rho_2$ have entered.

In summary, then, the problem of finding the optimal LSD of a given rank-4 ρ is reduced to solving (either one of) these eigenstate equations, because as soon as $\rho_1 = \rho_2^{\mathsf{T}}$ is known, the pure state ϱ of the optimal LSD is available. Further we note that $\frac{1}{4}(1-\mathcal{S})^2$ is the smallest eigenvalue of $\rho^{\mathsf{T}_1}\rho^{\mathsf{T}_2}$ and that $1 - \mathcal{S} = \mathcal{C}$ relates it to $\mathcal{C}$, the concurrence of ρ [which is to say that here both "=" signs hold in (2.75)] ; see [37] for the technical details justifying these two assertions.

To complete the argument, we must convince ourselves that (2.106) in conjuction with (2.99), etc. are not only necessarily obeyed by the ingredients of the optimal LSD, but are indeed sufficient to determine it. Suppose, then, that we have solved these equations and thus identified ϱ_{s}, ϱ, and $\mathcal{S}$, and assume that there is another LSD with $\lambda = \mathcal{S} + \epsilon > \mathcal{S}$. So, the first decomposition in

$$\rho = \mathcal{S}\varrho_{\mathrm{s}} + (1-\mathcal{S})\varrho = (\mathcal{S}+\epsilon)\rho_{\mathrm{sep}}^{(\epsilon)} + (1-\mathcal{S}-\epsilon)\rho_{\mathrm{pure}}^{(\epsilon)} \qquad (2.107)$$

is obtained from (2.106) and (2.99), and the second is better ($\epsilon > 0$) by assumption. Since there are no locally optimal LSDs, there must then be a continuum of "better" LSDs, such that $\rho_{\mathrm{sep}}^{(\epsilon)} \to \varrho_{\mathrm{s}}$ and $\rho_{\mathrm{pure}}^{(\epsilon)} \to \varrho$ as $\epsilon \to 0$. Upon setting

$$\rho_{\mathrm{pure}}^{(\epsilon)} = \varrho + \mathrm{i}\epsilon\left[\varrho, G\right] + O(\epsilon^2) \qquad (2.108)$$

with some hermitian generator G and recalling (2.90), (2.97), and (2.99), we conclude that

$$
\begin{aligned}
(\mathcal{S}+\epsilon)\,\mathrm{tr}\left\{\rho_{\mathrm{sep}}^{(\epsilon)\mathsf{T}_1}\rho_1\right\} \\
= \mathcal{S}\,\mathrm{tr}\left\{\varrho_{\mathrm{s}}^{\mathsf{T}_1}\rho_1\right\} + \epsilon\,\mathrm{tr}\left\{\varrho^{\mathsf{T}_1}\rho_1\right\} - (1-\mathcal{S})\epsilon\,\mathrm{tr}\left\{\mathrm{i}[\varrho, G]^{\mathsf{T}_1}\rho_1\right\} + O(\epsilon^2) \\
= -\tfrac{1}{2}\epsilon\overline{q} + O(\epsilon^2) \,.
\end{aligned}
\qquad (2.109)
$$

But $\overline{q} > 0$, so that this right-hand side is negative for sufficiently small positive ϵ values. Accordingly, $\rho_{\mathrm{sep}}^{(\epsilon)}$ is not separable after all, and we arrive at a contradiction.

All of this is, however, only true if the separable remainder ϱ_{s} of the optimal LSD is of rank 4, about which one has no prior knowledge. Nevertheless, we can accept $\varrho_{\mathrm{s}} > 0$ as a working hypothesis, determine ϱ in accordance with (2.98), and then check whether the resulting $\varrho_{\mathrm{s}} = \mathcal{S}^{-1}\left[\rho - (1-\mathcal{S})\varrho\right]$ is truly positive and separable. If it is, we have found the optimal LSD with ϱ_{s} of full rank and ϱ a Bell state. Otherwise, we learn that the actual ϱ_{s} is of rank 3.

2.6.2.2 Separable part has reduced rank

Then there is a pure state ρ_0 associated with the null eigenvalue of ϱ_s,

$$\varrho_s \rho_0 = \rho_0 \varrho_s = 0 \,. \tag{2.110}$$

But ρ_0 is not orthogonal to ϱ,

$$\text{tr}\{\varrho \rho_0\} > 0 \,, \tag{2.111}$$

because ρ is of rank 4.

In the limit $\epsilon \to 0$, we must have $\rho_{\text{pure}}^{(\epsilon)} \to \rho_0$ in the positivity criterion (2.95) and, as before, $\overline{\rho}_{\text{pure}}^{(\epsilon)} \to \rho_1$ in the separability criterion (2.96). Looking at the terms linear in ϵ, we find that

$$\begin{aligned} \text{either (i)} \quad & \text{tr}\{\text{i}[G,\varrho]\rho_0\} \geq 0 \quad \text{and} \quad \text{tr}\{\text{i}[G,\varrho]^{\text{T}_1}\rho_1\} \leq 0 \\ \text{or (ii)} \quad & \text{tr}\{\text{i}[G,\varrho]\rho_0\} \leq 0 \quad \text{and} \quad \text{tr}\{\text{i}[G,\varrho]^{\text{T}_1}\rho_1\} \geq 0 \end{aligned} \tag{2.112}$$

must be the case. There is no other possibility: if, for instance, both traces were negative, so that $\epsilon > 0$ would be all right with the generator G considered, then $\epsilon \to -\epsilon$ (or, equivalently, $G \to -G$) would lead to the contradictory situation in which neither (2.95) nor (2.96) holds.

Again, any hermitian G is permitted in (2.112), which implies

$$[\rho_1^{\text{T}_1}, \varrho] = -y[\rho_0, \varrho] \quad \text{with} \quad y \geq 0 \,. \tag{2.113}$$

We conclude that ϱ is an eigenvalue of $\rho_1^{\text{T}_1} + y\rho_0 = \rho_2^{\text{T}_2} + y\rho_0$,

$$\left(\rho_1^{\text{T}_1} + y\rho_0\right)\varrho = -\mu\varrho \,, \tag{2.114}$$

which turns into (2.98) for $y = 0$, $\mu = \frac{1}{2}\overline{q}$. This feeds the expectation that $\mu \geq 0$, and this is indeed the case, as Wellens and Kuś have shown [37] with a continuity argument that exploits the known optimal LSD of a state of the form (2.68) [24] and an observation by Karnas and Lewenstein [41], namely that there is a pure product state orthogonal to ρ_0 with its partial transpose orthogonal to ρ_1 (or ρ_2 if the second qubit is transposed).

As it stands, the eigenstate equation (2.114) is of little use because it involves ρ_0, which is unknown as yet. We determine it with an argument from the Lewenstein–Sanpera paper [14].*

---

*Incidentally, Karnas and Lewenstein [41] note that this argument was also used by Jaynes in his work on the foundations of statistical mechanics [42, 43].

For any positive operator A, we note this application of the Cauchy–Schwarz inequality (actually a particular case of a more general statement):

$$\mathrm{tr}\{A\varrho\} = \mathrm{tr}\big\{\big(\rho^{\frac{1}{2}}A\rho^{\frac{1}{2}}\big)\big(\rho^{-\frac{1}{2}}\varrho\rho^{-\frac{1}{2}}\big)\big\} \leq \mathrm{tr}\{A\rho\}\mathrm{tr}\{\rho^{-1}\varrho\} , \qquad (2.115)$$

or

$$\mathrm{tr}\big\{\big(\mathrm{tr}\,\{\rho^{-1}\varrho\}\,\rho - \varrho\big)A\big\} \geq 0$$

$$\text{with “=” only for } A \propto \rho^{-1}\varrho\rho^{-1} , \qquad (2.116)$$

so that $\mathrm{tr}\,\{\rho^{-1}\varrho\}\,\rho - \varrho$ is a positive operator of rank 3. This operator is a linear combination of ϱ_{s} and ϱ and, as a consequence of (2.111), it must be just a multiple of ϱ_{s} because if it had a nonzero contribution from ϱ it would be either nonpositive or of rank 4. Consistency then requires that

$$\mathrm{tr}\,\{\rho^{-1}\varrho\}\,(1 - \mathcal{S}) = 1 , \qquad (2.117)$$

and the “=” case of (2.116) tells us that

$$\rho_0 = \frac{\rho^{-1}\varrho\rho^{-1}}{\mathrm{tr}\,\{\rho^{-2}\varrho\}} . \qquad (2.118)$$

So, with $y = \nu\,\mathrm{tr}\,\{\rho^{-2}\varrho\}\,(1 - \mathcal{S})$, the eigenstate equation (2.114) becomes

$$\big(\rho_1^{\mathrm{T_1}} + \nu\rho^{-1}\big)\varrho = -\mu\varrho . \qquad (2.119)$$

We supplement it with the eigenstate equation for ρ_1,

$$\big(\mathrm{tr}\,\{\rho^{-1}\varrho\}\,\rho - \varrho\big)^{\mathrm{T_1}}\rho_1 = 0 , \qquad (2.120)$$

which is (2.90) after expressing ϱ_{s} in terms of ρ and ϱ with the aid of (2.117) and (2.71). The equation pair (2.119), (2.120) determines ϱ and ρ_1 for the given ρ. As Wellens and Kuś remark, there may be several solutions, but there is only one with $\mu, \nu \geq 0$ that gives a positive and separable ϱ_{s}; see [37] for the technical details justifying this uniqueness assertion.

Here, too, one can show that these equations are not only necessary but sufficient to determine the optimal LSD. We do not reproduce the argument though. It exploits the insights gained when demonstrating that $\mu \geq 0$ in (2.114) and is technically more involved than the argument given above for the case that ϱ_{s} is of rank 4. For details please consult the paper by Wellens and Kuś [37].

Presently we are not aware of an analytical method for solving (2.119), (2.120). But they certainly enable one to design a rapidly converging numerical scheme.

2.6.2.3 Summary

We summarize these lessons about optimal LSDs of 2-qubit states taught by the recent papers by Karnas and Lewenstein [41] and by Wellens and Kuś [37] as follows.

For a 2-qubit state ρ of full rank, $\rho > 0$, the optimal LSD has either (1) a separable state ϱ_s of rank 4 or (2) one of rank 3; in case (1) the pure state ϱ is a Bell state. There is an analytical procedure to decide which is the actual situation, and if (1) is the case, then the optimal LSD is available analytically. In case (2), however, we do not have, as yet, an analytical method to determine the optimal LSD.

Therefore, the $x \to 0$ limit mentioned at the beginning of Section 2.6.2 cannot be carried out analytically for a rank-3 state ρ, because we are surely dealing with case (2) for sufficiently small x values.

By contrast, states of rank 2 are not critical. Their optimal LSDs are known explicitly and reported in Section 2.7.3.

2.7   Examples

In this section we present a couple of special examples for which the optimal LSD is known. These optimal LSDs were first reported in [24] in a very concise manner; here we give all the details necessary for understanding how the results were obtained.

The examples have a feature in common that helps enormously, namely that the search can be limited to a one-parametric set of LSDs.

2.7.1   Self-transposed states

If both Pauli vectors vanish, $s^{\downarrow} = 0$ and $\vec{t} = 0$, the 2-qubit state ρ is equal to its total transpose ρ^{T}; it is *self-transposed*. In view of (2.39), all Bell states are of this kind, and so are all convex sums of Bell states. The converse is also true. All self-transposed states are convex sums of Bell states. Indeed, the eigenstates of a self-transposed ρ are Bell states.

Eigenvalues of			Eigenvalues of ρ	Monotonic order for	
$\sigma_1\tau_1$	$\sigma_2\tau_2$	$\sigma_3\tau_3$		"≥ 0"	"< 0"
1	1	-1	$\frac{1}{4}[1 \pm (c_1 + c_2 - c_3)]$	r_1	r_4
1	-1	1	$\frac{1}{4}[1 \pm (c_1 - c_2 + c_3)]$	r_2	r_3
-1	1	1	$\frac{1}{4}[1 \mp (c_1 - c_2 - c_3)]$	r_3	r_2
-1	-1	-1	$\frac{1}{4}[1 \mp (c_1 + c_2 + c_3)]$	r_4	r_1

Table 2.3 Eigenvalues of self-transposed 2-qubit states. The upper and lower signs correspond to the two cases of (2.121). The last two columns report the respective ordered assignment required by the standardized spectral decomposition (2.69) for the two cases.

Put differently, the pure states $\rho_{\text{pure}}^{(k)}$ in the spectral decomposition (2.69) are Bell states if $\rho = \rho^{\mathsf{T}}$.[*]

To be more explicit about these matters, we invoke the generic form of a self-transposed state,

$$\rho = \rho^{\mathsf{T}} = \frac{1}{4}\left(1 + \vec{\sigma}\cdot\overleftrightarrow{C}\cdot\tau^{\downarrow}\right) = \frac{1}{4}\left[1 \pm (c_1\sigma_1\tau_1 + c_2\sigma_2\tau_2 + c_3\sigma_3\tau_3)\right], \quad (2.121)$$

where c_1, c_2, c_3 are the characteristic values of $\overleftrightarrow{C}$ and, as in (2.28), the upper sign applies for $\det\{\overleftrightarrow{C}\} \geq 0$, the lower for $\det\{\overleftrightarrow{C}\} < 0$. Since $\sigma_3\tau_3 = -(\sigma_1\tau_1)(\sigma_2\tau_2)$, the eigenstates of ρ are the common eigenstates of $\sigma_1\tau_1$, $\sigma_2\tau_2$, and $\sigma_3\tau_3$, the four Bell states

$$\rho_{\text{Bell}}^{(++-)} = \tfrac{1}{4}(1 + \sigma_1\tau_1 + \sigma_2\tau_2 - \sigma_3\tau_3),$$

$$\rho_{\text{Bell}}^{(+-+)} = \tfrac{1}{4}(1 + \sigma_1\tau_1 - \sigma_2\tau_2 + \sigma_3\tau_3),$$

$$\rho_{\text{Bell}}^{(-++)} = \tfrac{1}{4}(1 - \sigma_1\tau_1 + \sigma_2\tau_2 + \sigma_3\tau_3),$$

$$\rho_{\text{Bell}}^{(---)} = \tfrac{1}{4}(1 - \sigma_1\tau_1 - \sigma_2\tau_2 - \sigma_3\tau_3), \quad (2.122)$$

which are mutually orthogonal. The superscripts refer to the ± 1 eigenvalues of $\sigma_1\tau_1$, $\sigma_2\tau_2$, $\sigma_3\tau_3$. The corresponding eigenvalues of ρ are given in Table 2.3. The positivity of ρ imposes the restriction $r_4 \geq 0$, that is:

$$c_1 + c_2 \leq 1 - c_3 \quad \text{for} \quad \det\{\overleftrightarrow{C}\} \geq 0,$$

$$\text{and} \quad c_1 + c_2 \leq 1 + c_3 \quad \text{for} \quad \det\{\overleftrightarrow{C}\} < 0, \quad (2.123)$$

---

[*]For this reason, self-transposed states are sometimes called "Bell-diagonal states."

which require that the triplet (c_1, c_2, c_3) — which is not a 3-vector — is inside the tetrahedron that R. and M. Horodecki speak of in [44].

In the "≥ 0" case, then, the largest eigenvalue of ρ cannot exceed $\frac{1}{2}$,

$$\det\left\{\overleftrightarrow{C}\right\} \geq 0 \; : \quad r_1 = \tfrac{1}{4}(1 + c_1 + c_2 - c_3) \leq \tfrac{1}{2} - \tfrac{1}{2}c_3 \,, \qquad (2.124)$$

so that $\rho^{\mathsf{T}_1} = \rho^{\mathsf{T}_2} = \frac{1}{2} - \rho \geq 0$ here, and the Peres–Horodeccy criterion (2.63) says that ρ is separable. In the "< 0" case, however, we have

$$\det\left\{\overleftrightarrow{C}\right\} < 0 \; : \quad r_1 = \tfrac{1}{4}(1 + c_1 + c_2 + c_3) = \frac{1}{4} + \frac{1}{4}\mathrm{Sp}\left\{\left|\overleftrightarrow{C}\right|\right\} \,, \quad (2.125)$$

so that $r_1 > \frac{1}{2}$ and $\mathrm{tr}\{\rho^{\mathsf{T}_1}\rho^{(---)}_{\mathrm{Bell}}\} = \frac{1}{2} - r_1 < 0$ if $\mathrm{Sp}\left\{\left|\overleftrightarrow{C}\right|\right\} > 1$. Accordingly, the self-transposed state (2.121) is

$$\text{separable if } \det\left\{\overleftrightarrow{C}\right\} \geq 0 \text{ or } \mathrm{Sp}\left\{\left|\overleftrightarrow{C}\right|\right\} \leq 1 \,;$$
$$\text{nonseparable if } \det\left\{\overleftrightarrow{C}\right\} < 0 \text{ and } \mathrm{Sp}\left\{\left|\overleftrightarrow{C}\right|\right\} > 1 \,. \qquad (2.126)$$

Rather than exploiting the Peres–Horodeccy criterion, we could have arrived at this observation by evaluating the Hill–Wootters concurrence of ρ. Here $\left|\sqrt{\rho^{\mathsf{T}}}\sqrt{\rho}\right| = \rho$, and so the h_k's of (2.23) are identical with the r_k's of Table 2.3, and we get

$$\mathcal{C} = \max\{0, r_1 - r_2 - r_3 - r_4\} = \max\{0, 2r_1 - 1\}$$
$$= \begin{cases} 0 & \text{if } \det\left\{\overleftrightarrow{C}\right\} \geq 0 \text{ or } \mathrm{Sp}\left\{\left|\overleftrightarrow{C}\right|\right\} \leq 1 \,; \\ \frac{1}{2}\mathrm{Sp}\left\{\left|\overleftrightarrow{C}\right|\right\} - \frac{1}{2} & \text{if } \det\left\{\overleftrightarrow{C}\right\} < 0 \text{ and } \mathrm{Sp}\left\{\left|\overleftrightarrow{C}\right|\right\} > 1 \,. \end{cases}$$
$$(2.127)$$

The search for the optimal LSD of a nonseparable self-transposed state is facilitated by the intrinsic symmetry of (2.121). In particular, ρ is invariant under the local transformations effected by the unitary operators $\sigma_1\tau_1$, $\sigma_2\tau_2$, $\sigma_3\tau_3$; it is also invariant under the swapping transformation (2.86). Therefore, the pure state ϱ of the optimal LSD must also be invariant under these transformations, and so it must be one of the Bell states in (2.122), namely the one associated with the largest eigenvalue

r_1, that is $\rho_{\text{Bell}}^{(---)}$. Thus we arrive at

$$
S = \begin{cases} 1 & \text{if } \det\left\{\overrightarrow{C}\right\} \geq 0 \text{ or } \text{Sp}\left\{\left|\overrightarrow{C}\right|\right\} \leq 1 \\ \dfrac{3}{2} - \dfrac{1}{2}\text{Sp}\left\{\left|\overrightarrow{C}\right|\right\} & \text{if } \det\left\{\overrightarrow{C}\right\} < 0 \text{ and } \text{Sp}\left\{\left|\overrightarrow{C}\right|\right\} > 1 \end{cases} = 1 - C
$$

(2.128)

so that self-transposed states have two "=" signs in (2.75).

We remark that the concavity (2.74) — applied, for $x = \frac{1}{2}$, to $\rho_1 = \rho$ and $\rho_2 = \rho^{\mathsf{T}}$ with an *arbitrary* 2-qubit state ρ — tells us that

$$
S\left(\tfrac{1}{2}(\rho + \rho^{\mathsf{T}})\right) \geq \tfrac{1}{2}S(\rho) + \tfrac{1}{2}S(\rho^{\mathsf{T}}) = S(\rho) . \tag{2.129}
$$

Since $\frac{1}{2}(\rho + \rho^{\mathsf{T}})$ is self-transposed by construction, the S value of (2.128), evaluated for the cross dyadic $\overrightarrow{C}$ of the arbitrary ρ considered, sets an upper bound on $S(\rho)$.

Another remark concerns the degenerate situation of $c_1 = c_2 = c_3 = |x| \geq 0$. Then the self-transposed state (2.121) is of the form (2.68) with ρ_{pure} a Bell state. If we write it in the form (2.36), we have

$$
\rho = \rho^{\mathsf{T}} = \frac{1}{4}\left(1 - x\vec{\sigma} \cdot \overrightarrow{O}_{en} \cdot \tau^{\downarrow}\right) \quad \text{with} \quad -\tfrac{1}{3} \leq x \leq 1 . \tag{2.130}
$$

These states are commonly called "Werner states" [45]. For $x \leq \frac{1}{3}$ they are separable, and for $x > \frac{1}{3}$ their degree of separability is $\frac{3}{2}(1 - x)$. Note that a violation of the Clauser–Horne inequality would require $x > 1/\sqrt{2}$, so that Werner states with $\frac{1}{3} < x \leq 1/\sqrt{2}$ are not separable and obey the Clauser–Horne inequality.

2.7.2 Generalized Werner states

States of the form (2.68) with ρ_{pure} not a Bell state represent a simple generalization of the Werner states.[*] Their generic form is

$$
\rho = \tfrac{1}{4}\left[1 + xp(\sigma_1 - \tau_1) - x\sigma_1\tau_1 - xq(\sigma_2\tau_2 + \sigma_3\tau_3)\right] . \tag{2.131}
$$

The corresponding values of A_2, A_1, A_0 are obtained from the pure-state values — that is: $A_2 = 6$, $A_1 = 8$, $A_0 = 3$ for $x = 1$ — by multiplying

---

[*]In [24], as an homage to 19th century mathematical physicists, we coined the term "generalized Werner states of the first kind" for the self-transposed states (2.121), and we referred to the states (2.68) as those "of the second kind." There is no commonly accepted terminology, and perhaps there is no need for it either.

with x^2, x^3, and x^4, respectively, because the convex sum (2.68) just
amounts to multiplying $K = 1 - 4\rho_{\text{pure}}$ by x. So,

$$A_2 = 6x^2 , \quad A_1 = 8x^3 , \quad A_0 = 3x^4 \qquad (2.132)$$

for the generalized Werner states (2.131), the particular values of p and
q being irrelevant here. They are crucial in

$$\det\left\{\overleftrightarrow{C}\right\} = -x^3 q^2 , \qquad \det\left\{\overleftrightarrow{C} - s^{\downarrow}\overrightarrow{t}\right\} = -x^3 q^2 (1 - xp^2) , \quad (2.133)$$

however, which we can use in (2.67) to establish that the state (2.131)
is separable if $(1 + 2q)x \leq 1$, and nonseparable otherwise. Equivalently,
we can determine the h_k's of (2.23),

$$h_1 = h_2 = \tfrac{1}{4}(1 - x) ; \quad \left.\begin{matrix} h_3 \\ h_4 \end{matrix}\right\} = \frac{1}{4}\sqrt{(1 + x)^2 - 4x^2 p^2} \mp \frac{1}{2}xq , \quad (2.134)$$

and then the concurrence

$$C = \max\left\{0, \tfrac{1}{2}(1 + 2q)x - \tfrac{1}{2}\right\} , \qquad (2.135)$$

with the same conclusion concerning the separability of ρ. Or, perhaps
simplest, we recall (2.37) and note that the smallest eigenvalue of ρ^{T_1} is

$$\tfrac{1}{4}(1 - x) - \tfrac{1}{2}xq = \tfrac{1}{4} - \tfrac{1}{4}(1 + 2q)x \qquad (2.136)$$

and learn once more that the sign of $(1 + 2q)x - 1$ decides whether a
generalized Werner state (2.131) is separable or not.

The local unitary transformations generated by $\sigma_1 + \tau_1$, the only local
generator in (2.50), leave the state (2.131) unchanged. When looking for
the optimal LSD, it is therefore sufficient to consider only those LSDs in
which the separable and pure parts commute with $\sigma_1 + \tau_1$ as well. This
is, in particular, the case for (2.68), which is one of the many LSDs of ρ.

Accordingly, the pure state of the optimal LSD must be of the form

$$\varrho = \tfrac{1}{4}\left[1 + p_0(\sigma_1 - \tau_1) - \sigma_1\tau_1 - q_0(\sigma_2\tau_2 + \sigma_3\tau_3)\right] \qquad (2.137)$$

with $p_0^2 + q_0^2 = 1$, and the separable part must have the structure

$$\varrho_s = \tfrac{1}{4}\left[1 + s(\sigma_1 - \tau_1) - c_1\sigma_1\tau_1 - c_2(\sigma_2\tau_2 + \sigma_3\tau_3)\right] . \qquad (2.138)$$

Now, this ϱ_s is positive if $c_1 \leq 1$ and $2\sqrt{c_2^2 + s^2} \leq 1 + c_1$, and it is
separable if $c_1 + 2|c_2| \leq 1$ and $2|s| \leq 1 + c_1$. Accordingly, the restrictions

$$c_1 + 2|c_2| \leq 1 \quad \text{and} \quad 2\sqrt{c_2^2 + s^2} \leq 1 + c_1 \qquad (2.139)$$

apply. Since ϱ_s is barely separable with respect to ϱ, its partial transpose has at least one zero eigenvalue, so that the equal sign has to hold in the left inequality of (2.139). Further, since q and p are positive by convention, it will suffice to consider nonnegative values for q_0, p_0.

Writing then $c \equiv c_2 \geq 0$, $c_1 = 1 - 2c$, the right inequality of (2.139) insists on

$$s^2 \leq 1 - 2c , \qquad (2.140)$$

and (2.72) is equivalent to the set of equations

$$xp = \mathcal{S}s + (1 - \mathcal{S})p_0 ,$$
$$x = \mathcal{S}c_1 + (1 - \mathcal{S}) = 1 - 2\mathcal{S}c ,$$
$$xq = \mathcal{S}c_2 + (1 - \mathcal{S})q_0 = \mathcal{S}c + (1 - \mathcal{S})q_0 . \qquad (2.141)$$

These are three equations for four unknowns: s, c, $q_0 = \sqrt{1 - p_0^2}$, and $\mathcal{S}$. Each solution consistent with (2.140) would give one LSD (with a barely separable ρ_{sep}), but we are only interested in the optimal LSD, the solution with the largest value of $\mathcal{S}$.

It is expedient to regard q_0, the concurrence of ϱ, as the basic parameter and the others as functions of q_0. Equations (2.141) supply

$$s = p_0 + \frac{xp - p_0}{\mathcal{S}} \quad \text{and} \quad c = \frac{1 - x}{2\mathcal{S}} \qquad (2.142)$$

with

$$\mathcal{S} = 1 - \frac{(1 + 2q)x - 1}{2q_0} . \qquad (2.143)$$

This tells us that the largest value of q_0 allowed by (2.140) specifies the optimal LSD. In conjunction with (2.135) it states further that the equal sign holds in the first inequality of (2.75); in the second inequality it will hold whenever $q_0 = 1$.

Having expressed s and c in terms of q_0, we note that (2.140) restricts q_0 by

$$\frac{1 + x - 2xpp_0}{q_0} \leq \left(qx - \frac{1 - x}{2}\right) + \left(qx - \frac{1 - x}{2}\right)^{-1} (x - x^2p^2) , \quad (2.144)$$

which gives $q_0 > q$ for $x < 1$ and $q_0 \to q$ in the limit $x \to 1$. We are now ready to ask the crucial question. Under which circumstances is (2.144) obeyed by $q_0 = 1$? In other words, under which circumstances is ϱ a Bell state? Well, for $q_0 = 1$, $p_0 = 0$ inequality (2.144) reads

$$2(2qx - 1)^2 \geq 5x^2 - 2x - 1$$
$$\text{or} \quad \left(\tfrac{3}{2}x^{-1} - 2q + \tfrac{1}{2}\right)^2 \geq 2(1 - q)(2 + q) , \qquad (2.145)$$

and one verifies easily that there are cases, such as $x = q = 0.8$, for which these equivalent inequalities are violated, so that $q_0 = 1$ is not possible and the largest q_0 value consistent with (2.144) must be really less than 1. More explicitly, the optimal LSD has

$$q_0 = 1 \quad \text{if} \quad x \leq \frac{3}{2} \left[2q - \frac{1}{2} + \sqrt{2(1-q)(2+q)} \right]^{-1} ,$$

$$q_0 < 1 \quad \text{if} \quad x > \frac{3}{2} \left[2q - \frac{1}{2} + \sqrt{2(1-q)(2+q)} \right]^{-1} . \qquad (2.146)$$

In the latter case, one finds q_0 from the quadratic equation obtained by equating the two sides of (2.144).

A graphical summary of these observations is presented in Figure 2.1. The two solid lines show where the equal sign holds in (2.145); it is the border line between the two cases of (2.146). The crosses mark the point with $x = \frac{1}{5}(\sqrt{6} + 1)$, $q = \frac{1}{2}(\sqrt{6} - 1)$, $\mathcal{S} = \frac{1}{10}(9 - \sqrt{6})$, where $q_0 = 1$, $2qx = 1$, and $\mathcal{S} = 1 - \frac{1}{2}x$. The top figure displays the lines of constant q_0. The solid line separates the regions $q_0 = 1$ from $q_0 < 1$. The dash-dotted line separates the separable states ($\mathcal{S} = 1$) from the nonseparable ones ($\mathcal{S} < 1$). The dashed lines indicate where $q_0 = 0.9, 0.8, 0.6, 0.4$, and 0.2, respectively. The bottom figure shows $\mathcal{S}$ as a function of x. To the left of the solid line one has $q_0 = 1$ and $\mathcal{S}$ depends linearly on x; to the right, one has $q_0 < 0$ and $\mathcal{S}$ is a nonlinear function of x. The dash-dotted line corresponds to the standard Werner states ($q = 1$); the dashed lines indicate $q = 0.8, 0.6, 0.4, 0.2$, and 0.1, respectively.

If the equal sign holds in (2.144), then it holds in (2.140), which is to say that ϱ_s is of rank 3. The two cases of (2.146) thus illustrate the observation of Karnas and Lewenstein [41] that a state of full rank, here exemplified by the generalized Werner state (2.131), has an optimal LSD with either ϱ_s of rank 4 and ϱ a Bell state (the $q_0 = 1$ case) or with ϱ_s of rank 3 (the $q_0 < 1$ case).

2.7.3   States of rank 2

States for which the equal signs hold in the first and the second inequality of (2.60), but not in the third, are states of rank 2. When we write their eigenvalues as

$$r_1 = \tfrac{1}{2}(1 + x) , \quad r_2 = \tfrac{1}{2}(1 - x) , \quad r_3 = r_4 = 0 \quad \text{with} \quad 0 \leq x < 1 ,$$
$$(2.147)$$

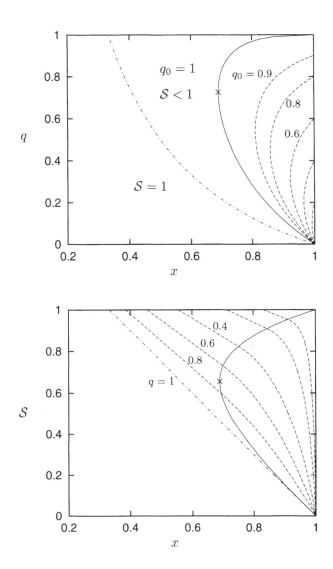

FIGURE 2.1
Optimal LSDs of generalized Werner states. **Top:** Lines of constant q_0 in the x, q plane. **Bottom:** Degree of separability $\mathcal{S}$ as a function of x for various values of q. Both plots cover $\frac{1}{5} \le x \le 1$, the more interesting part of the full range $-\frac{1}{3} \le x \le 1$.

the coefficients of the characteristic polynomial (2.56) are

$$A_2 = 4x^2 + 2 , \quad A_1 = 8x^2 , \quad A_0 = 4x^2 - 1 . \qquad (2.148)$$

In terms of the parameters x_1, x_2, x_3 in the generic form (2.47) of rank-2 states ρ_{rk2}, the value of x is given by

$$x = \sqrt{x_1^2 + x_2^2 + x_3^2} . \qquad (2.149)$$

Not unexpectedly, x is the distance from the center of the unit sphere $x_1^2 + x_2^2 + x_3^2 \leq 1$. The surface of the sphere is composed of the pure states on the boundary of this subspace of rank-2 states.

In view of (2.148), the (2.67) version of the Peres–Horodeccy criterion requires

$$\det\left\{ \overleftrightarrow{C} - s^{\downarrow}\vec{t} \right\} \geq 0 \quad \text{and} \quad \det\left\{ \overleftrightarrow{C} \right\} \geq 0 \qquad (2.150)$$

for a separable rank-2 state. The first determinant,

$$\det\left\{ \overleftrightarrow{C} - s^{\downarrow}\vec{t} \right\} = -\left[\left(z_2 - z_1 x_3\right)^2 + \left(1 - x_1^2 - x_3^2\right)\left(z_1^2 - z_2^2\right) \right]$$
$$\times \left[\left(z_2 - z_1 x_3\right)^2 + x_2^2\left(z_1^2 - z_2^2\right) \right] , \qquad (2.151)$$

is nonpositive and vanishes only if (recall that $z_1 \geq z_2 \geq 0$ by convention)

$$z_1 = 0 \quad \text{or} \quad z_1 x_3 = z_2 \quad \text{and} \quad x_2 = 0 , \qquad (2.152)$$

and then

$$\det\left\{ \overleftrightarrow{C} \right\} = x_3(z_1 - z_2 x_3)(z_2 - z_1 x_3)$$
$$- z_1 x_2^2(z_1 - z_2 x_3) - z_2 x_1^2(z_2 - z_1 x_3) \qquad (2.153)$$

vanishes too. This confirms the observations made after (2.49) that for $z_1 = 0$ all ρ_{rk2}'s of (2.47) are separable, whereas for $z_1 > 0$ the pure separable states are specified by $x_1 = \pm\sqrt{1 - (z_2/z_1)^2}$, $x_2 = 0$, $x_3 = z_2/z_1$.

More generally, for $z_1 > 0$, the separable states have

$$-\sqrt{1 - (z_2/z_1)^2} \leq x_1 \leq \sqrt{1 - (z_2/z_1)^2} , \quad x_2 = 0 , \quad x_3 = z_2/z_1 . \qquad (2.154)$$

In the geometrical picture associated with the sphere $x \leq 1$, these separable states constitute the straight line parallel to the x_1 axis that intersects the x_3 axis at $x_3 = z_2/z_1$. In the limiting situation of $z_2 = z_1$, the line degenerates into a single point, and then there is only one separable state and only one LSD for each $\rho_{\text{rk}2}$; it obtains for $z_2/z_1 = 1$ and $y = 0$ in (2.159) and (2.161) below.

With this visualization it is clear that, for $z_1 > z_2 \geq 0$, there is a one-parametric set of LSDs for each nonseparable rank-2 state. The given state corresponds to point $X = (x_1, x_2, x_3)$ inside the sphere. Pick any point $X_{\text{sep}} = (y, 0, z_2/z_1)$ on the line of separable states, and find the pure-state surface point $X_{\text{pure}} = (\sin\vartheta\cos\varphi, \sin\vartheta\sin\varphi, \cos\vartheta)$ on the ray from X_{sep} through X. With λ determined by

$$X = \lambda X_{\text{sep}} + (1 - \lambda) X_{\text{pure}} \qquad (2.155)$$

this construction produces one LSD for each X_{sep}. More explicitly, we need to solve the set of equations

$$x_1 = \lambda y + (1 - \lambda)\sin\vartheta\cos\varphi ,$$
$$x_2 = (1 - \lambda)\sin\vartheta\sin\varphi ,$$
$$x_3 = \lambda z_2/z_1 + (1 - \lambda)\cos\vartheta , \qquad (2.156)$$

where we regard x_1, x_2, x_3 as given and the unknowns $\vartheta, \varphi, \lambda$ as functions of y. After eliminating ϑ and φ, we get a single equation that relates λ to y,

$$(1 - \lambda)^2 = (x_1 - \lambda y)^2 + x_2^2 + (x_3 - \lambda z_2/z_1)^2 . \qquad (2.157)$$

For each y with $y^2 \leq 1 - (z_2/z_1)^2$ there is one λ in the range $0 \leq \lambda \leq 1$, and the degree of separability is the largest one of them.

This optimum is either at one of the boundaries of the y range or inside the range. In the first case, ϱ_s is a pure state, that is, it is of reduced rank; in the second case, ϱ_s is of rank 2. The inequality

$$x_2^2\sqrt{1 - (z_2/z_1)^2} \geq \left[(1 + x_3)\sqrt{1 - z_2/z_1} - |x_1|\sqrt{1 + z_2/z_1}\right]$$
$$\times \left[(1 - x_3)\sqrt{1 + z_2/z_1} - |x_1|\sqrt{1 - z_2/z_1}\right]$$
$$(2.158)$$

decides which one is the actual case. If it is obeyed, we have the "boundary case" with $y = \pm\sqrt{1 - (z_2/z_1)^2}$ for $x_1 \gtrless 0$ and

$$S = \frac{(1 - x^2)/2}{1 - x_3 z_2/z_1 - |x_1|\sqrt{1 - (z_2/z_1)^2}} . \qquad (2.159)$$

If (2.158) is violated, we have the "inside case" with

$$S = \frac{1}{1 - (z_2/z_1)^2}\left(1 - x_3\, z_2/z_1 - \sqrt{(x_3 - z_2/z_1)^2 + x_2^2\left[1 - (z_2/z_1)^2\right]}\,\right)$$
(2.160)

and $y = x_1/S$. These are supplemented by

$$X_{\text{pure}} = \left(x_1 - yS, x_2, x_3 - (z_2/z_1)S\right)/(1 - S)\,,$$
(2.161)

which tells us the coordinates of the surface point that specifies ϱ, the pure state of the optimal LSD. Tersely, if (2.158) is obeyed, then X_{sep} is the endpoint closest to X on the $x_2 = 0, x_3 = z_2/z_1$ line of X_{sep}'s; if it is violated, then X_{pure} is the $x_1 = 0$ point closest to X on the circumference of the circular cross section that the sphere $x \leq 1$ has with the plane defined by the point X and the line of X_{sep}'s. These matters are illustrated in Figure 2.2.

Note that this exemplifies the pairing property (2.85). The pure states with $x_1 = 0$ are paired with all separable states; those with $x_1 \gtrless 0$ are paired only with the product states specified by $x_1 = \pm\sqrt{1 - (z_2/z_1)^2}$, $x_2 = 0$, $x_3 = z_2/z_1$, respectively.

In the caption to Figure 2.2, there is a remark about the geometrical significance of the concurrence of rank-2 states. Its calculation is straightforward, but a bit tedious (it helps to observe that $\rho^{\mathsf{T}}\rho$ and $\Sigma_0\rho^{\mathsf{T}}\Sigma_0\rho$ have the same eigenvalues), with the outcome

$$C = \sqrt{\left(z_1^2 - z_2^2\right)x_2^2 + \left(z_2 - z_1 x_3\right)^2}\,.$$
(2.162)

Geometrically speaking, this says that the points X of states with common concurrence C constitute the surface of an elliptical cylinder, specified by

$$\left(x_1, x_2, x_3\right) = \left(x_1, \left[C/(z_1^2 - z_2^2)^{\frac{1}{2}}\right]\cos\varphi, z_2/z_1 + \left(C/z_1\right)\sin\varphi\right)$$
(2.163)

with arbitrary (real) values for x_1 and φ.

Now note that we can write the concurrence also as

$$C = z_1 R d\,,$$
(2.164)

where

$$R = \left(\frac{\left[1 - (z_2/z_1)^2\right]x_2^2 + (x_3 - z_2/z_1)^2}{x_2^2 + (x_3 - z_2/z_1)^2}\right)^{1/2}$$
(2.165)

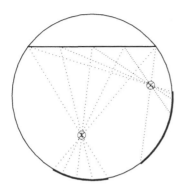 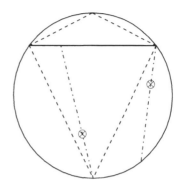

FIGURE 2.2

Geometry of the optimal LSDs of 2-qubit states of rank 2. The $x_2 = 0$, $x_3 = z_2/z_1$ line of X_{sep}s (thick horizontal line) and the point $X = (x_1, x_2, x_3)$ corresponding to the given nonseparable state (indicated by $\otimes$ for two different examples) define a plane that has a circular cross section with the sphere $x_1^2 + x_2^2 + x_3^2 \leq 1$. Vertical lines have constant x_1; horizontal lines have constant x_2 and x_3. The direction normal to the plane is $(0, z_2 - z_1 x_3, x_2) \propto (0, -\sin\theta, \cos\theta)$ where θ is the tilt angle relative to the plane $x_3 = 0$. The center of the circle is at $(z_2/z_1)\cos\theta (0, -\sin\theta, \cos\theta)$; its radius is $\sqrt{1 - [(z_2/z_1)\cos\theta]^2}$. The concurrence of a rank-2 state is proportional to the distance of its cross $\otimes$ from the thick line of separable states. **Left:** Different LSDs for the two exemplary states are indicated by the dotted lines that connect a separable state with a pure state through the $\otimes$s. The point $\otimes$ divides the lines in proportions of $(1 - \lambda) : \lambda$. The respective sets of pure states are marked by arcs just outside the circle. **Right:** The two optimal LSDs are indicated by the dash-dotted lines. Inequality (2.158) is violated inside the kite-shaped area bounded by the dashed lines.

is the radius of the circle in Figure 2.2 and

$$d = \left[x_2^2 + (x_3 - z_2/z_1)^2\right]^{1/2} \tag{2.166}$$

is the Hilbert–Schmidt distance between ρ and the nearest separable state.* In Figure 2.2 this is simply the Euclidean distance from the point $\otimes$ that specifies ρ to the thick line that represents the separa-

---

*That is, the nearest one among the separable states in the rank-2 subspace under consideration. There could be other separable states with an even smaller distance, but they cannot show up in a LSD of ρ_{rk2}.

ble states. * Since $\otimes$ divides the line representing the optimal LSD in proportions of $(1 - \mathcal{S}) : \mathcal{S}$, the first inequality in (2.75) is actually an equality for all rank-2 states, as it is also for the self-transposed states of Section 2.7.1 and the generalized Werner states of Section 2.7.2. This can hardly be accidental, but we don't know why the states of these three kinds are particular in this respect. There is an obvious challenge here. Find the conditions on ρ under which the equal sign holds.

The various LSDs on the left of Figure 2.2 involve pure states with a variety of concurrences. In view of the geometrical significance of the concurrence just noted, we observe that the pure state ϱ of the optimal LSD is the one with the largest concurrence. The same is clearly true for the self-transposed states and also for those LSDs of the generalized Werner states that we had to consider to find the optimal LSD. We surmise that

> the concurrence of the pure state in any LSD
> (2.71) cannot exceed the concurrence of ϱ, the
> pure state of the optimal LSD (2.72)   (2.167)

is generally true, but presently we cannot demonstrate the case.

2.8   Acknowledgments

We are grateful for many very helpful discussions with H.-J. Briegel, Č. Brukner, I. Cirac, M. Kuś, M. Lewenstein, and T. Wellens. BGE would like to thank H. Rauch and G. Badurek at the Atominstitut in Vienna, where part of this work was done, for the hospitality they provided, and the Technical University of Vienna for financial support. BGE would also like to thank M. O. Scully and the physics faculty at Texas A&M University, where another part of this work was done, for their hospitality and financial support. NM would like to thank the Egyptian government for granting a fellowship.

---

*The nearest separable state may or may not be equal to ϱ_s, the one that shows up in the optimal LSD. Recall the digression just before Section 2.6.1.

References

[1] S. Dürr, T. Nonn, and G. Rempe, *Nature* (London) **395** (1998) 33–37.

[2] S. Dürr, T. Nonn, and G. Rempe, *Phys. Rev. Lett.* **81** (1998) 5705–5709.

[3] P. G. Kwiat, P. D. D. Schwindt, and B.-G. Englert, What does a quantum eraser really erase?, in: *Mysteries, Puzzles, and Paradoxes in Quantum Mechanics*, edited by R. Bonifacio (AIP Conference Proceedings, vol. 461, Springer, New York 1999), pp. 69–80.

[4] P. D. D. Schwindt, P. G. Kwiat, and B.-G. Englert, *Phys. Rev. A* **60** (1999) 4285–4290.

[5] G. Badurek, R. J. Buchelt, B.-G. Englert, and H. Rauch, *Nucl. Instrum. Meth. A* **440** (2000) 562–567.

[6] B.-G. Englert, C. Kurtsiefer, and H. Weinfurter, *Phys. Rev. A* **63** (2001) art. 032301 (10 pages).

[7] A. Beige, B.-G. Englert, C. Kurtsiefer, and H. Weinfurter, LANL preprint `quant-ph/0101066` (2001).

[8] W. K. Wootters and B. D. Fields, *Ann. Phys.* (NY) **191** (1989) 363–381.

[9] Č. Brukner and A. Zeilinger, *Phys. Rev. Lett.* **83** (1999) 3354–3357.

[10] S. Hill and W. K. Wootters, *Phys. Rev. Lett.* **78** (1997) 5022–5025.

[11] W. K. Wootters, *Phys. Rev. Lett.* **80** (1998) 2245–2248.

[12] A. Peres, *Phys. Rev. Lett.* **77** (1996) 1413–1415.

[13] M. Horodecki, P. Horodecki, and R. Horodecki, *Phys. Lett. A* **223** (1996) 1–8.

[14] M. Lewenstein and A. Sanpera, *Phys. Rev. Lett.* **80** (1998) 2261–2264.

[15] M. Horodecki, P. Horodecki, and R. Horodecki, *Phys. Rev. Lett.* **80** (1998) 5239–5242.

[16] M. Lewenstein, B. Kraus, J. I. Cirac, and P. Horodecki, *Phys. Rev. A* **62** (2000) art. 052310 (16 pages).

[17] M. Lewenstein, B. Kraus, P. Horodecki, and J. I. Cirac, *Phys. Rev. A* **63** (2001) art. 044304 (4 pages).

[18] S. Wu, X. Chen, and Y. Zhang, *Phys. Lett. A* **275** (2000) 244–249.

[19] B. Kraus, J. I. Cirac, S. Karnas, and M. Lewenstein, *Phys. Rev. A* **61** (2000) art. 062302 (10 pages).

[20] T. Eggeling and R. F. Werner, *Phys. Rev. A* **63** (2001) art. 042111 (15 pages).

[21] S. Karnas and M. Lewenstein, *Phys. Rev. A* **64** (2001) art. 042313 (11 pages).

[22] A. Acín, D. Bruß, M. Lewenstein, and A. Sanpera, *Phys. Rev. Lett.* **87** (2001) art. 040401 (4 pages).

[23] V. M. Kendon, K. Nemoto, and W. J. Munro, LANL preprint `quant-ph/0106023` (2001).

[24] B.-G. Englert and N. Metwally, *J. Mod. Opt.* **47** (2000) 2221–2231. We thank Taylor & Francis (http://www.tandf.co.uk) for the kind permission to reproduce Table 1 (here: Table 2.1).

[25] B.-G. Englert and N. Metwally, *Appl. Phys. B* **72** (2001) 35–42. Note that, owing to absent-minded proof reading, the explicit expressions given for A_1 and A_0 in equations (29) are incorrect.

[26] M. Kuś and K. Życzkowski, *Phys. Rev. A* **63** (2001) art. 032307 (13 pages).

[27] A. Sanpera, R. Tarrach, and G. Vidal, *Phys. Rev. A* **58** (1998) 826–830.

[28] A. F. Abouraddy, B. E. A. Saleh, A. V. Sergienko, and M. C. Teich, *Phys. Rev. A* **64** (2001) art. 050101 (4 pages).

[29] L. Hardy, *Phys. Rev. Lett.* **71** (1993) 1665–1668.

[30] Y. Makhlin, LANL preprint `quant-ph/0002045` (2000).

[31] M. Grassl, M. Rötteler, and T. Beth, *Phys. Rev. A* **58** (1998) 1833–1839.

[32] N. J. Cerf, C. Adami, and R. M. Gingrich, *Phys. Rev. A* **60** (1999) 898–909.

[33] J. S. Bell, *Physics* **1** (1964) 195–200.

[34] J. F. Clauser and M. A. Horne, *Phys. Rev. D* **10** (1974) 526–535.

[35] R. Horodecki, M. Horodecki, and P. Horodecki, *Phys. Lett. A* **222** (1996) 21–25.

[36] J. Eisert and H.-J. Briegel, *Phys. Rev. A* **64** (2001) art. 022306 (4 pages).

[37] T. Wellens and M. Kuś, *Phys. Rev. A* **64** (2001) art. 052302 (10 pages).

[38] V. Vedral, M. B. Plenio, M. A. Rippin, and P. L. Knight, *Phys. Rev. Lett.* **78** (1997) 2275–2279.

[39] V. Vedral and M. B. Plenio, *Phys. Rev. A* **57** (1998) 1619–1633.

[40] F. Verstraete, J. Dehaene, and B. De Moor, LANL preprint `quant-ph/0107155` (2001).

[41] S. Karnas and M. Lewenstein, *J. Phys. A: Math. Gen.* **34** (2001) 6919–6937.

[42] E. T. Jaynes, *Phys. Rev.* **106** (1957) 620–630; reprinted in [46].

[43] E. T. Jaynes, *Phys. Rev.* **108** (1957) 171–190; reprinted in [46].

[44] R. Horodecki and M. Horodecki, *Phys. Rev. A* **54** (1996) 1838–1843.

[45] R. F. Werner, *Phys. Rev. A* **40** (1989) 4277–4281.

[46] E. T. Jaynes, *Papers on Probability, Statistics, and Statistical Physics*, edited by R. D. Rosenkrantz (Reidel, Dordrecht [1]1983; Kluwer, Dordrecht [2]1989).

Chapter 3

Invariants for multiple qubits: the case of 3 qubits

David A. Meyer* and Noland Wallach

Abstract   The problem of quantifying entanglement in multiparticle quantum systems can be addressed using techniques from the invariant theory of Lie groups. We briefly review this theory, and then develop these techniques for application to entanglement of multiple qubits.

3.1   Introduction

In quantum mechanics the state of a closed system is most completely described by a unit vector in a complex Hilbert space. (Such a state is *pure* in physics terminology.) For many systems, e.g., those characterizable as consisting of multiple particles, the Hilbert space has a natural decomposition into tensor factors. The standard model of quantum computation presumes an ability to implement unitary transformations which decompose into polynomially (in the number of factors, each of

*This work was supported in part by the National Security Agency (NSA) and Advanced Research and Development Activity (ARDA) under Army Research Office (ARO) contracts DAAG55-98-1-0376 and DAAD19-01-1-0520.

which is of no more than some constant dimension) many unitary transformations acting nontrivially on only one or two factors [1]. Such a
decomposition of states and operations makes possible the exponential
reduction in complexity of, for example, the quantum Fourier transform
relative to even the fast classical Fourier transform [2,3], and suggests,
more generally, that quantum computation may be more powerful than
classical computation.

It has been recognized, of course, since the famous paper of Einstein,
Podolsky and Rosen [4], that the quantum description of a multiparticle
system differs greatly from any classical description which decomposes in
the same way. Bohm distilled their two particle example to its essence by
considering pairs of spin-$\frac{1}{2}$ particles, i.e., systems described by elements
of the Hilbert space $\mathbb{C}^2 \otimes \mathbb{C}^2$ [5]. For this example, Bell's Theorem
specifies exactly the limits of any classical description [6]; these limits
are maximally exceeded by the "singlet" state $(|01\rangle - |10\rangle)/\sqrt{2}$. (Here,
and subsequently, we use Dirac notation $|\cdot\rangle$ to denote elements of Hilbert
space, $\langle\cdot|$ for dual elements, use $|0\rangle$ and $|1\rangle$ as a basis for $\mathbb{C}^2$, and write
$|01\rangle = |0\rangle \otimes |1\rangle \in \mathbb{C}^2 \otimes \mathbb{C}^2$.) In fact, they are exceeded equally by
any state obtained from the singlet state by unitary transformations
which decompose in the same way as the Hilbert space, i.e., elements of
$U(2) \times U(2)$. According to the terminology introduced by Schrödinger
[7], these states are equally *entangled*.

More precisely, an element of a Hilbert space with a specified tensor
product decomposition, $V = V_1 \otimes \cdots \otimes V_n$, is *not entangled* if and only
if it can be written as a product $v_1 \otimes \cdots \otimes v_n$ with $v_i \in V_i$. A *measure of
entanglement* is a function $f : V = V_1 \otimes \cdots \otimes V_n \to \mathbb{C}$ that is invariant
under $U(V_1) \times \cdots \times U(V_n)$.

In keeping with our interest in quantum computation (and because
they are easiest), in this paper we will consider only cases when each
of the tensor factors is two dimensional, i.e., a qubit. The situation
considered by Bohm [5], for example, is a pair of qubits. From a general
state $v \in \mathbb{C}^2 \otimes \mathbb{C}^2$, familiar constructions in physics are the *density
matrix*

$$\rho = v \otimes v^* \in (\mathbb{C}^2 \otimes \mathbb{C}^2) \otimes (\mathbb{C}^2 \otimes \mathbb{C}^2)^*,$$

and the *reduced density matrix* $\tilde{\rho} = \mathrm{Tr}_2\rho$. In a basis, $\rho = \sum \rho_{ijkl}|ij\rangle\langle kl|$
and $\tilde{\rho} = \sum \rho_{ijkj}|i\rangle\langle k|$. Notice that the cyclic property of trace im-
plies that $\tilde{\rho}$ is invariant under $I \times U(2)$. The usual analysis continues
by observing that the eigenvalues λ_i of $\tilde{\rho}$ are therefore invariant under
$U(2) \times U(2)$, and constructing the *entropy*, $-\sum \lambda_i \log \lambda_i$, to quantify
the entanglement of the bipartite state v from which ρ and $\tilde{\rho}$ were con-

structed. The eigenvalues of $\tilde{\rho}$ are the solutions of the characteristic equation $0 = \lambda^2 - (\text{Tr}\tilde{\rho})\lambda + \det \tilde{\rho}$; the functions $\text{Tr}\tilde{\rho} = 1$ and $\det \tilde{\rho}$ contain the same information as the eigenvalues. These functions are, in fact, invariants of $U(2) \times U(2)$, which are *polynomials* in the coefficients of $v = \sum v_{ij}|ij\rangle$ and $v^* = \sum \langle ij|\bar{v}_{ij}$; explicitly, $\det \tilde{\rho} = \det[v_{ij}] \det[\bar{v}_{ij}]$.

As others have noted [8–12], identifying such measures of entanglement is thus a problem in the invariant theory of Lie groups. In Section 3.2 we provide a brief introduction to the techniques of this theory, emphasizing the role of polynomial invariants. It is relatively straightforward to apply these techniques to small numbers of qubits; we do so for 1 and 2 qubits in Section 3.3, reproducing the result of the computation in the previous paragraph. Until recently, there was little understanding of entanglement for more than two factors, but this approach applies in principle to any number of factors. In Sections 3.4 and 3.5 we analyze the case of 3 qubits, obtaining a particularly nice set of generators and relations for the ring of invariants. These results are equivalent to, although in a different form and derived differently than the results of Grassl, Rötteler and Beth [10,13]. In particular, our approach has implications for entanglement invariants of 4 qubits; we sketch these in Section 3.6.

3.2 Invariants for compact Lie groups

Let W be a k dimensional vector space over K (the real numbers, $\mathbb{R}$, or the complex numbers, $\mathbb{C}$). Then a mapping $f : W \to \mathbb{C}$ is said to be a *polynomial* function if there exists a polynomial, φ, over $\mathbb{C}$ in variables $x_1, \ldots, x_k$ such that $f(\sum x_i w_i) = \varphi(x_1, \ldots, x_k)$. We will use the notation $\mathcal{P}(W)$ for the algebra of polynomials on W. We will also write $\mathcal{P}^d(W)$ for the space of polynomials of degree d. If V is a vector space over $\mathbb{C}$ but we are looking at V as a vector space over $\mathbb{R}$ then we will use the notation $\mathcal{P}_\mathbb{R}(V)$ and $\mathcal{P}^d_\mathbb{R}(V)$. Let G be a compact Lie group and let (π, V) be a finite dimensional unitary representation of G with V an n-dimensional complex Hilbert space. A function $f : V \to \mathbb{C}$ is said to be a *polynomial G-invariant* if $f(\pi(g)v) = f(v)$ for all $g \in G$ and $v \in V$ and $f \in \mathcal{P}_\mathbb{R}(V)$. We will write $\mathcal{P}_\mathbb{R}(V)^G$ for the G-invariant polynomials and $\mathcal{P}^d_\mathbb{R}(V)^G$ for the ones of degree d. The key reason for considering this class of invariants is that it is essentially the smallest algebra that separates the orbits.

THEOREM 3.1

If $v, w \in V$ then $f(v) = f(w)$ for all $f \in \mathcal{P}_{\mathbb{R}}(V)^G$ if and only if there exists $g \in G$ such that $\pi(g)v = w$.

PROOF   The function $v \mapsto |v|^2$ is an element of $\mathcal{P}_{\mathbb{R}}(V)^G$. Thus we may replace V by its unit sphere, $S(V)$, without any loss of generality. Since $S(V)$ is compact the Stone–Weierstrauss theorem implies that $\mathcal{P}_{\mathbb{R}}(V)$ is dense in the space of continuous functions on $S(V)$ in the uniform topology. Suppose that $\pi(G)v \cap \pi(G)w = \emptyset$. Urysohn's theorem implies that since $\pi(G)v$ and $\pi(G)w$ are compact there exists a real valued continuous function f on V such that $f(\pi(G)v) = \{1\}$ and $f(\pi(G)w) = \{0\}$. Let dg denote invariant measure on $\pi(G)$ normalized so that $\int_G dg = 1$. If g is a continuous function on V define $f^{\#}(z) = \int_G f(\pi(g)z)dg$. Then $f^{\#}(\pi(G)v) = \{1\}$ and $f^{\#}(\pi(G)w) = \{0\}$. Let $\phi \in \mathcal{P}_{\mathbb{R}}(V)$ be real valued and such that $|f^{\#}(z) - \phi(z)| < \frac{1}{4}$ for $z \in S(V)$. Then

$$\left| f^{\#}(v) - \int_G \phi(\pi(g)v)dg \right| = \left| \int_G (f^{\#}(v) - \phi(\pi(g)v))dg \right|$$

$$= \left| \int_G (f^{\#}(\pi(g)v) - \phi(\pi(g)v))dg \right|$$

$$\leq \int_G |f^{\#}(\pi(g)v) - \phi(\pi(g)v)|dg$$

$$\leq \frac{1}{4}.$$

Thus $\phi^{\#}(\pi(g)v) \geq \frac{3}{4}$ and $\phi^{\#}(\pi(g)w) \leq \frac{1}{4}$ for all $g \in G$. This implies the result. ∎

If we had used complex polynomials the analogous result would have been in general false.

We will set $\mathcal{P}(V)^G$ equal to $\mathcal{P}_{\mathbb{R}}(V)^G \cap \mathcal{P}(V)$. Let $\overline{V}$ denote V with the opposite complex structure. In other words we replace our choice, i, for $\sqrt{-1}$ by $-i$. If f is a function on V then we set $gf(x) = f(\pi(g)^{-1}x)$ for $g \in G$ and $x \in V$.

PROPOSITION 3.1

As a representation of G, $\mathcal{P}^d_{\mathbb{R}}(V)$ is equivalent with
$$\bigoplus_{0 \leq k \leq d} \mathcal{P}^{d-k}(V) \otimes \mathcal{P}^k(\overline{V}).$$

PROOF   Let $v_1, \ldots, v_n$ be a basis of V. Then a vector in V can be written as $\sum z_i v_i$ with $z_i \in \mathbb{C}$. Hence an element of $\mathcal{P}^d_{\mathbb{R}}(V)$ can be written as a polynomial of degree d in $z_1, \ldots, z_n, \overline{z}_1, \ldots, \overline{z}_n$. Such a polynomial is written as a sum of products of polynomials of degree $d - k$ in $z_1, \ldots, z_n$ and degree k in $\overline{z}_1, \ldots, \overline{z}_n$ for $0 \leq k \leq d$. The result now follows since the action of G on $\mathcal{P}^k(\overline{V})$ is equivalent to the action of G on $\overline{\mathcal{P}^k(V)}$ obtained by restricting the action of G on $\mathcal{P}^d_{\mathbb{R}}(V)$.   ∎

Let $\widehat{G}$ denote the set of equivalence classes of irreducible (necessarily finite dimensional) unitary representations of G (here a representation is always assumed to be strongly continuous). If (σ, W) is a finite dimensional unitary representation of G then W splits into a direct sum of irreducible subrepresentations. If $\gamma \in \widehat{G}$ then we denote by $W(\gamma)$ the sum of all irreducible invariant subspaces of W that are in the class of γ. Then $\dim W(\gamma) = m_W(\gamma) d(\gamma)$, where $d(\gamma)$ is the dimension of any member of γ and $m_W(\gamma)$ is the multiplicity of γ in W. If $\gamma \in \widehat{G}$ then we denote by $\overline{\gamma}$ the class of a representation of G that is dual (or complex conjugate) to an element of γ. If $W = \mathcal{P}^k(V)$ as above then we set $m_{\mathcal{P}^k(V)}(\gamma) = m_{V,k}(\gamma) = m_k(\gamma)$ (if V is understood). Then clearly, $m_{\overline{V},k}(\overline{\gamma}) = m_{V,k}(\gamma)$.

The formal power series $h_V(q,t) = \sum_{i,j} q^i t^j \dim\big(\mathcal{P}^i(V) \otimes \mathcal{P}^j(\overline{V})\big)^G$ is called the bigraded Hilbert series of the polynomial invariants. Also, $h_V(q,q)$ is the usual Hilbert series of the polynomial G-invariants in V.

PROPOSITION 3.2

We have
$$h_V(q,t) = \sum_{i,j} q^i t^j m_i(\gamma) m_j(\gamma).$$

PROOF   We note that $\big(\mathcal{P}^i(V)(\gamma) \otimes \mathcal{P}^j(\overline{V})(\tau)\big)^G$ is zero if $\gamma \neq \overline{\tau}$ and has dimension $m_i(\gamma) m_j(\gamma)$ if $\gamma = \overline{\tau}$. This implies the result.   ∎

The above result indicates that we should define the q-multiplicity of

γ in $\mathcal{P}(V)$ to be the formal power series $m(q, \gamma) = \sum_j q^j m_j(\gamma)$. Then we have:

LEMMA 3.1

With the notation above,

$$h(q, t) = \sum_{\gamma \in \widehat{G}} m(q, \gamma) m(t, \gamma).$$

In the next sections we will describe the implications of these results to qubits, i.e., the case when $V = \bigotimes^k \mathbb{C}^2$ and G is a product of k copies of $K = SU(2)$ (or $U(2)$) acting by, e.g., $(g_1, g_2, g_3)(v_1 \otimes v_2 \otimes v_3) = g_1 v_1 \otimes g_2 v_2 \otimes g_3 v_3$. Note that product (i.e., unentangled) states form a single orbit of the group G. This indicates (in light of Theorem 3.1) that the G-invariant polynomials on $\bigotimes^k \mathbb{C}^2$ are measures of entanglement.

If $K = SU(2)$ then the irreducible unitary representations are parameterized by their spin, which is a nonnegative half integer, s; that is, $\widehat{K}$ is $\{s \in \mathbb{Z}/2 \mid s \geq 0\} = (\mathbb{Z}/2)_{\geq 0}$. We fix an element in the class of s, F^s and observe that $\dim F^s = 2s + 1$. We choose $F^{\frac{1}{2}} = \mathbb{C}^2$. The corresponding parameterization of the irreducible unitary representations of G is $\left((\mathbb{Z}/2)_{\geq 0}\right)^k = \{(s_1, \ldots, s_k) \mid s_i \in (\mathbb{Z}/2)_{\geq 0}\}$. We choose $F^{\mathbf{s}} = F^{s_1} \otimes \cdots \otimes F^{s_k}$ as a representative of $\mathbf{s} = (s_1, \ldots, s_k)$.

3.3   The simplest cases

Before we get to more serious undertakings we will demonstrate our technique in the easiest cases.

Example 3.1

$k = 1$. Then we may take $F^s = \mathcal{P}^{2s}(\mathbb{C}^2)$. Thus $m_k(s) = \delta_{2k,s}$. This implies that

$$h_{\mathbb{C}^2}(q, t) = \sum_j (qt)^j = \frac{1}{1 - qt}.$$

From this we see (the well-known fact) that all of the polynomial invariants of the action of $SU(2)$ on $\mathbb{C}^2$ are polynomials in $v \mapsto |v|^2$.   □

Example 3.2

$k = 2$. In this case one has

$$\mathcal{P}^k(\mathbb{C}^2 \otimes \mathbb{C}^2) = F^{(\frac{k}{2},\frac{k}{2})} \oplus F^{(\frac{k}{2}-1,\frac{k}{2}-1)} \oplus \cdots \oplus \begin{cases} F^{(0,0)} & \text{if } k \text{ is even} \\ F^{(\frac{1}{2},\frac{1}{2})} & \text{otherwise} \end{cases}.$$

From this we see that

$$m\big(q, (\tfrac{k}{2}, \tfrac{k}{2})\big) = \sum_{j \geq 0} q^{k+2j} = \frac{q^k}{1 - q^2}.$$

This yields

$$\begin{aligned} h(q,t) &= \sum_k m\big(q, (\tfrac{k}{2}, \tfrac{k}{2})\big) m\big(t, (\tfrac{k}{2}, \tfrac{k}{2})\big) \\ &= \sum_k \frac{q^k t^k}{(1 - q^2)(1 - t^2)} \\ &= \frac{1}{(1 - q^2)(1 - qt)(1 - t^2)}. \end{aligned}$$

This immediately imples that the invariants of the action of $SU(2) \times SU(2)$ on $\mathbb{C}^2 \otimes \mathbb{C}^2$ are polynomials in three invariants. The invariant corresponding to tq is $v \mapsto |v|^2$ and there is a "new" invariant defined as follows: if $v = \sum v_{ij} |ij\rangle$ then $f(v) = \det[v_{ij}]$. Notice that this is an element of $\mathcal{P}(\mathbb{C}^2 \otimes \mathbb{C}^2)^G$ (and in fact generates the algebra). The invariant f corresponds to q^2 and the complex conjugate of f corresponds to t^2. ☐

In the above example we note that we could also have looked at the action of $U(2) \times U(2)$ on $\mathbb{C}^2 \otimes \mathbb{C}^2$. This is the same as the action of $S^1 \times G$ on $\mathbb{C}^2 \otimes \mathbb{C}^2$ via $(z, u, v)(x \otimes y) = z(ux \otimes vy)$. We note that $f(zv) = z^2 f(v)$. Thus the polynomial invariants of the action of $S^1 \times G$ are polynomials in $|v|^2$ and $|f(v)|^2$. These are exactly the invariants we described in the Introduction, namely $\operatorname{Tr} \tilde{\rho} = 1$ and $\det \tilde{\rho}$, respectively.

Although these examples are very simple, they illustrate an interesting feature of all such examples which will be useful subsequently. The representation $V = \bigotimes^k \mathbb{C}^2$ is equivalent with its complex conjugate. We are thus looking at the diagonal action of G on $\bigotimes^k \mathbb{C}^2 \oplus \bigotimes^k \mathbb{C}^2$. This can be interpreted as the action of G on $(\bigotimes^k \mathbb{C}^2) \otimes \mathbb{C}^2$ via $g \otimes I$. There is therefore a full $GL(2, \mathbb{C})$ acting by $I \otimes g$ that commutes with the action of G. This implies that $\mathcal{P}_{\mathbb{R}}(V)^G$ is naturally a representation space for $GL(2, \mathbb{C})$. The action of the Lie algebra of this group can be described in terms of polarization operators. Choose an orthonormal basis $\{v_i\}$ of V; let z_i be the corresponding linear coordinates; and set $x = \sum z_i \frac{\partial}{\partial \overline{z}_i}$, $y = \sum \overline{z}_i \frac{\partial}{\partial z_i}$ and $h = \sum z_i \frac{\partial}{\partial z_i} - \sum \overline{z}_i \frac{\partial}{\partial \overline{z}_i}$. Then $[x, y] = h$, $[h, x] = 2x$, $[h, y] = -2y$. In the case when $k = 2$ we note that the invariants are generated by f, yf, and $y^2 f$. The span of this space is the spin-1 representation of $SL(2, \mathbb{C})$. We also note that we may restrict this action to $SU(2)$ and thereby we have made a partial decomposition of the case of 3 qubits. Indeed, we have (using the classical theory of spherical harmonics):

$$m_{\otimes^3 \mathbb{C}^2}\left(q, (0, 0, \frac{k}{2})\right) = \begin{cases} 0 & \text{if } k \text{ is odd;} \\ \frac{q^{2k}}{1 - q^4} & \text{if } k \text{ is even.} \end{cases}$$

We will describe the full decomposition in the next section. Note that the above formula implies that the q-multiplicity of the trivial representation in the case of 3 qubits is $(1 - q^4)^{-1}$. We will now give a formula for an invariant of degree 4 which of necessity must generate all complex analytic polynomials for 3 qubits. Let

$$\left(\sum v_{ij} |ij\rangle, \sum w_{kl} |kl\rangle\right) = \sum \epsilon_{ik} \epsilon_{jl} v_{ij} w_{kl}$$

with $\epsilon_{ik} = 0$ if $i = k$; 1 if $i < k$; and -1 if $i > k$. Then $(\cdot, \cdot)$ defines a complex bilinear symmetric form on $\mathbb{C}^2 \otimes \mathbb{C}^2$ that is invariant under the action of $SU(2) \times SU(2)$. If $v \in \mathbb{C}^2 \otimes \mathbb{C}^2 \otimes \mathbb{C}^2$ then we write $v = v_0 \otimes |0\rangle + v_1 \otimes |1\rangle$. The desired invariant of degree 4 is given by

$$f(v) = \det\left[(v_i, v_j)\right].$$

For example, if $v = (|000\rangle + |111\rangle)/\sqrt{2}$ then $v_0 = |00\rangle/\sqrt{2}$ and $v_1 = |11\rangle/\sqrt{2}$ so $(v_0, v_0) = 0$, $(v_1, v_1) = 0$, and $(v_0, v_1) = \frac{1}{2}$. Hence

$$f(v) = \det\begin{bmatrix} 0 & \frac{1}{2} \\ \frac{1}{2} & 0 \end{bmatrix} = -\frac{1}{4}.$$

In particular, f is not the zero polynomial so

$$\mathcal{P}(\mathbb{C}^2 \otimes \mathbb{C}^2 \otimes \mathbb{C}^2)^{SU(2) \times SU(2) \times SU(2)}$$

is the algebra of polynomials in f.

3.4   The case of 3 qubits

In this section we will study the invariant polynomials under the action of $G = SU(2) \times SU(2) \times SU(2)$ acting on $\bigotimes^3 \mathbb{C}^2$. This means that we should be studying the real analytic (not complex analytic) polynomials on $V = \bigotimes^3 \mathbb{C}^2$. We will look at two cases. The first is the invariant theory for G and the second is that for $S^1 \times G$ acting via

$$(t, u_1, u_2, u_3)(v_1 \otimes v_2 \otimes v_3) = t(u_1 v_1 \otimes u_2 v_2 \otimes u_3 v_3),$$

the obvious action of $U(2) \times U(2) \times U(2)$. Both are a consequence of the decomposition of the space of complex analytic polynomials on V. If $\varsigma = (a, b, c)$ with $a, b, c \in (\mathbb{Z}/2)_{\geq 0}$ then set $m(\varsigma) = 2\min\{a, b, c\}$ and $n(\varsigma) = 2(a + b + c) - 2m(\varsigma)$. We note that if we write $\varsigma = a(\frac{1}{2}, \frac{1}{2}, \frac{1}{2}) + (b_1, b_2, b_3)$ with $a, b_i \geq 0$ and $b_1 b_2 b_3 = 0$ then $m(\varsigma) = a$ and $n(\varsigma) = a + 2(b_1 + b_2 + b_3)$. The following decomposition of the complex analytic polynomials on V under the action of G is taken from [14].

THEOREM 3.2
The algebra of G-invariants in the complex analytic polynomials on V consists of the polynomials in the invariant f described at the end of Section 3.3. Let Y denote the variety of all $v \in V$ such that $f(v) = 0$ and let $A^n(Y)$ denote the restriction of the space of polynomials of degree n to Y. Then $A^n(Y)$ decomposes into the multiplicity free direct sum of the representations with highest weight ς satisfying the following conditions:

$$n - n(\varsigma) \equiv 0 \ mod \ 2 \quad and \quad m(\varsigma) \geq \frac{n - n(\varsigma)}{2} \geq 0.$$

This result has as an immediate corollary (notation as in the discussion at the beginning of this section):

COROLLARY 3.1
We have

$$m(q, \varsigma) = \frac{q^{n(\varsigma)}(1 + q^2 + \cdots + q^{2m(\varsigma)})}{1 - q^4}$$

$$= \frac{q^{a + 2(b_1 + b_2 + b_3)}(1 + q^2 + \cdots + q^{2a})}{1 - q^4}$$

$$= q^{2(b_1+b_2+b_3)} q^a \frac{1 - q^{2a+2}}{(1 - q^2)(1 - q^4)}.$$

We are now ready to give the bigraded Hilbert series of the invariants in this case.

PROPOSITION 3.3
We have

$$h(q, t) = \frac{(1 + (qt)^2)(1 + (qt)^2 + (qt)^4)}{(1 - qt)(1 - q^4)(1 - q^3t)(1 - q^2t^2)(1 - qt^3)(1 - t^4)(1 - (qt)^3)}.$$

PROOF   Lemma 3.1 and the material above imply that (in the sums below $b_1 b_2 b_3 = 0$ means that we allow all possibilities of $b_i \geq 0$ where at least one of the b_i is 0):

$$h(q, t) = \sum_{\zeta} m(q, \zeta) m(t, \zeta) = \frac{1}{(1 - q^4)(1 - t^4)}$$

$$\sum_{b_1 b_2 b_3 = 0} (qt)^{2(b_1+b_2+b_3)} \sum_{a \geq 0} (qt)^a \frac{1 - q^{2a+2}}{1 - q^2} \cdot \frac{1 - t^{2a+2}}{1 - t^2}.$$

We note that we have in the sense of formal sums

$$\sum_{b_1 b_2 b_3 = 0} x^{b_1+b_2+b_3} = \frac{1}{(1 - x)^3} - \frac{x^3}{(1 - x)^3} = \frac{1 - x^3}{(1 - x)^3}$$

and

$$\sum_{a \geq 0} (qt)^a \frac{1 - q^{2a+2}}{1 - q^2} \cdot \frac{1 - t^{2a+2}}{1 - t^2} = \frac{(1 - q^2)(1 - t^2)(1 - (qt)^4)}{(1 - (qt)^3)(1 - q^3t)(1 - q^2t^2)(1 - qt^3)}.$$

If we make the obvious substitution the result follows.   ∎

Before we do any analysis of this formula we will look at the ordinary Hilbert series of the polynomial invariants for the above action of $S^1 \times G$ (there is no extra information in the bigraded Hilbert series since it would be a series in qt).

PROPOSITION 3.4
The Hilbert series for the polynomial invariants for the action of $S^1 \times G$

on $\bigotimes^3 \mathbb{C}^2$ is

$$h(q) = \frac{1 + q^{12}}{(1 - q^2)(1 - q^4)^3(1 - q^6)(1 - q^8)}.$$

PROOF   In this case we have that if $m(q, \zeta)$ is as above for G and if

$$m(q, \zeta) = \sum_{n \geq 0} a_n(\zeta) q^n$$

then

$$h(q) = \sum_{n \geq 0} q^{2n} \sum_{\zeta} a_n(\zeta)^2.$$

The argument for this is somewhat complicated. We will make some observations that follow from our formula for $m(q, \zeta)$. Define the non-negative integers $a_{n,m}$ by

$$\frac{1 - q^{m+1}}{(1 - q)(1 - q^2)} = \sum_{n \geq 0} a_{n,m} q^n.$$

If

$$w_m(q) = \sum_{n \geq 0} a_{n,m}^2 q^n$$

then if we set

$$g(q) = \frac{1 - q^6}{(1 - q^2)^3} \sum_{m \geq 0} q^m w_m(q^2),$$

we have $h(q) = g(q^2)$. This leaves the calculation of the series $w_m(q)$. The formula depends on the parity of m: if $k \geq 0$ then

$$w_{2k}(q) = \frac{1 + 2(q^2 + \cdots + q^{2k}) - (2k + 1)q^{2k+1}}{(1 - q)(1 - q^2)}$$

and

$$w_{2k+1}(q) = \frac{1 + 2(q^2 + \cdots + q^{2k}) - (2k + 1)q^{2k+2}}{(1 - q)(1 - q^2)}.$$

We write $b_m(q) = (1 - q^2)(1 - q^4)w_m(q^2)$. Then

$$g(q) = \frac{1 - q^6}{(1 - q^4)(1 - q^2)^4} \sum_{m \geq 0} q^m b_m(q).$$

Now

$$\sum_{m \geq 0} q^m b_m(q) = \sum_{k \geq 0} q^{2k} \big(1 + 2(q^4 + \cdots + q^{4k}) - (2k+1)q^{4k+2} \big)$$

$$+ \sum_{k \geq 0} q^{2k+1} \big(1 + 2(q^4 + \cdots + q^{4k}) - (2k+1)q^{4k+4} \big).$$

This expression can be written

$$\frac{1}{1-q} + 2(1+q) \sum_{k \geq 0} q^{2k}(q^4 + \cdots + q^{4k}) - (1+q^3)q^2 \sum_{k \geq 0} (2k+1)q^{6k}$$

$$= \frac{1}{1-q} + 2(1+q)q^4 \sum_{k \geq 0} q^{2k} \frac{1 - q^{4k}}{1 - q^4} + \frac{(1+q^3)q^2}{1-q^6} - \frac{2(1+q^3)q^2}{(1-q^6)^2}$$

$$= \frac{1}{1-q} + \frac{2(1+q)q^4}{(1-q^2)(1-q^4)} - \frac{2(1+q)q^4}{(1-q^4)(1-q^6)}$$

$$+ \frac{(1+q^3)q^2}{1-q^6} - \frac{2(1+q^3)q^2}{(1-q^6)^2}$$

$$= \frac{(1+q^6)(1+q)}{(1-q^6)(1-q^3)}.$$

We therefore have

$$g(q) = \frac{1 + q^6}{(1-q)(1-q^2)^3(1-q^3)(1-q^4)}.$$

Hence

$$h(q) = g(q^2) = \frac{1 + q^{12}}{(1-q^2)(1-q^4)^3(1-q^6)(1-q^8)}.$$

∎

Our next task is to write out a basic set of invariants. This will be done in the next section.

3.5   A basic set of invariants for 3 qubits

In this section we construct a set of invariants for the action of $G = SU(2) \times SU(2) \times SU(2)$ on $V = \mathbb{C}^2 \otimes \mathbb{C}^2 \otimes \mathbb{C}^2$. We will first do this abstractly and then give more concrete formulae for the invariants which are necessary for our proof that they are, in fact, basic. We define an inner product $\langle \cdot, \cdot \rangle$ on $S(V)$ (the symmetric algebra on V) which is the restriction of the usual inner product on the tensor algebra:

$$\langle v_1 \otimes \cdots \otimes v_k | w_1 \otimes \cdots \otimes w_l \rangle = \langle v_1 | w_1 \rangle \cdots \langle v_k | w_k \rangle \, \delta_{k,l}.$$

As usual we will write $v^k = v_1 \otimes \cdots \otimes v_k$ with $v_i = v$ for all $1 \le i \le k$. Then $S(V)$ is the span of the v^k for $v \in V$ and $k \in \mathbb{Z}_{\ge 0}$. We will write $S^k(V)$ for the span of the elements v^k with $v \in V$. Since the representation of G on V is self dual the results we described for the decomposition of the (complex analytic) polynomial functions on V also describe the decomposition of $S(V)$. Thus we have that as representations of G:

$$S^1(V) = V = F^{(\frac{1}{2}, \frac{1}{2}, \frac{1}{2})}$$

$$S^2(V) = F^{(1,0,0)} \oplus F^{(0,1,0)} \oplus F^{(0,0,1)} \oplus F^{(1,1,1)}$$

$$S^3(V) = F^{(\frac{1}{2}, \frac{1}{2}, \frac{1}{2})} \oplus F^{(\frac{1}{2}, \frac{1}{2}, \frac{3}{2})} \oplus F^{(\frac{1}{2}, \frac{3}{2}, \frac{1}{2})} \oplus F^{(\frac{3}{2}, \frac{1}{2}, \frac{1}{2})} \oplus F^{(\frac{3}{2}, \frac{3}{2}, \frac{3}{2})}$$

$$S^4(V) = F^{(0,0,0)} \oplus F^{(2,0,0)} \oplus F^{(0,2,0)} \oplus F^{(0,0,2)} \oplus F^{(1,1,0)} \oplus F^{(1,0,1)}$$
$$\oplus F^{(0,1,1)} \oplus F^{(1,1,1)} \oplus F^{(1,1,2)} \oplus F^{(1,2,1)} \oplus F^{(2,1,1)} \oplus F^{(2,2,2)}.$$

The bigraded formula above implies that there is one invariant of bidegree $(1,1)$, four linearly independent invariants of bidegree $(2,2)$, and one each of bidegrees $(4,0)$, $(3,1)$, $(1,3)$ and $(0,4)$. In bidegree $(3,3)$ there is a 1-dimensional space of invariants that cannot be a subspace of the algebra generated by the ones of lower degree. We assert that the nine dimensional space of invariants obtained from these observations generates the algebra of invariants. We will now give our first description of the desired invariants. Let P_ζ denote the projection onto the F^ζ constituents in each of the symmetric powers described above. Then there is only one $(1,1)$ invariant up to a scalar multiple and that must be $|v|^2$. In bidegree $(2,2)$ the following invariants span: $|v|^4$ and $\langle P_\zeta(v^2), v^2 \rangle$ for $\zeta \in \{(1,0,0), (0,1,0), (0,0,1), (1,1,1)\}$. It is clear that $\sum_\zeta \langle P_\zeta(v^2), v^2 \rangle = |v|^4$. Thus we can choose $\psi_j(v) = \langle P_{\varepsilon_j}(v^2), v^2 \rangle$ with $\varepsilon_1 = (1,0,0)$, $\varepsilon_2 = (0,1,0)$ and $\varepsilon_3 = (0,0,1)$. Up to a scalar multiple the only invariant of bidegree $(3,1)$ is obtained as follows. The above decomposition of $S^3(V)$ implies that there exists a unique (up to a scalar multiple) intertwining operator

$$T : V \to S^3(V)$$

(that is, $T(gv) = gT(v)$ for $g \in G$). We set $\psi_4(v) = \langle v^3, T(v) \rangle$ and $\psi_5(v) = \langle T(v), v^3 \rangle$. It is clear that up to a scalar multiple the only

$(4,0)$ invariant is our original one, f, and the one of bidegree $(0,4)$ is its complex conjugate. Finally we set $\psi_6(v) = \langle P_{(\frac{1}{2}, \frac{1}{2}, \frac{1}{2})}(v^3), v^3 \rangle$.

Our next task is to give more concrete descriptions of ψ_j, $1 \leq j \leq 6$. For this we must use a bit more of the structure of the representation of G on $V = \mathbb{C}^2 \otimes \mathbb{C}^2 \otimes \mathbb{C}^2$. We observe that there is a symplectic structure over $\mathbb{C}$. Indeed, if we write

$$
\begin{aligned}
v &= v(x, y) \\
&= x_1|000\rangle + x_2|011\rangle + x_3|101\rangle + x_4|110\rangle \\
&\quad + y_1|111\rangle + y_2|100\rangle + y_3|010\rangle + y_4|001\rangle,
\end{aligned}
$$

then the symplectic structure is given by

$$
\omega\big(v(x, y), v(s, t)\big) = \sum x_i t_i - \sum y_i s_i.
$$

We therefore have a Poisson bracket on the polynomial functions on V given by

$$
\begin{aligned}
\{g, h\}\big(v(x, y)\big) = \sum \frac{\partial g}{\partial x_i}\big(v(x, y)\big) \frac{\partial h}{\partial y_i}\big(v(x, y)\big) \\
- \sum \frac{\partial g}{\partial y_i}\big(v(x, y)\big) \frac{\partial h}{\partial x_i}\big(v(x, y)\big).
\end{aligned}
$$

Since the action of G is symplectic it follows that the action of its Lie algebra on polynomials is given by a Poisson bracket with quadratic elements. The complexified Lie algebra of G is a direct sum of three copies of $\mathfrak{sl}(2, \mathbb{C})$. We will now write out the corresponding polynomials. Note that the Lie algebra of $\mathfrak{sl}(2, \mathbb{C})$ has basis $\{e, f, h\}$ with

$$
e = \begin{bmatrix} 0 & 1 \\ 0 & 0 \end{bmatrix}, \quad f = \begin{bmatrix} 0 & 0 \\ 1 & 0 \end{bmatrix}, \quad h = \begin{bmatrix} 1 & 0 \\ 0 & -1 \end{bmatrix}.
$$

So the three sets of polynomials are:

$$
\begin{aligned}
e_1 &= x_1 x_2 - y_3 y_4, & f_1 &= x_3 x_4 - y_1 y_2, \\
e_2 &= x_1 x_3 - y_2 y_4, & f_2 &= x_2 x_4 - y_1 y_3, \\
e_3 &= x_1 x_4 - y_2 y_3, & f_3 &= x_2 x_3 - y_1 y_4, \\
h_1 &= -x_1 y_1 - x_2 y_2 + x_3 y_3 + x_4 y_4; \\
h_2 &= -x_1 y_1 + x_2 y_2 - x_3 y_3 + x_4 y_4; \\
h_3 &= -x_1 y_1 + x_2 y_2 + x_3 y_3 - x_4 y_4.
\end{aligned}
$$

The elements $\frac{1}{2}h_i^2 + 2e_i f_i$ are all the same, and up to scalar multiple equal to the polynomial f above. One can check that $\{e_i, f\} = \{f_i, f\} = \{h_i, f\} = 0$ for $i \in \{1, 2, 3\}$ directly.

We now need notation for the two copies of V that come into the study of the previous section. Let $z_i = x_i$ and $z_{4+i} = y_i$ for $1 \le i \le 4$. We note that the symplectic basis used above is also orthonormal. Thus if we think of the second copy as the conjugate space using the same basis, the action of an element of K is by its conjugate and thus by the transpose inverse relative to the above basis. It is convenient to think of the z_j as $s_j + it_j$ with s_j and t_j real, and introduce new variables w_j with $w_j = -s_{j+4} + it_{j+4}$ and $w_{j+4} = s_j - it_j$ for $1 \le j \le 4$. We now note that in this context the polynomials on $V \oplus V$ (using the coordinates z_j and w_j) admit polarization operators. We set

$$D_{w,z} = \sum w_i \frac{\partial}{\partial z_i} \quad \text{and} \quad D_{z,w} = \sum z_i \frac{\partial}{\partial w_i}.$$

One checks that

$$H = [D_{w,z}, D_{z,w}] = \sum w_i \frac{\partial}{\partial w_i} - \sum z_i \frac{\partial}{\partial z_i}$$

to see that we have yet another Lie algebra isomorphic with the Lie algebra $\mathfrak{sl}(2, \mathbb{C})$ viz:

$$e \longmapsto D_{z,w}, \quad f \longmapsto D_{w,z}, \quad h \longmapsto H.$$

The action of this Lie algebra commutes with the action of K on the polynomials in the two copies of V. We now write the operators analogous to the e_i, f_i and h_i in terms of the coordinates w_i. They become:

$$E_1 = w_1 w_2 - w_7 w_8, \quad F_1 = w_3 w_4 - w_5 w_6,$$
$$E_2 = w_1 w_3 - w_6 w_8, \quad F_2 = w_2 w_4 - w_5 w_7,$$
$$E_3 = w_1 w_4 - w_6 w_7, \quad F_3 = w_2 w_3 - w_5 w_8,$$
$$H_1 = -w_1 w_5 - w_2 w_6 + w_3 w_7 + w_4 w_8;$$
$$H_2 = -w_1 w_5 + w_2 w_6 - w_3 w_7 + w_4 w_8;$$
$$H_3 = -w_1 w_5 + w_2 w_6 + w_3 w_7 - w_4 w_8.$$

We can now write down formulae for our invariants:

$$|v|^2 = \sum z_i w_{i+4} - \sum z_{i+4} w_i$$
$$\psi_i = \frac{h_i H_i}{2} + e_i F_i + f_i E_i, \quad 1 \le i \le 3$$

$$f = \frac{h_i^2}{2} + 2e_i f_i \quad \text{(for any } i)$$

$$g = \bar{f} = \frac{H_i^2}{2} + 2E_i F_i \quad \text{(for any } i)$$

$$\psi_4 = D_{w,x} f$$

$$\psi_5 = D_{z,w} g \propto D_{w,z}^3 f$$

$$\psi_6 = \sum \frac{\partial f}{\partial z_i} \frac{\partial g}{\partial w_i}.$$

Note that up to a scalar multiple

$$D_{w,z}^2 f = 2(\psi_1 + \psi_2 + \psi_3) + |v|^4.$$

The following five elements:

$$f, \ D_{w,z} f, \ D_{w,z}^2 f, \ D_{w,z}^3 f, \ D_{w,z}^4 f$$

span a representation space for the fourth action of $\mathfrak{sl}(2,\mathbb{C})$, equivalent with the 5-dimensional irreducible representation. We denote those elements by u_1, u_2, u_3, u_4, u_5. We also observe that it is well known that the algebra of invariants in the polynomials in $V \oplus V$ under the action of the four copies of $\mathfrak{sl}(2,\mathbb{C})$ is a polynomial ring in 4 generators of respective degrees $2, 4, 4, 6$. A calculation shows that an element $a_1\psi_1 + a_2\psi_2 + a_3\psi_3$ is invariant under all three actions if and only if $a_1 + a_2 + a_3 = 0$. One can check that ψ_6 is not of the form

$$|v|^2 (a_1\psi_1 + a_2\psi_2 + a_3\psi_3 + a_4|v|^2)$$

for any choice of a_j, $1 \leq j \leq 4$. One can also show that the algebra of $\mathfrak{sl}(2,\mathbb{C})$ invariants in the polynomials on the 5-dimensional irreducible representation is a polynomial ring in two invariants, α_1 and α_2, of degrees 2 and 3.

PROPOSITION 3.5
The algebra generated by $|v|^2$, u_i $(1 \leq i \leq 5)$, and ψ_6 is isomorphic with the polynomial algebra in seven variables.

This result has been proved with the help of the computer algebra package Maple as follows. Form the matrix with entries

$$A_{i,j} = \frac{\partial u_i}{\partial z_j} \quad \text{and} \quad A_{i,j+8} = \frac{\partial u_i}{\partial w_j}$$

for $2 \leq i \leq 6$, $1 \leq j \leq 8$ and

$$A_{1,j} = \frac{\partial |v|^2}{\partial z_j}, \quad A_{1,j+8} = \frac{\partial |v|^2}{\partial w_j}, \quad A_{7,j} = \frac{\partial \psi_6}{\partial z_j}, \quad A_{7,j+8} = \frac{\partial \psi_6}{\partial w_j}.$$

Substitute "random" values for the z_i and w_i and then use Gaussian elimination to find a nonzero 7×7 minor (e.g., use C_{ij} with $i \in \{1,2,3,4,5,6,7\}$ and $j \in \{1,2,3,4,5,6,9\}$). Thus if $f_1 = |v|^2$, $f_{i+1} = u_i$ for $1 \leq i \leq 5$, and $f_7 = \psi_7$ then

$$\mathrm{d}f_1 \wedge \mathrm{d}f_2 \wedge \mathrm{d}f_3 \wedge \mathrm{d}f_4 \wedge \mathrm{d}f_5 \wedge \mathrm{d}f_6 \wedge \mathrm{d}f_7$$

is nonzero on an open dense subset of $\mathbb{C}^{16}$. This clearly implies that if h is a polynomial in 7 indeterminates and $h(f_1, f_2, f_3, f_4, f_5, f_6, f_7)$ is identically 0 then h is identically 0.

We are finally ready to give the main result on invariants:

THEOREM 3.3
The algebra of G-invariants is generated by $|v|^2, u_1, \ldots, u_5, \psi_6, \psi_1 - \psi_2,$
$\psi_2 - \psi_3$.

PROOF We note that the general theory of symmetric pairs, applied to $SO(4,4)$ (a reference for this theory, used throughout this proof, can be found in [15, Section 12.4]), implies that the algebra, I, of invariants under G annihilated by both D_{zw} and D_{wz}, is a polynomial ring in generators of degrees $2, 4, 4, 6$. We already know that the elements $|v|^2$, $\psi_1 - \psi_2$ and $\psi_2 - \psi_3$ have these additional properties. We also note that $D_{zw}(\psi_6 + \frac{|v|^2 u_3}{6}) = D_{wz}(\psi_6 + \frac{|v|^2 u_3}{6}) = 0$. Thus if we take $\alpha_1 = |v|^2$, $\alpha_2 = \psi_1 - \psi_2$, $\alpha_4 = \psi_2 - \psi_3$ and $\alpha_4 = \psi_6 + \frac{|v|^2 u_3}{6}$ then these give algebraically independent generators of the algebra I. The general theory also implies that the algebra of all polynomials on $V \oplus V$ is a free I-module under multiplication.

We also note that the same theory for the symmetric pair $(SL(3, \mathbb{R}),$ $SO(3))$ implies that $J = \mathbb{C}[u_1, \ldots, u_5] \cap I$ is a polynomial algebra in two generators β_1 and β_2 of respective degrees 8 and 12 with

$$\beta_1 = 2f D_{wz}^4 f + (D_{wz}^2 f)^2 - 2D_{wz} f D_{wz}^3 f$$
$$\beta_2 = 2(D_{wz}^2 f)^3 - 6D_{wz} f D_{wz}^2 f D_{wz}^3 f + 9f(D_{wz}^3 f)^2$$
$$- 12f D_{wz}^2 f D_{wz}^4 f + 6(D_{wz} f)^2 D_{wz}^4 f.$$

Furthermore, $\mathbb{C}[u_1, \ldots, u_5]$ is a free J-module under multiplication and
the algebra $\mathbb{C}[\alpha_1, \alpha_2, \alpha_3, \alpha_4]$ is a free $L = \mathbb{C}[\alpha_1, \beta_1, \beta_2, \alpha_4]$ module under
multiplication. Thus since the algebra of all polynomials is free as a
$\mathbb{C}[\alpha_1, \alpha_2, \alpha_3, \alpha_4]$-module under multiplication and the algebra $\mathbb{C}[\alpha_1, u_1,$
$\ldots, u_5, \alpha_4]$ is a module direct summand, it is thus free as an L-module
under multiplication. This implies that the algebra $\mathbb{C}[\alpha_1, \alpha_2, \alpha_3, \alpha_4,$
$u_1, u_2, u_3, u_4, u_5]$ is isomorphic with

$$\mathbb{C}[\alpha_1, u_1, \ldots, u_5, \alpha_4] \bigotimes_L \mathbb{C}[\alpha_1, \alpha_2, \alpha_3, \alpha_4],$$

which has Hilbert series

$$\frac{(1 - q^8)(1 - q^{12})(1 - q^2)(1 - q^6)}{(1 - q^4)^5(1 - q^2)(1 - q^6)(1 - q^2)(1 - q^4)^2(1 - q^6)}$$
$$= \frac{(1 - q^8)(1 - q^{12})}{(1 - q^4)^5(1 - q^2)(1 - q^6)(1 - q^4)^2}$$
$$= \frac{(1 + q^4)(1 + q^4 + q^8)}{(1 - q^2)(1 - q^4)^5(1 - q^6)}.$$

This agrees with the Hilbert series that we calculated for the invariants
in the previous section. ∎

Before we go on to the invariants for $U(2) \times U(2) \times U(2)$ a few observations about the invariants are in order. If $v \in V$ then we can construct
three pairs of vectors from v. Write $v = \sum x_i|i\rangle$. Let $\alpha = (x_0, x_1, x_2, x_3)$,
$\beta = (x_4, x_5, x_6, x_7)$, $\gamma = (x_0, x_1, x_4, x_5)$, $\delta = (x_2, x_3, x_6, x_7)$, $\mu = (x_0, x_2,$
$x_4, x_6)$ and $\nu = (x_1, x_3, x_5, x_7)$. Then the pairs are (α, β), (γ, δ) and
(μ, ν). In the first pair we are looking at whether or not the first (most
significant) bit (of i) is 0, for the next the second bit and for the last the
last bit. If u, v are vectors in $\mathbb{C}^4$ then we define $\Delta(u, v)$ to be the sum
of the absolute value squared of the 2×2 minors of the matrix

$$\begin{bmatrix} u_1 & u_2 & u_3 & u_4 \\ v_1 & v_2 & v_3 & v_4 \end{bmatrix}.$$

Then the invariants ψ_1, ψ_2 and ψ_3 are up to scalar multiples $\Delta(\alpha, \beta)$,
$\Delta(\gamma, \delta)$ and $\Delta(\mu, \nu)$. Thus $\sum \psi_j$ is up to a scalar multiple the invariant
Q defined for arbitrarily many qubits in [16].

3.6 Some implications for other representations

In this section we will show how the results in the previous sections apply to other compact Lie groups. We first observe (as in Section 3.2) that we may look upon the results as the analysis of the action of $G = SU(2) \times SU(2) \times SU(2)$ on $\mathbb{C}^2 \otimes \mathbb{C}^2 \otimes \mathbb{C}^2 \otimes \mathbb{C}^2$ via $g \otimes I$. The group $U(2)$ acting only on the last factor commutes with the action of G. We will now rewrite the formula in Proposition 3.3 to take into account the total homogeneity. In other words we write $q = qx$ and $t = qx^{-1}$. The formula now becomes

$$\frac{(1 + q^4)(1 + q^4 + q^8)}{(1 - q^2)(1 - q^4 x^4)(1 - q^4 x^2)(1 - q^4)(1 - q^4 x^{-2})(1 - q^4 x^{-4})(1 - q^6)}.$$

We note that the variable x can be thought of as the parameter of the circle subgroup, T, of all

$$\begin{bmatrix} x & 0 \\ 0 & x^{-1} \end{bmatrix}$$

in the $SU(2)$ acting on the fourth variable, and the formula above is just the q-character for the action on the G-invariants. In [17] the q-multiplicity formulae for the action of $SU(2)$ on the spin-2 (5-dimensional) representation was determined. Set $W = F^2$. Then $m_{F^2}(q, k) = 0$ if k is not an integer. If $k = 2l$ is an even integer we have

$$(1 - q^2)(1 - q^3) m_W(q, 2l) = q^l + q^{l+1} + \cdots + q^{2l} = q^l \frac{1 - q^{l+1}}{1 - q}.$$

If $k = 2l + 3$ then we have

$$m_W(q, 2l + 3) = q^3 m_W(q, 2l).$$

One can prove this by observing that these formulae satisfy

$$\frac{1}{(1 - qx^4)(1 - qx^2)(1 - q)(1 - qx^{-2})(1 - qx^{-4})}$$
$$= \sum_{k \geq 0} m_W(q, k) \frac{x^{2k+1} - x^{-2k-1}}{x - x^{-1}}.$$

We first note that this gives an alternate proof of Proposition 3.3 since that proposition describes the Hilbert series of the invariants for the action of T on the polynomial invariants. Note that there is a shift $q \to q^4$. Thus since every repesentation F^k with k an integer has a T-fixed vector, we see that the Hilbert series for the action of $S^1 \times G$ as in Section 3.3 is

$$\frac{(1 + q^4)(1 + q^4 + q^8)}{(1 - q^2)(1 - q^6)} \sum_{k \geq 0} m_W(q^4, k).$$

We leave it to the reader to check that this formula agrees with Proposition 3.3. We also note that, on the other hand, Proposition 3.3 can be used to derive information about the series $m_W(q, k)$.

More seriously, we note that our formulae give information about 4 qubits:

COROLLARY 3.2

Let $SU(2) \times SU(2) \times SU(2) \times SU(2)$ act on $U = \mathbb{C}^2 \otimes \mathbb{C}^2 \otimes \mathbb{C}^2 \otimes \mathbb{C}^2$ as above. Then

$$m_U\big(q, (0, 0, 0, k)\big) = \frac{(1 + q^4)(1 + q^4 + q^8)}{(1 - q^2)(1 - q^6)} m_{F^2}(q^4, k).$$

References

[1] M. H. Freedman, Poly-locality in quantum computing, `quant-ph/0001077`.

[2] P. W. Shor, Algorithms for quantum computation: discrete logarithms and factoring, in S. Goldwasser, Ed., *Proceedings of the 35th Symposium on Foundations of Computer Science*, Santa Fe, NM, 20–22 November 1994 (Los Alamitos, CA: IEEE Computer Society Press 1994) 124–134;
P. W. Shor, Polynomial-time algorithms for prime factorization and discrete logarithms on a quantum computer, *SIAM J. Comp.* **26** (1997) 1484–1509.

[3] D. Coppersmith, An approximate Fourier transform useful in quantum factoring, *IBM Research Report RC 19642* (12 July 1994).

[4] A. Einstein, B. Podolsky and N. Rosen, Can quantum-mechanical description of physical reality be considered complete?, *Phys. Rev.* **47** (1935) 777–780.

[5] D. Bohm, *Quantum Theory* (New York: Prentice-Hall 1951).

[6] J. S. Bell, On the Einstein-Podolsky-Rosen paradox, *Physics* **1** (1964) 195–200.

[7] E. Schrödinger, Die gegenwärtige Situation in der Quantenmechanik, *Naturwissenschaften* **23** (1935) 807–812; 823–828; 844–849.

[8] E. M. Rains, Polynomial invariants of quantum codes, *IEEE Trans. Inform. Theory* **46** (2000) 54–59.

[9] N. Linden and S. Popescu, On multi-particle entanglement, *Fortsch. Phys.* **46** (1998) 567–578.

[10] M. Grassl, M. Rötteler and T. Beth, Computing local invariants of quantum-bit systems, *Phys. Rev. A* **58** (1998) 1833–1839.

[11] A. Sudbery, On local invariants of pure three-qubit states, *J. Phys. A: Math. Gen.* **34** (2001) 643–652.

[12] J.-L. Brylinski and R. Brylinski, Invariant polynomial functions on k qudits, `quant-ph/0010101`. In Chapter 11 of this book.

[13] M. Grassl, Description of multi-particle entanglement through polynomial invariants, talk presented at the *Workshop on Complexity, Computation and the Physics of Information*, Isaac Newton Institute for Mathematical Sciences, Cambridge, UK, 22 July 1999;
`http://iaks-www.ira.uka.de/home/grassl/paper/CCP.ps.gz`.

[14] B. Gross and N. Wallach, On quaternionic discrete series representations, and their continuations, *J. Reine Angew. Math.* **481** (1996) 73–123.

[15] R. Goodman and N. Wallach, *Representations and Invariants of the Classical Groups, Encyclopedia of Mathematics and its Applications* **68** (Cambridge, UK: Cambridge University Press 1998).

[16] D. A. Meyer and N. Wallach, Global entanglement in multiparticle systems, `quant-ph/0108104`.

[17] N. Wallach and J. Willenbring, On some q-analogs of a theorem of Kostant-Rallis, *Canad. J. Math.* 52 (2000) 438–448.

Universality of Quantum Gates

Chapter 4

Universal quantum gates

Jean-Luc Brylinski* and Ranee Brylinski

Abstract   In this paper we study universality for quantum gates acting on qudits. Qudits are states in a Hilbert space of dimension d where d can be any integer ≥ 2. We determine which 2-qudit gates V have the properties: (i) the collection of all 1-qudit gates together with V produces all n-qudit gates up to arbitrary precision, or (ii) the collection of all 1-qudit gates together with V produces all n-qudit gates exactly. We show that (i) and (ii) are equivalent conditions on V, and they hold if and only if V is not a primitive gate. Here we say V is primitive if it transforms any decomposable tensor into a decomposable tensor. We discuss some applications and also relations with work of other authors.

4.1   Statements of main results

We determine which 2-qudit gates V have the property that all 1-qudit gates together with V form a universal collection, in either the approximate sense or the exact sense. Here d is an arbitrary integer ≥ 2. Our results are new for the case of qubits, i.e., $d = 2$ (which for

*Research supported in part by NSF Grant No. DMS-9803593

many is the case of primary interest). We treat the case $d > 2$ as well because it is of independent interest and requires no additional work.

Since Deutsch [3] found a universal gate (on 3 qubits), universal gates for qubits have been extensively studied. We mention in particular the papers [1, 2, 4, 5, 6] which will be further discussed in Section 4.2.

First we set up some notations. A *qudit* is a (normalized) state in the Hilbert space $\mathbb{C}^d$. An *n-qudit* is a state in the tensor product Hilbert space $H = (\mathbb{C}^d)^{\otimes n} = \mathbb{C}^d \otimes \cdots \otimes \mathbb{C}^d$. The *computational basis* of H is the orthonormal basis given by the d^n classical n-qudits

$$|i_1 i_2 \cdots i_n\rangle = |i_1\rangle \otimes |i_2\rangle \otimes \cdots \otimes |i_n\rangle \qquad (4.1)$$

where $0 \leq i_j \leq d - 1$. The general state in H is a superposition

$$|\psi\rangle = \sum \psi_{i_1 i_2 \cdots i_n} |i_1 i_2 \cdots i_n\rangle \qquad (4.2)$$

where $||\psi||^2 = \sum |\psi_{i_1 i_2 \cdots i_n}|^2 = 1$. We say ψ is *decomposable* when it can be written as a tensor product $|x_1 \cdots x_n\rangle = |x_1\rangle \otimes |x_2\rangle \otimes \cdots \otimes |x_n\rangle$ of qudits.

A quantum gate on n-qudits is a unitary operator $L : (\mathbb{C}^d)^{\otimes n} \to (\mathbb{C}^d)^{\otimes n}$. These gates form the unitary group $U((\mathbb{C}^d)^{\otimes n}) = U(d^n)$. A sequence $L_1, \ldots, L_k$ of gates constitutes a quantum circuit on n-qudits. The output of that circuit is the product gate $L_1 \cdots L_k$. In practice, one wants to build circuits out of gates L_i which are *local* in that they operate on only a small number of qudits, typically 1, 2 or 3.

We can produce local gates in the following way. A 1-qudit gate A gives rise to n different n-qudit gates $A(1), \cdots, A(n)$ obtained by making A act on the individual tensor slots. So

$$A(l)|x_1 \cdots x_l \cdots x_n\rangle = |x_1\rangle \otimes \cdots \otimes |x_{l-1}\rangle \otimes A|x_l\rangle \otimes |x_{l+1}\rangle \otimes \cdots \otimes |x_n\rangle \quad (4.3)$$

Similarly, for a 2-qudit gate B, we have $n(n-1)$ different n-qudit gates $B(p,q)$ obtained by making B act on pairs of slots. For $B = \sum S_i \otimes T_i$ we have $B(p,q) = \sum S_i(p)T_i(q)$.

A basic problem in quantum computation is to find collections of gates that are universal in the following sense.

DEFINITION 4.1   *A collection of 1-qudit gates A_i and 2-qudit gates B_j is called* universal *if, for each $n \geq 2$, every n-qudit gate can be approximated with arbitrary accuracy by a circuit made up of the n-qudit gates produced by the A_i and B_j.*

We also have the stronger notion, which we call exact universality.

DEFINITION 4.2   *A collection of 1-qudit gates A_i and 2-qudit gates B_j is called* exactly universal *if, for each $n \geq 2$, every n-qudit gate can be obtained exactly by a circuit made up of the n-qudit gates produced by the A_i and B_j.*

In mathematical terms, universality means that the n-qudit gates produced by the A_i and B_j generate a dense subgroup of $U(d^n)$, while exact universality means that these gates generate the full group $U(d^n)$.

Note that a finite collection of 1-qudit and 2-qudit gates can be universal, but it can never be exactly universal, as the group it generates is countable, while $U(d^n)$ is uncountable.

We now state our main result. We introduce the following terminology. A 2-qudit gate V is *primitive* if V maps decomposables to decomposables, i.e., if $|x\rangle$ and $|y\rangle$ are qudits then we can find qudits $|u\rangle$ and $|v\rangle$ such that $V|xy\rangle = |uv\rangle$. We say V is *imprimitive* when V is not primitive. Let $P : (\mathbb{C}^d)^{\otimes 2} \to (\mathbb{C}^d)^{\otimes 2}$ denote the 2-qudit gate such that $P|xy\rangle = |yx\rangle$.

THEOREM 4.1
Suppose we are given a 2-qudit gate V. Then the following are equivalent:

(i) The collection of all 1-qudit gates A together with V is universal.

(ii) The collection of all 1-qudit gates A together with V is exactly universal.

(iii) V is imprimitive.

We prove Theorem 4.1 in Sections 4.3-4.7. The implications (ii)$\Rightarrow$(i)$\Rightarrow$(iii) are easy. The hard part is showing (iii)$\Rightarrow$(ii). In Section 4.9 we give a variant of Theorem 4.1.

In Section 4.8 we characterize primitive gates in the following way.

THEOREM 4.2
V is primitive if and only if $V = S \otimes T$ or $V = (S \otimes T)P$ for some 1-qudit gates S and T. Thus V acts by $V|xy\rangle = S|x\rangle \otimes T|y\rangle$ or by $V|xy\rangle = S|y\rangle \otimes T|x\rangle$.

COROLLARY 4.1
Almost every 2-qudit gate is imprimitive. In fact the imprimitive gates form a connected open dense subset of $U(d^2)$.

For the proofs, we use Lie group theory, including some representation theory for compact groups. For exact universality, we also use some real algebraic geometry (in proving Lemma 4.1). Our methods can be used to prove a variety of results on universality and exact universality. We illustrate this is in Section 4.9.

We thank Goong Chen and Martin Rötteler for useful discussions, and for asking us questions that led, respectively, to the results in Section 4.9 and the results on exact universality.

Part of this work was carried out while both authors were Professeurs Invités at the CPT and IML of the Université de la Méditerranée in Marseille, France. They are grateful to the Université de la Méditerranée for its hospitality. Also part of this work was done at the Mini-Symposium on Quantum Computation at College Station on May 4–6, 2001, where RKB lectured on it. We thank Andreas Klappenecker for organizing this symposium.

4.2 Examples and relations to works of other authors

In this section, we give examples of primitive and imprimitive gates.

PROPOSITION 4.1
Suppose a 2-qudit gate V is diagonal in the computational basis with $V|jk\rangle = e^{i\theta_{jk}}|jk\rangle$. Then V is primitive if and only if for all j, k, p, q we have

$$\theta_{jk} + \theta_{pq} \equiv \theta_{jq} + \theta_{pk} \pmod{2\pi} \tag{4.4}$$

PROOF We apply V to the decomposable tensor

$$|\psi\rangle = (|j\rangle + |p\rangle) \otimes (|k\rangle + |q\rangle)$$

We put $\alpha_{jk} = e^{i\theta_{jk}}$. If V is primitive then the result

$$V|\psi\rangle = \alpha_{jk}|jk\rangle + \alpha_{jq}|jq\rangle + \alpha_{pk}|pk\rangle + \alpha_{pq}|pq\rangle$$

must be decomposable. Thus $\alpha_{jk}\alpha_{pq} - \alpha_{jq}\alpha_{pk}$ vanishes, which amounts to (4.4). Conversely, if (4.4) holds, we can solve for scalars β_j and γ_k such that $\alpha_{jk} = \beta_j\gamma_k$. Then $V = B \otimes C$ where $B|j\rangle = \beta_j|j\rangle$ and $C|k\rangle = \gamma_k|k\rangle$. ∎

For example, if all θ_{jk} are zero except that $\theta_{00} \not\equiv 0 \pmod{2\pi}$, then V is imprimitive. In the case $d = 2$, (4.4) reduces to the condition $\theta_{00} + \theta_{11} \equiv \theta_{01} + \theta_{10} \pmod{2\pi}$ found in [6].

In another direction, consider the generalized CNOT gate X given by $X|ij\rangle = |i, i \oplus j\rangle$ where $i \oplus j$ denotes addition of integers modulo d. For $d = 2$, X is the standard CNOT gate. Then X is imprimitive because X transforms the decomposable tensor $(|0\rangle + |1\rangle) \otimes |0\rangle)$ into the indecomposable tensor $|00\rangle + |11\rangle$. Therefore the collection of all 1-qudit gates together with X is exactly universal. This was already proven when $d = 2$ in [2].

Here is another kind of controlled gate. Take some 1-qudit gate U and define a 2-qudit gate X_U by $X_U|0k\rangle = |0\rangle \otimes U|k\rangle$ and, for $j \neq 0$, $X_U|jk\rangle = |jk\rangle$. Then X_U is primitive if and only if U is a scalar operator, i.e., $U|x\rangle = e^{i\theta}|x\rangle$. Indeed, for any $j \neq 0$ we have

$$X_U(|jx\rangle + |0x\rangle) = |j\rangle \otimes |x\rangle + |0\rangle \otimes U|x\rangle$$

This must be decomposable if X_U is primitive. This can only happen if $U|x\rangle = e^{i\theta}|x\rangle$. Since $|x\rangle$ is arbitrary, we see that $e^{i\theta}$ is independent of $|x\rangle$. Thus U is a scalar operator. This construction yields many imprimitive gates that have finite order.

Another point of view is to consider a 2-qudit gate V just by itself. This is interesting because almost any V is universal. This was proven in [6] and (for $d = 2$) in [4]. More precisely, these authors found finitely many open conditions on gates (e.g., the closure of the subgroup generated by the gate is a maximal torus in $U(d^2)$) which automatically imply the gate is universal by itself. In particular, all their gates have infinite order. We will call V an *IU gate* (individually universal) if V is universal by itself.

By Theorem 4.1, IU gates are imprimitive. There are many gates that are imprimitive but not IU: for instance, imprimitive gates that are diagonal in the computational basis.

4.3   Proof of theorem 4.1 (outline)

We will end up focusing on 2-qudits, and so we put $G = U(d^2)$. We define H to be the subgroup of G generated by the 2-qudit gates $A(1)$ and $A(2)$ for $A \in U(d)$. Let F be the subgroup of G generated by H, V and $V(2,1)$.

(ii)$\Rightarrow$(i): obvious

(i)$\Rightarrow$(iii): Suppose V is primitive. We will show that universality fails for $n = 2$, i.e., F is not dense in G. Clearly F lies in the set L of primitive gates. But (a) L is a closed subgroup of G and (b) $L \neq G$. Indeed (a) follows easily from the definition of primitive since the decomposable tensors in $(\mathbb{C}^d)^{\otimes 2}$ form a closed subset. Also (b) is true because we already exhibited in Section 4.2 some 2-qudit gates that are imprimitive. So L is not dense in G. So F is not dense in G.

(iii)$\Rightarrow$(ii): This takes more work. Here is an outline. The details are given in Section 4.4 (first step), Section 4.5 (second step), Section 4.6 (fourth step) and Section 4.7 (fifth step).

First step: We give a general abstract result, Lemma 4.1, which says that if k closed connected subgroups of a compact group $\mathcal{G}$ generate a dense subgroup of $\mathcal{G}$, they must in fact generate $\mathcal{G}$.

Second step: Using Lemma 4.1 we reduce the problem to $n = 2$.

Third step: H is the set of 2-qudit gates of the form $S \otimes T$. So H is a closed connected Lie subgroup of G. Lemma 4.1 suggests that we look for a closed connected subgroup H' of G such that

$$H \text{ and } H' \text{ generate a dense subgroup of } G \qquad (4.5)$$

The trick is to find a nice way to choose H'. We introduce the subgroup $H' = VHV^{-1}$; this is clearly closed and connected. The next two steps of the proof are devoted to showing our group H' satisfies (4.5).

Fourth step: We will use the Lie algebras $\mathfrak{g} = Lie\ G$, $\mathfrak{h} = Lie\ H$ and $\mathfrak{h}' = Lie\ H'$. Showing (4.5) amounts to showing that $\mathfrak{h}$ and $\mathfrak{h}'$ generate $\mathfrak{g}$ as a Lie algebra. Let $\mathfrak{z}$ be the Lie subalgebra generated by $\mathfrak{h}$ and $\mathfrak{h}'$. So $\mathfrak{h} \subseteq \mathfrak{z} \subseteq \mathfrak{g}$. Using some representation theory, we show abstractly in Lemma 4.2 that there is no Lie algebra strictly in between $\mathfrak{h}$ and $\mathfrak{g}$. Thus $\mathfrak{z} = \mathfrak{h}$ or $\mathfrak{z} = \mathfrak{g}$.

Fifth step: We need to rule out $\mathfrak{z} = \mathfrak{h}$. Clearly $\mathfrak{z} = \mathfrak{h} \Leftrightarrow \mathfrak{h}' = \mathfrak{h} \Leftrightarrow H' = H \Leftrightarrow V$ normalizes H. But we prove in Proposition 4.2 that the normalizer of H is the set of primitive gates. So V cannot normalize H. Thus $\mathfrak{z} \neq \mathfrak{h}$.

Sixth step: Thus $\mathfrak{z} = \mathfrak{g}$. This proves (4.5). Now (4.5) and Lemma 4.1 imply that H and H' generate G. So *a fortiori*, H and V generate G.

REMARK 4.1   (i) We actually proved something stronger than exact universality, namely that H and V generate G.

(ii) To prove (iii)$\Rightarrow$(i) directly, there is no need for H' or Lemma 4.1. We can simply work with F. The problem is to show that F is dense in G, which amounts to showing that $\mathfrak{f} = \mathfrak{g}$ where $\mathfrak{f}$ is the Lie algebra of the closure $\overline{F}$ of F in G. Clearly $\mathfrak{h} \subseteq \mathfrak{f} \subseteq \mathfrak{g}$ Then we use the same two results, Lemma 4.2 and Proposition 4.2, to show, respectively, that (a) $\mathfrak{f} = \mathfrak{h}$ or $\mathfrak{f} = \mathfrak{g}$ and (b) $\mathfrak{f} = \mathfrak{h}$ does not happen.   ∎

4.4   First step: from universality to exact universality

Our bridge from universality to exact universality is

LEMMA 4.1

Let G be a compact Lie group. If $\mathcal{H}_1, \ldots, \mathcal{H}_k$ are closed connected subgroups and they generate a dense subgroup of G, then in fact they generate G.

PROOF   We can take $k = 2$ since the general case easily reduces to this. Consider the subset $\Sigma = \mathcal{H}_1\mathcal{H}_2$ of G and its n-fold products $\Sigma^n = \Sigma \cdots \Sigma$. Then $\Sigma, \Sigma^2, \ldots$ is an increasing sequence of subsets whose union, call it Σ^∞, is dense in G. We want to show that there exists m such that $\Sigma^m = G$.

To begin with, we observe that Σ^n is compact and connected. This follows as Σ^n is the image of the continuous multiplication map μ from the compact connected set $(\mathcal{H}_1 \times \mathcal{H}_2)^{\times n}$ into G. So Σ^∞ is connected. So G is connected.

In fact we can conclude much more using μ. For G has an additional structure compatible with its Lie group structure: G is a smooth irreducible real algebraic variety. (In fact, we can faithfully represent G on some $\mathbb{C}^N$ and then G is an irreducible closed real algebraic subvariety of

the space of matrices of size N.) The subgroups $\mathcal{H}_1$ and $\mathcal{H}_2$ are closed irreducible subvarieties; to obtain irreducibility we use the connectedness
of $\mathcal{H}_1$ and $\mathcal{H}_2$.

Clearly μ is a morphism of irreducible real algebraic varieties. It follows using the Tarski–Seidenberg theorem that Σ^n is a semi-algebraic
set in $\mathcal{G}$ and its "algebraic closure" Z_n is irreducible. Here Z_n is the
unique smallest closed real algebraic subvariety of $\mathcal{G}$ which contains Σ^n.
So $Z_1, Z_2, \ldots$ is an increasing sequence of closed irreducible subvarieties
whose union is dense in $\mathcal{G}$. It follows, by dimension theory in algebraic
geometry, that $Z_p = \mathcal{G}$ for some p. Since Σ^p is semi-algebraic, the fact
$Z_p = \mathcal{G}$ implies that Σ^p contains an open neighborhood $\mathcal{O}$ of one of
its points g. (This is the payoff for introducing real algebraic geome-
try.) Now it follows that Σ^{2p+1} contains an open neighborhood $\mathcal{U}$ of the
identity. Indeed, we take $\mathcal{U} = \mathcal{O}g^{-1}$ and notice that g^{-1} lies in Σ^{p+1}.

We next claim that $\Sigma^\infty = \mathcal{G}$. First, Σ^∞ is open in $\mathcal{G}$; this follows since
Σ^{2p+1+k} contains the open neighborhood $\Omega_k = \mathcal{U}\Sigma^k$ of Σ^k. Second, Σ^∞
is clearly a subgroup of $\mathcal{G}$. But $\mathcal{G}$ is connected and so $\mathcal{G}$ has no open
subgroup other than itself. So $\Sigma^\infty = \mathcal{G}$.

The last paragraph shows that $\mathcal{G}$ is the union of the increasing se-
quence of open sets Ω_k. But $\mathcal{G}$ is compact, and so this forces $\mathcal{G} = \Omega_q$ for
some q. Hence $\mathcal{G} = \Sigma^{2p+1+q}$.   ∎

4.5   Second step: reduction to $n = 2$

THEOREM 4.3
The set of all 2-qudit gates is exactly universal.

PROOF   We will apply Lemma 4.1 to the $\binom{n}{2}$ subgroups $H(p,q) =$
$\{B(p,q) \mid B \in G\}$ of $U(d^n)$, indexed by pairs (p,q) with $p < q$. Each
$H(p,q)$ is a connected closed subgroup of $U(d^n)$. We need to show that
the $H(p,q)$ generate a dense subgroup of $U(d^n)$; this amounts to showing
that the Lie algebras of the $H(p,q)$ generate the Lie algebra of $U(d^n)$.
This was done by DiVincenzo in [5] . Although DiVincenzo only worked
in the case $d = 2$, his method easily extends to the case $d > 2$. Thus

Lemma 4.1 applies and tells us the $H(p,q)$ generate $U(d^n)$. ∎

For $d = 2$, Theorem 4.3 was already known by rather different methods. It was explained in [2] how to explicitly build any n-qudit gate out of the n-qudit gates produced by the 1-qubit gates A together with the CNOT gate.

Note that the third step has already been mentioned in the outline of the proof of Theorem 4.1 in Section 4.3.

4.6 Fourth Step: analyzing the Lie algebra $\mathfrak{g}$

LEMMA 4.2

There are no Lie algebras strictly in between $\mathfrak{h}$ and $\mathfrak{g}$.

PROOF   We will write elements of $G = U(d^2)$ and $\mathfrak{g} = \mathfrak{u}(d^2)$ as matrices of size d^2, by using the computational basis of $(\mathbb{C}^d)^{\otimes 2}$. Now H is the subgroup of G of matrices of the form $h_{S,T}$ where $h_{S,T} = S \otimes T$ is the Kronecker product of unitary matrices S and T of size d.

The main idea now is to study $\mathfrak{g}$ as a representation π of the Lie group $K = SU(d) \times SU(d)$ where (S,T) acts on $\mathfrak{g}$ by $\pi^{S,T}(\xi) = h_{S,T}\, \xi\, h_{S,T}^{-1}$. This is useful because if $\mathfrak{r}$ is a Lie subalgebra of $\mathfrak{g}$ and $\mathfrak{r}$ contains $\mathfrak{h}$, then the operators $\pi^{S,T}$ preserve $\mathfrak{r}$. So we get inclusions $\mathfrak{h} \subseteq \mathfrak{r} \subseteq \mathfrak{g}$ of representations of K. We will show that there is no representation of K strictly in between $\mathfrak{h}$ and $\mathfrak{g}$.

Now $\mathfrak{g}$ decomposes into a direct sum of irreducible representations of K. This follows formally since K is a compact Lie group. In fact it is useful for us to write down the decomposition explicitly.

To do this, we observe that each element of $\mathfrak{g}$ is a finite sum of Kronecker products $X \otimes Y$ where X lies in $\mathfrak{u}(d)$ and Y lies in $i\mathfrak{u}(d)$. Here $\mathfrak{u}(d) = Lie\ U(d)$ is the space of skew-hermitian matrices of size d. Then

$$\pi^{S,T}(X \otimes Y) = (SXS^{-1}) \otimes (TYT^{-1})$$

Thus $\mathfrak{g}$ identifies naturally with the tensor product $\mathfrak{u}(d) \otimes (i\mathfrak{u}(d))$, where Kronecker product corresponds to tensor product. The representation π then corresponds to the obvious tensor product representation of

K on $\mathfrak{u}(d) \otimes (i\mathfrak{u}(d))$. As a representation of $U(d)$ under conjugation, $\mathfrak{u}(d)$ decomposes into the direct sum of two irreducible representations: $\mathfrak{u}(d) = i\mathbb{R} I \oplus \mathfrak{su}(d)$, where I is the identity matrix and $\mathfrak{su}(d) = Lie\ SU(d)$ is the space of skew-hermitian matrices of trace 0. Thus we obtain the decomposition

$$\mathfrak{g} = (i\mathbb{R} I \oplus \mathfrak{su}(d)) \otimes (\mathbb{R} I \oplus i\,\mathfrak{su}(d)) = \mathfrak{p}_0 \oplus \mathfrak{p}_1 \oplus \mathfrak{p}_2 \oplus \mathfrak{p}_3 \qquad (4.6)$$

into four irreducible representations of K, where $\mathfrak{p}_0 = i\mathbb{R} I \otimes I$, $\mathfrak{p}_1 = \mathfrak{su}(d) \otimes I$, $\mathfrak{p}_2 = I \otimes \mathfrak{su}(d)$ and $\mathfrak{p}_3 = i\,\mathfrak{su}(d) \otimes \mathfrak{su}(d)$.

We recognize $\mathfrak{h} = \mathfrak{p}_0 \oplus \mathfrak{p}_1 \oplus \mathfrak{p}_2$; this follows since $\mathfrak{h}$ consists of matrices of the form $X \otimes I + I \otimes Y$ where X and Y lie in $\mathfrak{u}(d)$. Thus $\mathfrak{g} = \mathfrak{h} \oplus \mathfrak{p}_3$ and so there is no representation of K strictly in between $\mathfrak{h}$ and $\mathfrak{g}$. ∎

4.7   Fifth Step: the normalizer of H

We can now show

PROPOSITION 4.2
The normalizer of H in G is the group L of primitive gates.

PROOF   We showed in Section 4.3 in proving (i)⇒(iii) that L is a closed subgroup of G with L lying strictly in between H and G. It follows by Lemma 4.2 that the Lie algebra of L is $\mathfrak{h}$. Now, since H is a connected Lie group, it follows that L normalizes H.

For the converse, we return to our setup in the proof of Lemma 4.2. Let us write $X(1) = X \otimes I$ and $Y(2) = I \otimes Y$ for any matrices X and Y of size d. We identified $\mathfrak{h}$ as the set of matrices $X(1) + Y(2)$ of size d^2 where X and Y lie in $\mathfrak{u}(d)$.

Suppose $R \in G$ normalizes H. Then the conjugation action of R on $\mathfrak{g}$ preserves $\mathfrak{h}$. So given any $X, Y \in \mathfrak{u}(d)$, we have

$$R(X(1) + Y(2))R^{-1} = X'(1) + Y'(2) \qquad (4.7)$$

for some $X', Y' \in \mathfrak{u}(d)$. Then tr X + tr Y = tr X' + tr Y' where tr X denotes the trace of X. Consequently we can make X' and Y'

unique by requiring tr $X = $ tr X' and tr $Y = $ tr Y'. In particular, if $X, Y \in \mathfrak{su}(d)$, then $X', Y' \in \mathfrak{su}(d)$. In this way, R defines a linear endomorphism γ_R of $\mathfrak{su}(d) \oplus \mathfrak{su}(d)$ where $\gamma_R(X, Y) = (X', Y')$. Clearly γ_R is invertible. Moreover γ_R preserves the Lie algebra bracket: this follows using $[X(1) + Y(2), U(1) + V(2)] = [X, U](1) + [Y, V](2)$. Thus γ_R is a Lie algebra automorphism.

Any Lie algebra automorphism of $\mathfrak{su}(d) \oplus \mathfrak{su}(d)$ either preserves the two summands or permutes them. This is forced because $\mathfrak{su}(d)$ is a simple Lie algebra. So we have two cases: either γ_R preserves the summands so that $\gamma_R(X, 0) = (X', 0)$ and $\gamma_R(0, Y) = (0, Y')$, or γ_R permutes the summands so that $\gamma_R(X, 0) = (0, Y')$ and $\gamma_R(0, Y) = (X', 0)$. We call the first case the *straight* case, and the second the *cross-over* case.

Suppose we are in the straight case. Then

$$RX(1)R^{-1} = X'(1) \quad \text{and} \quad RY(2)R^{-1} = Y'(2) \qquad (4.8)$$

We want to show (4.8) implies that R is primitive. Consider a decomposable 2-qudit $|xy\rangle$. We want to show $R|xy\rangle$ is also decomposable. To do this, we introduce matrices X and Y in $\mathfrak{u}(d)$ as follows: $X = ip_x$ and $Y = ip_y$ where p_x is the matrix which orthogonally projects $\mathbb{C}^d$ onto the line $\mathbb{C}x$. Now (4.8) produces two matrices X' and Y' in $\mathfrak{u}(d)$. (Clearly (4.8) extends automatically to the case where X, Y, X', Y' lie in $\mathfrak{u}(d)$, since $\mathfrak{u}(d) = i\,\mathbb{R}I \oplus \mathfrak{su}(d)$.)

We claim that X' and Y' are also of the form $X' = ip_{x'}$ and $Y' = ip_{y'}$ for some qudits $|x'\rangle$ and $|y'\rangle$. The point is that both X' and Y' are skew-hermitian, have trace equal to i, and have rank equal to 1. To compute the rank of X', we observe that (4.8) implies $X(1)$ and $X'(1)$ have the same rank. But rank $X(1) = d(\text{rank } X) = d$ and rank $X'(1) = d(\text{rank } X')$. So rank $X' = 1$, and similarly rank $Y' = 1$. Now using (4.8) we find

$$RX(1)Y(2)R^{-1} = X'(1)Y'(2) = -p_{x'}(1)p_{y'}(2) \qquad (4.9)$$

Let us apply both sides of (4.9) to $R|xy\rangle$. The left hand side gives $-R|xy\rangle$. The right hand side must be of the form $-e^{i\theta}|x'y'\rangle$. So $R|xy\rangle = e^{i\theta}|x'y'\rangle$ is decomposable. Thus R is primitive.

This settles the straight case. In the cross-over case, the same reasoning applies to RP to show that RP is primitive. Since P itself is primitive, we again conclude that R is primitive. ∎

4.8   Proof of theorem 4.2

In this section, we use only the work from Sections 4.6-4.7. The following result combined with Proposition 4.2 gives Theorem 4.2.

PROPOSITION 4.3
The normalizer of H in G is the union of H and HP.

PROOF   We return to the proof of Proposition 4.2. Suppose R normalizes H and R satisfies (4.8). We showed not only $R|xy\rangle = e^{i\theta}|x'y'\rangle$ (where θ depends on x, y, x', y') but also x' depends only on x while y' depends only on y. Furthermore x and y determined x' and y' uniquely up to phase factors.

We now construct a 1-qudit gate S as follows: we fix choices of y and y' and then define S by $R|xy\rangle = S|x\rangle \otimes |y'\rangle$. If we change our choices of y and y', then this changes S only by an overall phase factor. Similarly, we construct a 1-qudit gate T by $R|xy\rangle = |x'\rangle \otimes T|y\rangle$ where this time we fixed choices of x and x'. Now, for each $|xy\rangle$, $R|xy\rangle$ coincides with $S|x\rangle \otimes T|y\rangle$ up to a phase factor which depends on $|xy\rangle$. It is easy to see that these phase factors are in fact all the same. Thus $R = e^{i\theta}S(1)T(2)$. So R belongs to H.

This finishes the case where (4.8) holds. In the cross-over case we have $RX(1)R^{-1} = Y'(2)$ and $RY(2)R^{-1} = X'(1)$. Then we conclude that RP lies in H. Thus every R normalizing H belongs to either H or HP. The converse is clear.   ∎

We note that $HP = PH$ since P normalizes H.

Using Theorem 4.2 we can derive explicit equations characterizing primitive gates. Let $V_{ij,kl}$ be the matrix coefficients of V in the computational basis.

COROLLARY 4.2
Let V be a 2-qudit gate. Then V is primitive if and only if V satisfies one of the following two conditions:

(i) $V_{ij,kl}V_{\bar{i}\bar{j},\bar{k}\bar{l}} = V_{i\bar{j},kl}V_{\bar{i}j,\bar{k}\bar{l}}.$

(ii) $V_{ij,kl}V_{\bar{i}\bar{j},\bar{k}\bar{l}} = V_{i\bar{j},kl}V_{\bar{i}j,kl}.$

PROOF By Theorem 4.2, the set of primitive gates is the union of H with HP. We will show that V belongs to H if and only if (i) holds, while V belongs to HP if and only if (ii) holds.

We can view V as an element of $M(d) \otimes M(d)$ where $M(d)$ is the space of matrices of size d. We recognize (i) as the classical set of quadratic equations which characterize when V is *decomposable* in the sense that we can find A and B in $M(d)$ such that $V = A \otimes B$ (so that $V|xy\rangle = A|x\rangle \otimes B|y\rangle$). But V is decomposable if and only if V belongs to H. Indeed, if $V = A \otimes B$ then, since V is unitary, it follows easily that $A = \lambda S$ and $B = \lambda^{-1}T$ where λ is a positive number and S and T are unitary. So $V = S \otimes T$. The converse is obvious. Therefore (i) characterizes when V belongs to H.

On the other hand, V belongs to HP if and only if VP belongs to H. But (i) holds for VP if and only if (ii) holds for V. So (ii) characterizes when V belongs to HP. ∎

REMARK 4.2 We have a different (and more direct) way of proving Theorem 4.2 using some projective complex algebraic geometry. The starting point is to observe that a primitive gate V induces a holomorphic automorphism of $\mathbb{CP}^{d-1} \times \mathbb{CP}^{d-1}$. ∎

Finally, we prove Corollary 4.1. The set of imprimitive gates is just $G \setminus L$. This is open in G since we proved L is closed. The rest requires using results on the topology of smooth manifolds. Since L is a closed submanifold of G with $L \neq G$, it follows that $G \setminus L$ is dense in G. Now connectedness of $G \setminus L$ follows as soon as we check that L has codimension at least two in G. This is the case because $\dim G = d^4$ and $\dim L = \dim H = 2d^2 - 1$ and so the codimension is $d^4 - 2d^2 + 1 \geq 9$.

4.9 A variant of theorem 4.1

In this section we consider, in response to a question of G. Chen, what happens to Theorem 4.1 when we require that the 1-qudit gates A are *special*, i.e., satisfy $\det A = 1$. We can prove an analog of (i)⇔(iii): given a 2-qudit gate V, the following are equivalent:

(i') The collection of all special 1-qudit gates A together with V is universal.

(iii') V is imprimitive and $\det V$ is not a root of unity.

We cannot get exact universality here because the determinants of the gates generated by $A(1), A(2), V, V(2,1)$ are constrained to all be powers of $\det V$. But these powers form only a dense subset of $U(1)$. So a certain set of determinants never appears.

We can get a full analog of Theorem 4.1 in the following way:

THEOREM 4.4

Suppose we are given a family X of 2-qudit gates Q_ϕ, indexed by angles ϕ modulo 2π, such that $\det Q_\phi = e^{i\phi}$. Then the following are equivalent:

(i) *The collection of all special 1-qudit gates A together with X is universal.*

(ii) *The collection of all special 1-qudit gates A together with X is exactly universal.*

(iii) *At least one Q_ϕ is imprimitive.*

PROOF   Each part runs parallel to the proof of Theorem 4.1. We define $H^\sharp$ to be the subgroup of $SU(d^2)$ generated by the gates $A(1)$ and $A(2)$ for A special; then $H^\sharp$ is the set of gates of the form $S \otimes T$ where S and T belong to $SU(d)$. Let $F^\sharp$ be the subgroup of $U(d^2)$ generated by $H^\sharp$ and all the gates Q_ϕ and $Q_\phi(2,1)$.

(ii) $\Rightarrow$(i) is obvious.

(i) $\Rightarrow$(iii): if (iii) fails, then $F^\sharp$ lies in the group of L of primitive gates. But L is not dense in G.

(iii)$\Rightarrow$(ii): We can take $n = 2$ as in the proof of Theorem 4.1. Pick some Q_ϕ which is not primitive, and put $V = Q_\phi$. Our aim is to show $F^\sharp = U(d^2)$.

We claim that $H^\sharp$ and $H^\flat$ generate $SU(d^2)$, where we put $H^\flat = V H^\sharp V^{-1}$. Clearly, $H^\sharp$ and $H^\flat$ are closed connected subgroups of $SU(d^2)$. So, by Lemma 4.1, proving the claim reduces to showing that $H^\sharp$ and $H^\flat$ generate a dense subgroup of $SU(d^2)$. This amounts to showing that the Lie algebras $\mathfrak{h}^\sharp = Lie\, H^\sharp$ and $\mathfrak{h}^\flat = Lie\, H^\flat$ generate $\mathfrak{g}$.

Let $\mathfrak{z}^\sharp$ be the Lie algebra generated by $\mathfrak{h}^\sharp$ and $\mathfrak{h}^\flat$. Then we have $\mathfrak{h}^\sharp \subseteq \mathfrak{z}^\sharp \subseteq \mathfrak{su}(d^2)$. As in the proof of Lemma 4.2, $\mathfrak{z}^\sharp$ must be a representation of K. So we return to the decomposition (4.6). We recognize that $\mathfrak{su}(d^2) = \mathfrak{p}_1 \oplus \mathfrak{p}_2 \oplus \mathfrak{p}_3$ while $\mathfrak{h}^\sharp = \mathfrak{p}_1 \oplus \mathfrak{p}_2$. Therefore $\mathfrak{su}(d^2) = \mathfrak{h}^\sharp \oplus \mathfrak{p}_3$. We conclude $\mathfrak{z}^\sharp = \mathfrak{h}^\sharp$ or $\mathfrak{z}^\sharp = \mathfrak{su}(d^2)$.

We want to rule out $\mathfrak{z}^\sharp = \mathfrak{h}^\sharp$. Clearly $\mathfrak{z}^\sharp = \mathfrak{h}^\sharp \Leftrightarrow \mathfrak{h}^\flat = \mathfrak{h}^\sharp \Leftrightarrow H^\flat = H^\sharp \Leftrightarrow V$ normalizes $H^\sharp$. But $H^\sharp$ and H have the same normalizer: this follows since H is the product of $H^\sharp$ with the scalar 2-qudit gates, and also $H^\sharp$ is the set of gates in H with determinant equal to 1. So Proposition 4.2 tells us that V cannot normalize $H^\sharp$. Thus $\mathfrak{z}^\sharp \neq \mathfrak{h}^\sharp$.

This proves our claim that $H^\sharp$ and $H^\flat$ generate $SU(d^2)$. Therefore $F^\sharp$ contains $SU(d^2)$. But also $F^\sharp$ contains a gate of each determinant $e^{i\phi}$. So $F^\sharp = U(d^2)$. ∎

Here is a concrete illustration that was suggested to us by G. Chen. We take $d = 2$ and consider the gates (written in the computational basis)

$$U_{\theta,\phi} = \begin{pmatrix} \cos\theta & -ie^{i\phi}\sin\theta \\ -ie^{-i\phi}\sin\theta & \cos\theta \end{pmatrix}, \quad Q_\phi = \begin{pmatrix} 1 & 0 & 0 & 0 \\ 0 & 1 & 0 & 0 \\ 0 & 0 & 1 & 0 \\ 0 & 0 & 0 & e^{i\phi} \end{pmatrix} \quad (4.10)$$

COROLLARY 4.3

The collection of gates $U_{\theta,\phi}$ and Q_ϕ (where θ and ϕ run through $\mathbb{R}$) is exactly universal.

PROOF It is known that the gates $U_{\theta,\phi}$ generate $SU(2)$. We can also see this directly using Lemma 4.1. Indeed, for each value of ϕ, the $U_{\theta,\phi}$ form a closed connected subgroup S_ϕ of $SU(2)$. Consider the two subgroups S_0 and $S_{\pi/2}$. It is easy to see that their Lie algebras generate $\mathfrak{su}(2)$. This means S_0 and $S_{\pi/2}$ generate a dense subgroup of $SU(2)$. So by Lemma 4.1, S_0 and $S_{\pi/2}$ generate $SU(2)$.

Obviously $\det Q_\phi = e^{i\phi}$, and so we get exact universality from Theorem 4.4 as soon as we check that some Q_ϕ is imprimitive. In fact, we saw in Section 4.2 that Q_ϕ is always imprimitive, except, of course, if Q_ϕ is the identity. ∎

References

[1] A. Barenco, A universal two bit gate for quantum computation, *Proc. Royal Soc. London* A, 449 (1995), 679–693.

[2] A. Barenco, C. H. Bennett, R. Cleve, D. P. DiVincenzo, N. Margolus, P. Shor, T. Sleator, J. A. Smolin and H. Weinfurter, Elementary gates for quantum computation, *Phys. Rev.* A, 52 (1995), 3457–3467.

[3] D. Deutsch, Quantum computational networks, *Proc. Royal Soc. London* A, 425 (1989), 73–90.

[4] D. Deutsch, A. Barenco and A. Ekert, Universality in quantum computation, *Proc. Royal Soc. London* A, 449 (1995), 669–677.

[5] D. P. DiVincenzo, Two-bit quantum gates are universal for quantum computation, *Phys. Rev.* A, 51 (1995), 1015-1022.

[6] S. Lloyd, Almost any quantum logic gate is universal, *Phys. Rev. Lett.*, 75 (1995), 346–349.

Quantum Search Algorithms

Chapter 5

From coupled pendulums to quantum search

Lov K. Grover* and Anirvan M. Sengupta

Abstract   Quantum search is a quantum mechanical technique for searching N possibilities in only $\sqrt{N}$ steps. There are several different perspectives from which one can get to the algorithm - Schrödinger's equation, antenna array, rotation in a two-dimensional Hilbert space, just to name a few. This paper gives a fresh perspective on the algorithm in terms of classical coupled oscillators. Consider N oscillators, one of which is of a different resonant frequency. We could identify which one this is by measuring the oscillation frequency of each oscillator, a procedure that would take about N cycles. We show how, by coupling the oscillators together in a very simple way, it is possible to identify the different one in only $\sqrt{N}$ cycles. In case there are multiple oscillators of a different frequency, we can estimate the number of these in a time which is the square-root of that required by the sampling method. An extension of this technique to the quantum case leads to the quantum search and some novel algorithms.

*Research was partly supported by NSA and ARO under contract No. DAAG55-98-C-0040.

5.1 Introduction

A single quantum oscillator has multiple modes of oscillation. For example a simple harmonic oscillator in quantum mechanics has infinite modes of oscillation; a spin 1/2 particle in a magnetic field has two modes of oscillation and is referred to as a *qubit*. Furthermore, it is in general in both of these simultaneously. This is in contrast to a classical oscillator that just has a single mode of oscillation. Therefore to obtain N modes of oscillation, we will need at least N classical oscillators, while if we use quantum oscillators, each of which has two modes of oscillation (qubits), we need just $\log_2 N$ oscillators.

Quantum computing algorithms, such as quantum search, make use of the fact that a quantum system is simultaneously in multiple states to carry out certain computations in parallel in the same hardware. To implement the actual quantum search algorithm one needs a quantum mechanical system where one can carry out certain elementary quantum mechanical operations in a controlled way; it is *not* possible to implement the algorithm on classical hardware. Yet, in this paper we show that a very similar algorithm works in a classical system. This is a consequence of the fact that the algorithm is essentially a resonance phenomenon and can thus be implemented in different ways. Even though the mathematics of the classical and quantum systems is very different, similar phenomena show up in both places. The main difference is that in a classical system the hardware required is proportional to N whereas in the quantum system, the hardware is only proportional to $\log_2 N$.

Yet for all the difference between quantum and classical algorithms, there is considerable similarity in the approach. In this paper we start by discussing a purely classical problem where the problem is to identify an oscillator that has a different resonant frequency from the rest of the system.

5.2 Classical analogy

The analysis and results of the following section hold for any system of classical oscillators, either mechanical or electrical. For concreteness we consider the oscillators to be pendulums.

The following is the problem. There are N pendulums - one of which is slightly shorter than the rest. The problem is to identify which one this is. By carefully coupling them together and letting them oscillate for $O(\sqrt{N})$ cycles, a substantial portion of the energy can be transferred to the shorter pendulum whose amplitude becomes very high. This is accomplished by a resonance phenomenon very similar to that in quantum search. Using this, it is possible to identify the different pendulum as described in Section 5.4.

There have been previous classical analogs of the quantum search algorithm using electromagnetic waves [4], [5], [6]. The contribution of this paper is to present the algorithm as a resonance phenomenon. As a result it can be implemented in a way very different from the original algorithm while obtaining the square-root speedup. The implementation of this paper is also simpler than other classical analogs where there are two parts to the implementation, a propagation part (which leads to the phase shift) and a coupling part. In the scheme of this paper both parts happen naturally in a single interaction.

5.3 N Coupled pendulums

We show that by suspending the N pendulums from a bigger pendulum (Figure 5.1) and adjusting the masses and lengths of the bigger pendulum appropriately, it is possible to achieve a coupling similar to that of the N states in the quantum search algorithm. To simplify notation, in the following analysis we make the first pendulum special while the rest of the $(N-1)$ of them are identical.

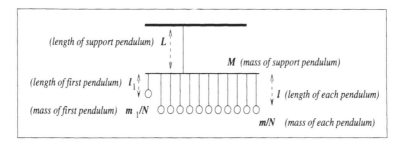

FIGURE 5.1
N pendulums are suspended from a single pendulum.

The Lagrangian of the system of Figure 5.1 is given by

$$L = \frac{1}{2}\left[M\dot{X}^2 - KX^2 + \frac{1}{N}(m_1\dot{x}_1{}^2 - k_1(x_1 - X)^2) \right.$$

$$\left. + \frac{1}{N}\sum_{j=2}^{N}(m\dot{x}_j{}^2 - k(x_j - X)^2) \right];$$

$$K \equiv \left(M + \frac{m}{N} \right)\frac{g}{L}, \quad k \equiv m\frac{g}{l}, \quad k_1 \equiv m_1\frac{g}{l_1} \qquad (5.1)$$

where X is the displacement of the support pendulum, x_j is the displacement of the jth pendulum hanging from the support; M, L are the mass and the length of the support pendulum, $\frac{m_1}{N}$, l_1 are the mass and length of the first pendulum and $\frac{m}{N}$, l are the mass and the length of each of the other pendulums (g is the acceleration due to gravity). It was probably simpler to keep the Lagrangian of (5.1) in terms of the m_i's, l_i's and g. However, as mentioned before, the framework of this paper applies to any system of oscillators, electrical or mechanical. In order to be able to quickly translate the results to other applications, we express the Lagrangian (5.1) in a more general notation in terms of the stiffnesses (k_i's).

Now we change variables so that we consider the center of mass mode $\bar{x}$ of pendulums $2, \ldots, N$, and other modes of excitation of the same pendulums orthogonal to the center of mass mode, which we denote by y_l, $l = 1, \ldots, (N-2)$. In terms of these variables, the Lagrangian, may be written as:

$$L = \frac{1}{2}\left[M\dot{X}^2 - KX^2 + \frac{1}{N}(m_1\dot{x}_1{}^2 - k_1(x_1 - X)^2) \right. \qquad (5.2)$$

$$\left. + \left(1 - \frac{1}{N}\right)(m\dot{\bar{x}}^2 - k(\bar{x} - X)^2) + \frac{1}{N}\sum_{l=1}^{N-2}(m\dot{y}_l^2 - ky_l^2) \right].$$

Note that the y's decouple from the rest of the variables. If we consider an initial condition where each y is zero, they will stay zero. Hence we can omit these variables and concentrate on the three crucial ones: $X, x_1, \bar{x}$. Defining $\xi \equiv \frac{1}{\sqrt{N}}x_1$, and ignoring some irrelevant $O(\frac{1}{N})$ terms, the reduced Lagrangian (without the y's) may be written as:

$$L_{\text{red}} \approx \frac{1}{2}\left[M\dot{X}^2 - KX^2 + m_1\dot{\xi}^2 - k_1\left(\xi - \frac{1}{\sqrt{N}}X\right)^2 \right.$$

$$\left. + m\dot{\bar{x}}^2 - k(\bar{x} - X)^2 \right]. \qquad (5.3)$$

The Lagrangian, L_{red}, represents two strongly coupled degrees of freedom, X and $\bar{x}$, and a variable ξ that is weakly coupled to others. We first solve the $X, \bar{x}$ system. This gives rise to two modes with frequencies that we denote by ω_a and ω_b. The natural frequency of the ξ degree of freedom, corresponding to the special pendulum, is $\omega_1 = \sqrt{\frac{k_1}{m_1}}$ (ignoring the $O\left(\frac{1}{\sqrt{N}}\right)$ coupling ξ has with the other modes). If ω_1 is arranged to be very close to either ω_a or ω_b, we get a resonant transfer of energy between the two weakly coupled systems. The number of cycles required for significant transfer of energy to the special pendulum varies inversely with the coupling and will be $O(\sqrt{N})$.*

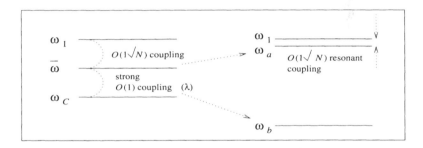

FIGURE 5.2
The center of mass mode ($\bar{\omega}$) and the coupling mode (ω_c) interact to produce two new modes (ω_a and ω_b); one of these (ω_a) is resonantly coupled to the oscillation mode of the different pendulum (ω_1) with an $O(1/\sqrt{N})$ coupling.

We next analyze the three-mode system defined by the reduced Lagrangian (5.3) by writing its equations of motion. The equations of motion can be written in matrix form as follows:

$$\widehat{M}\ddot{\vec{X}}(t) = -\widehat{K}\vec{X}(t), \tag{5.4}$$

---

*Clearly, when the deviation of the length of the pendulum approaches zero, there should be no energy transfer to this pendulum. Yet the previous analysis seems to suggest that the transfer time will be $O(\sqrt{N})$ cycles irrespective of the deviation. The reason for this becomes clear by examining the frequency diagram of Figure 5.2 when the deviation between ω_1 and $\bar{\omega}$ becomes zero. Then whatever value we choose for ω_c will result in an order 1 difference between ω_1 and ω_a, i.e., we will never be able to satisfy the resonance condition.

where the displacement vector $\vec{X}$, the mass matrix $\widehat{M}$, and the stiffness matrix $\widehat{K}$ are defined as follows: $\vec{X}(t) \equiv \begin{pmatrix} X(t) \\ \bar{x}(t) \\ \xi(t) \end{pmatrix}$, $\widehat{M} \equiv \begin{pmatrix} M & 0 & 0 \\ 0 & m & 0 \\ 0 & 0 & m_1 \end{pmatrix}$,

$$\widehat{K} \equiv \begin{pmatrix} K + k + \frac{k_1}{N} & -k & -\frac{k_1}{\sqrt{N}} \\ -k & k & 0 \\ -\frac{k_1}{\sqrt{N}} & 0 & k_1 \end{pmatrix}.$$

We solve (5.4) by assuming a solution with time dependence of each component of $\vec{X}(t)$ as $e^{i\rho t}$. Assume a solution for $\vec{X}(t)$ of the form $\begin{pmatrix} a \\ b \\ c \end{pmatrix} e^{i\rho t}$. Substituting in (5.4), we obtain $-\rho^2 \widehat{M} \begin{pmatrix} a \\ b \\ c \end{pmatrix} = \widehat{K} \begin{pmatrix} a \\ b \\ c \end{pmatrix}$. It

follows that ρ^2 is given by eigenvalues of the matrix $\widehat{M}^{-1}\widehat{K}$. Since $\widehat{M}$ is a diagonal matrix, the eigenvalues of the matrix $\widehat{M}^{-1}\widehat{K}$ are the same as the eigenvalues of $\widehat{M}^{\frac{1}{2}} \left(\widehat{M}^{-1}\widehat{K} \right) \widehat{M}^{-\frac{1}{2}}$. Therefore ρ^2 is given by the eigenvalues of the following matrix, Λ:

$$\Lambda \equiv \widehat{M}^{-\frac{1}{2}} \widehat{K} \widehat{M}^{-\frac{1}{2}} = \begin{pmatrix} \omega_c^2 & -\lambda & -\frac{k_1}{\sqrt{NMm_1}} \\ -\lambda & \bar{\omega}^2 & 0 \\ -\frac{k_1}{\sqrt{NMm_1}} & 0 & \omega^2 \end{pmatrix}. \tag{5.5}$$

Here $\omega_c^2 \equiv \frac{1}{M}\left(K + k + \frac{k_1}{N}\right)$ (ω_c corresponds to the frequency of the *coupling* degree of freedom, i.e., the frequency of the large pendulum), $\bar{\omega}^2 \equiv \frac{k}{m}$ ($\bar{\omega}$ is the frequency of the center of mass mode), $\omega^2 \equiv \frac{k_1}{m_1}$ (ω is the frequency of the deviating pendulum), $\lambda \equiv \frac{k}{\sqrt{Mm}}$ (λ is the coupling between the large pendulum and the center of mass mode).

Inspecting the matrix Λ makes it clear that the first two modes are strongly coupled, whereas the first mode is only weakly coupled to the third mode by a term of order $\frac{1}{\sqrt{N}}$. We can thus change basis so that the (1,2) block is diagonalized. The corresponding frequencies are given by the eigenvalues of the (1,2) block:

$$\omega_{1,2}^2 = \frac{1}{2}\left[\omega_c^2 + \bar{\omega}^2 \pm \sqrt{(\omega_c^2 - \bar{\omega}^2)^2 + 4\lambda^2} \right]. \tag{5.6}$$

In the rotated basis, each of the first two modes will have $O\left(\frac{1}{\sqrt{N}}\right)$ coupling with the third mode, and the matrix Λ gets transformed into a matrix Λ of the following form:

$$\tilde{\Lambda} \equiv \begin{pmatrix} \omega_1^2 & 0 & -\frac{\alpha}{\sqrt{N}} \\ 0 & \omega_2^2 & -\frac{\beta}{\sqrt{N}} \\ -\frac{\alpha}{\sqrt{N}} & -\frac{\beta}{\sqrt{N}} & \omega^2 \end{pmatrix} \; ; \; (\alpha, \beta \text{ are of order 1}). \tag{5.7}$$

We start this system by giving a push to the large support pendulum, delivering order 1 energy. This energy will initially be in the (1,2) subsystem. However, under the condition of resonance, in $O(\sqrt{N})$ cycles, the special pendulum will swing with an amplitude of order 1. All the other $(N-1)$ identical pendulums would move in lock step; their total energy would be order 1, but individual pendulums will have energy of $O(1/N)$, and their amplitudes would be $O\left(\frac{1}{\sqrt{N}}\right)$.

It must be noted that precise information about the difference oscillator is required in order to satisfy the resonance condition—we would have to know precisely how much longer or shorter this pendulum was as compared to the remaining pendulums. This would determine the value of M and L (the mass and length of the support pendulum from which the rest of the pendulums are suspended).

5.4 The algorithm

As described above, we have a means for transferring a large portion of the energy from the support pendulum into an aberrant pendulum, assuming we have precise information about the length of this pendulum but do not know which this is. This procedure can be used to identify which pendulum this is (as in the quantum search algorithm). In order to better define the problem, it is important to list some of the associated constraints.

5.4.1 Rules of the game

1. The system is started by giving a single push to the support pendulum.

2. We can redesign parameters and observe the motion of a constant number of pendulums.

3. Observations can only be resolved with a finite precision that is independent of N.

These constraints are meant to reflect realistic limitations on the system. Also, these constraints are what make the problem interesting. For example, if we could observe the system with arbitrary precision, then we could deduce the presence of the short pendulum just by observing the motion of any pendulum in only a constant number of cycles, even without any resonance. However, this demands a precision of $O(1/N)$.

5.4.2   Algorithm

The following procedure ascertains whether or not there is a special pendulum in the set that is connected to the support pendulum. Once we have a procedure for identifying the presence (or absence) of a desired item in a specified set, it is possible to identify precisely which one this is by $\log_2 N$ repetitions of the identification procedure in a binary search fashion.

Identification procedure:

> Select any one of the pendulums and shorten its length so that it is of the same length as the short pendulum (assuming it is not already a short pendulum). It is assumed that we know the length of the short pendulums. Set the system in motion by giving a push to the support. Observe the cyclic variation in amplitude of the shortened pendulum for $O(\sqrt{N})$ cycles.

In case the set of pendulums connected to the support originally had a short pendulum, then, including the one we had shortened, it will have two short pendulums. If it did not originally have a short pendulum, then it will have just one short pendulum. An analysis similar to the previous section shows that the resonant coupling transfers a large fraction of the energy to and from the short pendulums with a periodicity of $O(\sqrt{N/\tau})$ cycles, where τ is the number of short pendulums. Thus there will be a difference of a factor of $\sqrt{2}$ in the periodicity of the cyclic variation in amplitude, depending on whether there are one or two short pendulums. This periodicity is inferred from observation of the variations in amplitude of the shortened pendulum.

5.5 Towards quantum searching

Consider an N state system, whose Hamiltonian in a particular basis is known to be of the form:

$$H = \begin{bmatrix} 0 & 0 & \cdots & \cdots & 0 \\ 0 & \ddots & \cdots & \cdots & 0 \\ \vdots & \vdots & 1 & 0 & 0 \\ \vdots & \vdots & 0 & \ddots & 0 \\ 0 & 0 & 0 & 0 & 0 \end{bmatrix},$$

i.e., only one of the diagonal entries is nonzero, say the w^{th}, where the value of w is unknown, the challenge is to find out what w is. In Dirac notation, H may be written in the form $|w\rangle\langle w|$ where w is known to be a basis state. We are allowed to add on any additional term to the Hamiltonian (provided, of course, this does not depend on w) and let the system evolve in any way we choose. The question is as to how rapidly can we identify w?

One obvious method is to add an additional term to the Hamiltonian that is given by $|X\rangle\langle X| + |X\rangle\langle j| + |j\rangle\langle X|$ where X is an ancilla state and j is one of the N states. The total Hamiltonian hence becomes: $|w\rangle\langle w| + |X\rangle\langle X| + |X\rangle\langle j| + |j\rangle\langle X|$. If we start the system from the state X and let it evolve for a time of order 1 and then observe the system, the probability of finding the system in state j will depend strongly on whether or not j equals w. Thus by trying out each value of j in the range $(1, N)$ we can identify w. This method requires evolving the system for a total time of $O(N)$. Any obvious technique will need $O(N)$ time. However, by using an analogy with the quantum search algorithm, it is possible to devise a scheme to identify w that requires only $O(\sqrt{N})$ time [2].

The idea is to first add an additional term of $\frac{1}{N}(|1\rangle + \cdots + |N\rangle)(\langle 1| + \cdots + \langle N|)$ to the given Hamiltonian. Then start the system from the superposition $\frac{1}{\sqrt{N}}(|1\rangle + \cdots + |N\rangle)$, let it evolve for a time $O(\sqrt{N})$ and finally carry out an observation—with a high probability the state observed after this will be $|w\rangle$.

To simplify notation, assume that w is the first of the N states, i.e., $w = 1$. The total Hamiltonian then becomes:

$$H = \frac{1}{N}(|1\rangle + \cdots + |N\rangle)(\langle 1| + \cdots + \langle N|) + |1\rangle\langle 1|. \qquad (5.8)$$

Writing this in the subspace spanned by the two orthogonal states $|1\rangle$ and $|B\rangle \equiv \frac{1}{\sqrt{N-1}} \sum_{j=2}^{N} |j\rangle$, and leaving out terms of order $\frac{1}{N}$, the above Hamiltonian becomes

$$H \approx (|1\rangle\langle 1| + |B\rangle\langle B|) + \frac{1}{\sqrt{N}}(|1\rangle\langle B| + |B\rangle\langle 1|). \qquad (5.9)$$

Thus the quantum dynamics of the system is essentially that of two *degenerate* levels with mixing amplitude of $O(1/\sqrt{N})$. The initial state $1/\sqrt{N} \sum_{j=1}^{N} |j\rangle \approx |B\rangle$ "rotates" to $|1\rangle$ in a time inversely related to the mixing matrix element. Since this element is $O(1/\sqrt{N})$, the time required is $O(\sqrt{N})$. As we mention later, this technique is similar to the quantum search algorithm in that it consists of a rotation of the state vector in a two-dimensional vector space defined by $|w\rangle$ and $\frac{1}{\sqrt{N}}(|1\rangle + \cdots + |N\rangle)$.

5.6   The quantum search algorithm

The search problem is the following: a function $f(x)$, $x = 1, 2, \ldots, N$ is known to be nonzero at a single point x_0 and zero everywhere else. The problem is to identify x_0. Any classical scheme, whether probabilistic or deterministic, will need at least N evaluations of $f(x)$. By using the same resonance phenomenon as that described in Section 5.3 for classical pendulums, the quantum search algorithm is able to accomplish this in $O(\sqrt{N})$ steps.

The discrete quantum search algorithm is very similar to the analog scheme discussed in Section 5.5. The main difference is that instead of having the Hamiltonian be constant throughout, it is adjusted so that the item-specific portion acts separately from the mixing portion, i.e., there are alternate steps of $|w\rangle\langle w|$ and $\frac{1}{N}(|1\rangle + \cdots + |N\rangle)(\langle 1| + \cdots + \langle N|)$. This perspective is described in [3].

Assume that we have a quantum mechanical circuit that, when given a state x, can tell us whether or not it is the desired state. It is then possible to put a small amount of hardware around this circuit and redesign it so that if we input a superposition with a certain amplitude in a desired state, then the circuit selectively inverts the amplitude in this state. Note that this does not need advance knowledge of which desired state it is—it just needs the capability to infer whether or not a specified state is the desired state. Let the transformation induced by such a block be denoted by I_f. This is like the $|w\rangle\langle w|$ portion of the Hamiltonian of Section 5.5 in that it induces a phase shift of the desired state.

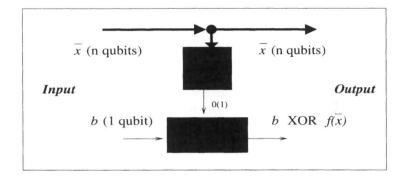

FIGURE 5.3
Given a circuit, one of whose outputs is a Boolean function $f(x)$, it is possible to synthesize a quantum mechanical circuit (shown above), which inverts the amplitude in the state where $f(x)$ evaluates to 1.

Denote the Walsh–Hadamard transformation by W, the state with all qubits in the 0 state by $|\bar{0}\rangle$ and the transformation that selectively inverts the amplitude in the $|\bar{0}\rangle$ state by $I_{\bar{0}}$. The quantum search algorithm consists of the following sequence of quantum mechanical transformations:

$$(W I_{\bar{0}} W)(I_f) \cdots O(\sqrt{N}) \text{ repetitions } \cdots (W I_{\bar{0}} W)(I_f)(W I_{\bar{0}} W)(I_f) W |\bar{0}\rangle.$$

The transformation $(W I_{\bar{0}} W)$ is the inversion about average transformation as described in [3]. It corresponds to the coupling portion of the Hamiltonian: $\frac{1}{N}(|1\rangle + \cdots + |N\rangle)(\langle 1| + \cdots + \langle N|)$. As mentioned earlier, I_f corresponds to the item-specific portion of the Hamiltonian: $|w\rangle\langle w|$. Indeed, it can be shown that by applying the respective Hamiltonians for a time π produces the indicated transformations [7].

Just as in $O(\sqrt{N})$ cycles the energy in the coupled pendulum case gets driven to the pendulum that has a higher frequency; similarly in quantum search, in $O(\sqrt{N})$ steps, the probability gets driven into the state that has a higher energy. In the coupled pendulum case, only a large fraction of the energy gets transferred into the desired state; in the quantum search case, with proper design, the entire probability can get transferred into the desired state. A simple observation then reveals which one this is.

5.7   Why does it take $O(\sqrt{N})$ cycles?

The quantum search algorithm has been rigorously proved to be the best possible algorithm for exhaustive search, i.e., no other algorithm can carry out an exhaustive search of N items in fewer than $O(\sqrt{N})$ steps. The proof for this is complicated and based on subtle properties of unitary transformations [15]. Fortunately, in the classical analog, there is a simple argument as to why it needs $O(\sqrt{N})$ cycles to transfer the energy to the desired pendulum.

As described in Section 5.3, the oscillation mode of the single pendulum is resonantly coupled to one of the two modes arising out of interaction of the center of mass mode (which has a mass $O(N)$ times that of the single pendulum) with the mode of the coupling pendulum (which too has a mass $O(N)$ times that of the single pendulum). The modes that arise out of the interaction between the coupling pendulum and the center of mass mode also behave as oscillators with a mass $O(N)$ times that of the single pendulum.

The question is as to how rapidly can we transfer energy from a pendulum of mass $O(N)$ to that of a pendulum with a mass of order 1 through a resonant coupling. Assume both pendulums to have an energy of order 1. Then the amplitude of the larger pendulum is $O(1/\sqrt{N})$ times that of the smaller pendulum. Since they have the same frequencies, the peak velocity of the larger pendulum is also $O(1/\sqrt{N})$ times that of the smaller pendulum.

Consider an elastic collision between a sphere of mass of N, traveling with a velocity of $O(1/\sqrt{N})$, with another sphere of unit mass traveling with a velocity less than 1. As shown in Figure 5.4, in the center of mass frame, the larger sphere is almost stationary and the smaller sphere bounces off the larger sphere. The speed of the smaller sphere stays unaltered and the velocity changes sign (in order to conserve kinetic energy). Translating back to the original frame, we see that the magnitude of the velocity of the smaller sphere has increased by $2/\sqrt{N}$. Therefore it will take $O(\sqrt{N})$ such interactions for the velocity of the smaller sphere to be able to rise from 0 to order 1 (or equivalently to transfer an energy of order 1).

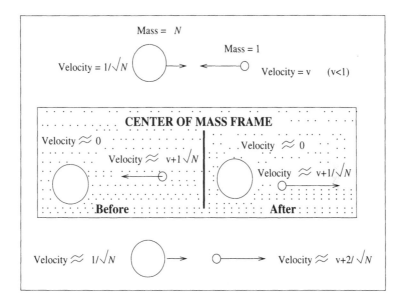

FIGURE 5.4
When a sphere of unit mass moving with unit velocity collides with a larger sphere of mass equal to N that is moving with a velocity of $1/\sqrt{N}$, the magnitude of the velocity of the smaller sphere can change by at most $2/\sqrt{N}$.

5.8 Applications and extensions

5.8.1 Counting

Estimating the number of occurrences is an important problem in statistics and computer science. One of the first extensions of the original quantum search algorithm was to the problem of counting where too it gave a square-root advantage over the best possible classical algorithm [8]. As might be anticipated, our classical analog too gives a square-root advantage over the standard estimation technique.

Consider the following variant of the previous pendulum problem. We are given N pendulums; a small fraction of them (say ϵ) are shorter than the rest. The problem is to estimate ϵ. The standard sampling technique is to pick a certain number of pendulums at random and measure their oscillation frequency. Since the probability of getting a shorter pendulum in each sample is ϵ, it will take about $\frac{1}{\epsilon}$ samples before we get a single occurrence of a shorter pendulum. Since it takes $O(1)$ cycles to estimate the oscillation frequency of a pendulum, it will take $O\left(\frac{1}{\epsilon}\right)$ cycles in all to be able to derive any reasonable estimate of ϵ. On the other hand, by extending the technique of the previous section, it is possible to estimate ϵ in only $O\left(\frac{1}{\sqrt{\epsilon}}\right)$ cycles.

The approach is to suspend all N pendulums from a single pendulum as in Section 5.3 thus coupling them. Now, as before, a resonant coupling is designed between the shorter pendulums and the rest of the system. The strength of this coupling is $O(\sqrt{\epsilon})$. This causes energy to flow back and forth from the shorter pendulums with a periodicity of $O\left(\frac{1}{\sqrt{\epsilon}}\right)$ cycles. Measuring this periodicity immediately yields ϵ.

The analysis of this case is very similar to that of Section 5.4. As before, there are three coupled modes: the mode of the large (coupling pendulum), the center of mass mode of the $N(1-\epsilon)$ longer pendulums and the center of mass mode of the $N\epsilon$ shorter pendulums. The first two modes interact with an $O(1)$ coupling; the third mode is weakly coupled to the first with an $O(\sqrt{\epsilon})$ coupling (note that this coupling to the third mode in the previous case is $O(\frac{1}{\sqrt{N}})$). The first two modes interact to produce two modes with frequencies ω_1 and ω_2. Denote the natural frequency of the shorter pendulum by ω. If either ω_1 or ω_2 is arranged to be very close to ω, we will have a resonant transfer of energy between the two weakly coupled systems which have an $O(\sqrt{\epsilon})$ coupling. The number of cycles required for significant transfer of energy to the shorter pendulum varies inversely with the coupling and will be $O\left(\frac{1}{\sqrt{\epsilon}}\right)$ cycles.

As in Section 5.4, we design the first pendulum to be a short pendulum. By following its amplitude for $O\left(\frac{1}{\sqrt{\epsilon}}\right)$ cycles, we will observe a cyclic variation. The length of this cycle will immediately identify ϵ.

5.8.2   Mechanical applications

Consider an application where we need to transfer energy to one of N (oscillator) subsystems. This can be accomplished by coupling the subsystems as described in this paper and making a slight perturbation to the subsystem into which we want the energy to flow. After $O(\sqrt{N})$ cycles, a large fraction of the energy will flow into the selected subsystem. Alternatively, if we want to transfer energy from one subsystem to another, this can be similarly accomplished by a two-step process. First, make a perturbation to the subsystem from which the energy is coming. If the system is now allowed to oscillate for $O(\sqrt{N})$ cycles, the energy transfers into the support structure. Now, if the perturbation is removed from the source subsystem and made in the destination subsystem, the energy will flow from the support into the destination subsystem. By proper design it is possible to accomplish a lossless transfer to energy from one to another subsystem. This type of scheme would be especially useful in an application where we need the flexibility of transferring energy from any one to any one of N components with minimal changes in hardware: a mechanical router.

5.8.3   Quantum mechanical applications

In quantum mechanical settings there are several applications where various modes of oscillation are coupled through the *center of mass mode*. For example, consider N atoms coupled resonantly to a photon mode in an optical cavity [9]. The atoms are trapped in the cavity by some kind of electromagnetic fields. The photon mode plays the role of the support pendulum through which the particles are coupled. Consider the basis state $|i\rangle$ to be the state where the photon excitation is localized on the ith atom. Due to the coupling there is a certain amplitude for the excitation to transfer to another atom. Since the atoms are close together in the cavity, this amplitude is the same between any two atoms. Therefore the Hamiltonian is of the form: $a \sum_i |i\rangle\langle i| + b \sum_{i,j} |i\rangle\langle j|$. This is exactly the kind of Hamiltonian that motivated the analysis of Section 5.5. A similar analysis applies in the case of an ion-trap [10] or in the case of Josephson junctions [11] coupled through a mutual inductance.

References

[1] L.K. Grover, Quantum mechanics helps in searching for a needle in a haystack, *Phys. Rev. Letters*, **78(2)** (1997), 325; also at

http://www.bell-labs.com/user/lkgrover/.

[2] E. Farhi and S. Gutmann, An analog analogue of a digital quantum computation, *Phys. Rev.* A **57** (1998), 2403.

[3] L.K. Grover, From Schrödinger's equation to the quantum search algorithm, *American Journal of Physics*, July 2001; also at http://www.bell-labs.com/user/lkgrover/.

[4] M.O. Scully and M.S. Zubairy, Quantum search protocol for an atomic array, *Physical Review* A **64** 022304 (2001).

[5] S. Lloyd, Quantum search without entanglement, *Phys. Rev.* A **61** (1999).

[6] P. Kwiat, et al., Grover's search algorithm: an optical approach, *J. Mod. Optics* **47** (2000), 257.

[7] C. Zalka, Grover's quantum searching is optimal, *Phys. Rev.* A **60** (1999), 2746.

[8] M. Boyer, G. Brassard, P. Hoeyer and A. Tapp, Tight bounds on quantum searching, *Fortsch. Phys. Lett.* **46** (1998), 493–506.

[9] A. Imamoglu, et al., *Physical Review Lett.* **83** (1999), 4204.

[10] J. Cirac and P. Zoller, *Physical Review Lett.* **74** (1995), 4091.

[11] Y. Makhlin, et al., *Nature*, **398** (1999), 305.

Chapter 6

Generalization of Grover's algorithm to multiobject search in quantum computing, Part I: continuous time and discrete time

Goong Chen,[*] **Stephen A. Fulling, and Jeesen Chen**

Abstract   L. K. Grover's search algorithm in quantum computing gives an optimal, quadratic speedup in the search for a single object in a large unsorted database. In this paper, we generalize Grover's algorithm in a Hilbert-space framework for both continuous and discrete time cases that isolates its geometrical essence to the case where more than one object satisfies the search criterion.

6.1   Introduction

A quantum computer (QC) is envisaged as a collection of 2-state "quantum bits," or *qubits* (e.g., spin 1/2 particles). Quantum com-

---

[*]Supported in part by Texas A&M University Interdisciplinary Research Grant IRI 99-22 and by DARPA QuIST grant F49620-01-1-0566.

putation does calculations on data densely coded in the entangled states that are the hallmark of quantum mechanics, potentially yielding unprecedented parallelism in computation, as P. Shor's work on factorization [13, 14] proved in 1994. Two years later, L. K. Grover [7] showed that for an unsorted database with N items in storage, it takes an average number of $\mathcal{O}(\sqrt{N})$ searches to locate a single desired object by his quantum search algorithm. If N is a very large number, this is a significant quadratic speedup over the exhaustive search algorithm in a classical computer, which requires an average number of $\frac{N+1}{2}$ searches (see Remark A.1 in Appendix). Even though Grover's algorithm is not exponentially fast (as Shor's is), it has been argued that the wide range of its applicability compensates for this [4]. Furthermore, the quantum speedup of the search algorithm is *indisputable*, whereas for factoring the nonexistence of competitively fast classical algorithms has not yet been proved [1, 2].

Grover's original papers [7, 8] deal with search for a single object. In practical applications, typically more than one item will satisfy the criterion used for searching. In the simplest generalization of Grover's algorithm, the number of "good" items is known in advance (and greater than one). Here we expound this generalization, along the lines of a treatment of the single-object case by Farhi and Gutmann [6], which makes the Hilbert-space geometry of the situation very clear.

The success of Grover's algorithm and its multiobject generalization is attributable to two main sources:

1. The notion of amplitude amplification

2. The dramatic reduction to invariant subspaces of low dimension for the unitary operators involved

Indeed, the second of these can be said to be responsible for the first. A proper geometrical formulation of the process shows that all the "action" takes place within a *two-dimensional, real* subspace of the Hilbert space of quantum states. Since the state vectors are normalized, the state is confined to a one-dimensional unit circle and (if moved at all) initially has nowhere to go except toward the place where the amplitude for the sought-for state is maximized. This accounts for the robustness of Grover's algorithm — that is, the fact that Grover's original choice of initial state and of the Walsh–Hadamard transformation can be replaced by (almost) any initial state and (almost) any unitary transformation [4, 9, 10].

The notion of amplitude amplification was emphasized in the original works [7]–[9] of Grover himself and in those of Boyer, Brassard, Høyer and Tapp [3] and Brassard, Høyer and Tapp [4]. (See also [1, 2].) Dimensional reduction is prominent in the papers by Farhi and Gutmann [6] and Jozsa [10]. We applied dimensional reduction to multiobject search independently of references [3] and [4] and later learned that the same conclusions about multiobject search (and more), in the *discrete time case,* had been obtained there in a different framework.

The rest of the paper is divided into three parts. In Section 6.2, we present the continuous-time version of the multiobject search algorithm, and in Section 6.3 the discrete-time version. In the Appendix, the computational complexity of the classical random multiobject search algorithm is analyzed and some relevant points on the literature are also made.

6.2 Continuous time quantum computing algorithm for multiobject search

Farhi and Gutmann [6] first considered quantum computation from a different point of view by regarding it as controlled Hamiltonian (continuous) time evolution of a system. This view is definitely proper, because quantum mechanical systems naturally evolve continuously in time. We inherit their point of view in this section.

Let an unsorted database consist of N objects $\{w_j \mid 1 \leq j \leq N\}$, and let f be an *oracle* (or Boolean) function such that

$$f(w_j) = \begin{cases} 1, j = 1, 2, \ldots, \ell, \\ 0, j = \ell + 1, \ell + 2, \ldots, N. \end{cases} \tag{6.1}$$

Here the ℓ elements $\{w_j \mid 1 \leq j \leq \ell\}$ are the desired objects of search. However, in general ℓ is *not* explicitly given. Note that the assignment of the "desired" objects to the first ℓ values of the index j is just a convention for the purposes of theoretical discussion; from the point of view of the user of the algorithm, the N objects are in random or unknown order, or, perhaps better, have no meaningful ordering whatever. Consider now a Hilbert space $\mathcal{H}$ of dimension N with an orthonormal basis $\mathcal{B} = \{|w_j\rangle \mid 1 \leq j \leq N\}$; each $|w_j\rangle$ is an eigenstate in the quantum computer representing the object w_j in the database. Denote

$L = \text{span}\{|w_j\rangle \mid 1 \leq j \leq \ell\}$. Here we have adopted the notation of *ket* $|\cdot\rangle$ and *bra* $\langle\cdot|$ in mathematical physics to denote, respectively, vectors and linear functionals in $\mathcal{H}$. Suppose we are given a Hamiltonian H in $\mathcal{H}$ and we are told that H has an eigenvalue $E \neq 0$ on the entire subspace L defined above and all the other eigenvalues are zero. The task is to find an eigenvector $|w_j\rangle$ in L that has eigenvalue E. The task is regarded as completed when a *measurement* of the system shows that it is in the state $|w_j\rangle$ for some j: $1 \leq j \leq \ell$.

Define a linear operator H_L, whose action on a basis element is given by

$$H_L|w_j\rangle = \frac{E}{2}(|w_j\rangle - (-1)^{f(w_j)}|w_j\rangle), \qquad j = 1, 2, \ldots, N. \qquad (6.2)$$

Note that here we have only utilized the knowledge of f; no knowledge of the desired search objects $\{|w_j\rangle \mid j = 1, 2, \ldots, \ell\}$ is required or utilized since it is assumed to be hidden in the oracle (black box). Nevertheless, since H_L is a linear operator, through *linear extension* we know that H_L is uniquely defined, and it necessarily has the following unique "explicit" representation

$$H_L = E \sum_{j=1}^{\ell} |w_j\rangle\langle w_j|. \qquad (6.3)$$

The explicitness of H_L in (6.3) is somewhat misleading. We need to emphasize that ℓ in (6.3) is *not explicitly* known or given since the only knowledge we have is f in (6.1). For the implementation of the algorithms here as well as in the next section, this does not constitute any problem, however, except that in most applications the determination of and the information about ℓ are important. This becomes a separate class of problems calling *counting* that is studied in [4], which we hope to expound further in a sequel. As in [6], we now add to H_L the "driving Hamiltonian"

$$H_D = E|s\rangle\langle s| \qquad (6.4)$$

for some (yet arbitrary) unit vector $|s\rangle \in \mathcal{H}$, $|s\rangle \notin L$. This gives the overall Hamiltonian as

$$H = H_L + H_D. \qquad (6.5)$$

Our quantum computer is governed by the Schrödinger equation

$$\begin{cases} i\frac{d}{dt}|\psi(t)\rangle = H|\psi(t)\rangle, \, t > 0, \\ |\psi(0)\rangle = |s\rangle, \end{cases} \qquad (6.6)$$

as a continuous-time, controlled controlled Hamiltonian system. Since $(6.6)_1$ is autonomous, the state of the system at time t is given by

$$|\psi(t)\rangle = e^{-iHt}|s\rangle, \qquad t \geq 0, \tag{6.7}$$

where e^{-iHt} is the exponential $N \times N$ (time evolution) matrix.

Define an augmented space $\widetilde{L}$ from L and $|s\rangle$:

$$\widetilde{L} = \mathrm{span}(L \cup \{|s\rangle\}). \tag{6.8}$$

Let $\widetilde{L}^{\perp}$ be the orthogonal complement of $\widetilde{L}$ satisfying $\widetilde{L} \oplus \widetilde{L}^{\perp} = \mathcal{H}$, where $\oplus$ denotes the orthogonal orthogonal direct sum. With respect to this orthogonal decomposition, we now have our first reduction of dimensionality below.

PROPOSITION 6.1

Fix $|s\rangle \in \mathcal{H}$, $|s\rangle \notin L$ in (6.6). Let $H, \widetilde{L}$ and $\widetilde{L}^{\perp}$ be defined as above. For any $|w\rangle \in \mathcal{H}$, write $|w\rangle = |v\rangle + |u\rangle \in \widetilde{L} \oplus \widetilde{L}^{\perp}$ according to the orthogonal direct sum. Then $H|w\rangle = H|v\rangle$ and $H|u\rangle = 0$. Consequently, the Hamiltonian H has an associated blockwise decomposition

$$H = \begin{bmatrix} H_{\widetilde{L}} & \vdots & 0_{12} \\ \cdots & \cdots & \cdots \\ 0_{21} & \vdots & 0_{22} \end{bmatrix} \tag{6.9}$$

where $H_{\widetilde{L}}$ is an invertible $(\ell + 1) \times (\ell + 1)$ matrix defined on $\widetilde{L}$ such that $H_{\widetilde{L}}|v\rangle = H|v\rangle$ for all $|v\rangle \in \widetilde{L}$, and $0_{12}, 0_{21}$ and 0_{22} are, respectively, $(\ell+1) \times (N-\ell-1)$, $(N-\ell-1) \times (\ell+1)$ and $(N-\ell-1) \times (N-\ell-1)$ zero matrices.

PROOF   Straightforward verification.   ∎

COROLLARY 6.1

Fix $|s\rangle \in \mathcal{H}, |s\rangle \notin L$ in (6.6). Let $H, \widetilde{L}$ and $\widetilde{L}^{\perp}$ be defined as above. Then the state of the solution of (6.6) at time $t, |\psi(t)\rangle$, has zero component in $\widetilde{L}^{\perp}$ for all $t > 0$.

PROOF   The action of the evolution dynamics e^{-iHt} on the invariant subspace $\widetilde{L}^{\perp}$ is, by (6.9), $e^{-i0_{22}t} = e^0 = I_{\widetilde{L}^{\perp}}$, the identity operator on

$\widetilde{L}^{\perp}$. Since the component of $|s\rangle$ in $\widetilde{L}^{\perp}$ is the zero vector, the action of $\boldsymbol{I}_{N-\ell-1}$ [the $(N-\ell-1)\times(N-\ell-1)$ identity matrix] on it remains zero for all $t > 0$.   ∎

By the properties obtained above, we need to fix our attention only on $H_{\widetilde{L}}$ defined on $\widetilde{L}$. By abuse of notation, we still write H instead of $H_{\widetilde{L}}$ on $\widetilde{L}$.

PROPOSITION 6.2

[Matrix representation of $\boldsymbol{H}$ on $\widetilde{L}$] Under the same assumptions as in Prop. 6.1, define $x_i = \langle s|w_i\rangle$ for $i = 1, 2, \ldots, \ell$. Let

$$|r\rangle = \frac{1}{C_r}\left(|s\rangle - \sum_{i=1}^{\ell} x_i|w_i\rangle\right), \qquad C_r \equiv \sqrt{1 - \sum_{i=1}^{\ell}|x_i|^2}. \qquad (6.10)$$

Then $\{|w_1\rangle, \ldots, |w_\ell\rangle, |r\rangle\}$ forms an orthonormal basis of $\widetilde{L}$ with respect to which H admits the matrix representation

$$H = E[H_{ij}]_{(\ell+1)\times(\ell+1)}; \qquad (6.11)$$

$$H_{ij} = \begin{cases} x_j\bar{x}_i, & 1 \le i,j \le \ell, i \ne j, \\ 1 + |x_j|^2, & 1 \le i,j \le \ell, i = j, \\ (\delta_{j,\ell+1}x_j + \delta_{i,\ell+1}\bar{x}_i)C_r, & i = \ell+1 \text{ or } j = \ell+1, i \ne j, \\ C_r^2, & i = j = \ell+1. \end{cases}$$

PROOF   Solve (6.10) for $|s\rangle$:

$$|s\rangle = \sum_{i=1}^{\ell} x_i|w_i\rangle + C_r|r\rangle. \qquad (6.12)$$

Substituting (6.12) into (6.3) and (6.4), we obtain (6.11) in bra-ket form. ∎

The exponential matrix function e^{-iHt} on $\widetilde{L}$ based upon the representation can be obtained, e.g., by

$$e^{-iHt} = \sum_{k=0}^{\infty} \frac{t^k}{k!}(-iH)^k \qquad (6.13)$$

or

$$e^{-iHt} = \frac{1}{2\pi i} \oint_C (\zeta \mathbf{I}_{\ell+1} + iH)^{-1} e^{\zeta t} d\zeta \qquad (6.14)$$

where C is a simple closed curve in $\mathbb{C}$ enclosing all the values ζ such that the $(\ell+1) \times (\ell+1)$ matrix $\zeta \mathbf{I}_{\ell+1} + iH$ is not invertible. However, since ℓ can be arbitrarily large, it is a highly nontrivial task (if possible at all) to calculate e^{-iHt} explicitly based on (6.13), (6.14) or any other known techniques.

It turns out that the above difficulty can be bypassed if $|s\rangle$ is chosen in an ingenious way that can effect another reduction of dimensionality. (Nevertheless, we again call attention to the fact that any choice of $|s\rangle$ must be independent of any knowledge of $|w_i\rangle$, for $i = 1, 2, \ldots, \ell$.) The choice by Grover [7, 8] is

$$|s\rangle = \frac{1}{\sqrt{N}} \sum_{j=1}^{N} |w_j\rangle, \qquad (6.15)$$

implementable and obtainable on the QC through an application of the Walsh–Hadamard transformation on all the qubits; $|s\rangle$ is a superposition state with the same amplitude in all eigenstates. With this choice of $|s\rangle$, we now have

$$x_i = x \equiv 1/\sqrt{N}, i = 1, 2, \ldots, \ell; \qquad (6.16)$$

$$C_r^2 = 1 - (\ell/N), \quad |r\rangle = \frac{1}{\sqrt{1 - (\ell/N)}} \left(|s\rangle - \frac{1}{\sqrt{N}} \sum_{i=1}^{\ell} |w_i\rangle \right)$$

in (6.11) and (6.10).

THEOREM 6.1

A two-dimensional invariant subspace for the Hamiltonian H. Let $|s\rangle$ be given as in (6.15). Denote

$$\mathcal{V} = \{|v\rangle \in \widetilde{L} \mid |v\rangle = a \sum_{i=1}^{\ell} |w_i\rangle + b|r\rangle; \quad a, b \in \mathbb{C}\}. \qquad (6.17)$$

Then $\mathcal{V}$ is a invariant two-dimensional subspace of H in $\widetilde{L}$ such that

(1) $r, s, \in \mathcal{V}$;

(2) $H(\mathcal{V}) = \mathcal{V}$.

PROOF   It is obvious that (1) holds. To see (2), we have

$$H|v\rangle \equiv H\left[a\sum_{i=1}^{\ell}|w_i\rangle + b|r\rangle\right]$$

$$= a\sum_{i=1}^{\ell}H|w_i\rangle + bH|r\rangle$$

$$= E\left\{a\sum_{i=1}^{\ell}\left[(1+x^2)|w_i\rangle + x^2\sum_{\substack{j=1\\j\neq i}}^{\ell}|w_j\rangle + x\sqrt{1-\ell x^2}|r\rangle\right]\right.$$

$$\left. +b\left[x\sqrt{1-\ell x^2}\sum_{j=1}^{\ell}|w_j\rangle + (1-\ell x^2)|r\rangle\right]\right\}$$

$$= E\left\{a\sum_{i=1}^{\ell}\left[|w_i\rangle + x^2\sum_{j=1}^{\ell}|w_j\rangle\right]\right.$$

$$\left. +bx\sqrt{1-\ell x^2}\sum_{j=1}^{\ell}|w_j\rangle + [a\ell x\sqrt{1-\ell x^2} + b(1-\ell x^2)]|r\rangle\right\}$$

$$= E\left\{(a + a\ell x^2 + bx\sqrt{1-\ell x^2})\sum_{i=1}^{\ell}|w_i\rangle \right. \tag{6.18}$$

$$\left. +[a\ell x\sqrt{1-\ell x^2} + b(1-\ell x^2)]|r\rangle\right\} \in \mathcal{V}.$$

∎

COROLLARY 6.2
 Define

$$|\widetilde{w}\rangle = \frac{1}{\sqrt{\ell}}\sum_{i=1}^{\ell}|w_i\rangle. \tag{6.19}$$

Then $\{|\widetilde{w}\rangle, |r\rangle\}$ *forms an orthonormal basis for* $\mathcal{V}$ *such that with respect to this basis,* H *admits the matrix representation*

$$H = E\begin{bmatrix} 1 + \frac{\ell}{N} & \frac{\sqrt{\ell(N-\ell)}}{N} \\ \frac{\sqrt{\ell(N-\ell)}}{N} & 1 - \frac{\ell}{N} \end{bmatrix} \tag{6.20}$$

with

$$e^{-iHt} = e^{-iEt} \begin{bmatrix} \cos(Eyt) - iy\sin(Eyt) & -\sqrt{1-y^2}\,i\sin(Eyt) \\ -\sqrt{1-y^2}\,i\sin(Eyt) & \cos(Eyt) + iy\sin(Eyt) \end{bmatrix}, \quad (6.21)$$

$$y \equiv \sqrt{\ell/N}.$$

PROOF   The representation (6.20) follows easily from (6.18). To calculate e^{-iHt}, write

$$H = E \begin{bmatrix} 1+y^2 & y\sqrt{1-y^2} \\ y\sqrt{1-y^2} & 1-y^2 \end{bmatrix}, \qquad y \equiv \sqrt{\ell/N},$$

analogous to [6, (8)], and apply (6.14) to obtain (6.21); or apply (6.13) and the properties of the $SU(2)$ generators commonly used in quantum mechanics [12, (XIII.84), p. 546].   ∎

Since $\mathcal{V}$ is invariant under H and H is a self-adjoint matrix, the orthogonal complement $\mathcal{V}^\perp$ of $\mathcal{V}$ is also an invariant subspace of H. The precise action of H or e^{iHt} on $\mathcal{V}^\perp$ does not seem to be describable in simple terms. However, this knowledge is not needed as we now have the explicit form of the solution available below.

COROLLARY 6.3
  The solution $\psi(t)$ of (6.6) given (6.15) is

$$\psi(t) = e^{-iEt}\{[y\cos(Eyt) - i\sin(Eyt)]|\widetilde{w}\rangle \qquad (6.22)$$
$$+ \sqrt{1-y^2}\cos(Eyt)|r\rangle\}, \quad t > 0,$$

where $y = \sqrt{\ell/N}$.

PROOF   Use (6.21) and (6.16).   ∎

Note that (6.22) has the same form as [6, (10)]. Since $|\widetilde{w}\rangle$ and $|r\rangle$ are unit, mutually orthogonal vectors, the probability of reaching the state $|\widetilde{w}\rangle$ at time t is

$$P(t) = \sin^2(Eyt) + y^2\cos^2(Eyt), \qquad y = \sqrt{\ell/N}. \qquad (6.23)$$

At time $T \equiv \pi/(2Ey)$, the probability is 1. By (6.19), if we make a measurement of $|\psi(t)\rangle$ at $t = T$, we will obtain any one of the eigenstates

$|w_i\rangle$, $i = 1, 2, \ldots, \ell$, with equal probability $1/\ell$. Therefore the task of search is completed (*with probability 1*), requiring a total time duration

$$T = \pi/(2Ey) = \frac{\pi}{2E}\sqrt{\frac{N}{\ell}}. \tag{6.24}$$

Formula (6.23) manifests the notion of *amplitude amplification* because the amplitude of $|\psi(t)\rangle$ along $|r\rangle$, $\sqrt{1 - y^2}\cos(Eyt)$, is steadily decreasing in magnitude as a function of $t \in [0, T]$, implying that $P(t)$ in (6.23) is increasing for $t \in [0, T]$.

Next, let us address the optimality of the search algorithm as given above. We assume that N/ℓ is a large number. We show that the above generalized Grover–Farhi–Gutmann algorithm for multiobject search is time-optimal within the order $\mathcal{O}(\sqrt{N/\ell})$. In contrast with the classical random search which requires on average $(N + 1)/(\ell + 1)$ searches (see (A.6) in the Appendix), again we see that there is a quadratic speedup.

The idea of proof follows by combining those in [3] and [6]. Let the Hamiltonian H_L be given as in (6.3). We wish to add a somewhat arbitrary (generally, time-dependent) driving Hamiltonian $H_D(t)$ to H_L so that the terminal state (at time $\widetilde{T}$) is in L (which, after a measurement, becomes one of the eigenstates $|w_1\rangle, |w_2\rangle, \ldots, |w_\ell\rangle$). Our objective is to find a lower bound on $\widetilde{T}$. Of course, $H_D(t)$ and $\widetilde{T}$ must be independent of L, since they are part of the algorithm prescribed for determining L.

Let $|w_I\rangle \in \mathcal{H}$ be an arbitrary initial state such that $\langle w_I|w_I\rangle = 1$. Let $\psi_L(t)$ denote the solution of the Schrödinger equation

$$\begin{cases} i\dfrac{d}{dt}|\psi_L(t)\rangle = (H_L + H_D(t))|\psi_L(t)\rangle, & 0 < t \leq \widetilde{T}, \\ |\psi_L(0)\rangle = |w_I\rangle, & \text{(initial condition)} \\ |\psi_L(\widetilde{T})\rangle = \displaystyle\sum_{i=1}^{\ell} \alpha_i|w_i\rangle \in L, & \text{(terminal condition)}. \end{cases} \tag{6.25}$$

Note that $\alpha_i \in \mathbb{C}$ for $i = 1, 2, \ldots, \ell$ and

$$\sum_{i=1}^{\ell} |\alpha_i|^2 = 1 \tag{6.26}$$

because the evolution process is unitary at any $t \in (0, \widetilde{T}]$. On the other hand, let $|\psi(t)\rangle$ evolve with $H_D(t)$:

$$\begin{cases} i\dfrac{d}{dt}|\psi(t)\rangle = H_D(t)|\psi(t)\rangle, \ 0 < t \leq \widetilde{T}, \\ |\psi(0)\rangle = |w_I\rangle. \end{cases} \tag{6.27}$$

LEMMA 6.1
Assume that $\widetilde{N} \equiv N/\ell$ is an integer. Let the orthonormal basis $\mathcal{B} = \{|w_i\rangle \mid 1 \leq i \leq N\}$ be grouped into $\widetilde{N}$ disjoint subsets

$$\mathcal{B} = \bigcup_{k=1}^{\widetilde{N}} B_k \tag{6.28}$$

where each $B_k \equiv \{|w_{i,k}\rangle \mid i = 1, 2, \ldots, \ell\}$ contains exactly ℓ orthonormal basis elements. Then we have

$$2\widetilde{N} - 2\sqrt{\widetilde{N}} \leq \sum_{k=1}^{\widetilde{N}} \langle \psi_{L_k}(\widetilde{T}) - \psi(\widetilde{T}) | \psi_{L_k}(\widetilde{T}) - \psi(\widetilde{T}) \rangle$$

$$\leq 2E\sqrt{\widetilde{N}}\,\widetilde{T}, \tag{6.29}$$

where $|\psi_{L_k}(t)\rangle$ is the solution $|\psi_L(t)\rangle$ of (6.25) with $L = L_k = \text{span } B_k$ in both the differential equation and the terminal condition. Consequently,

$$\widetilde{T} \geq (1 - \varepsilon_{\widetilde{N}})\frac{\widetilde{N}^{1/2}}{E} = \frac{1 - \varepsilon_{\widetilde{N}}}{E}\sqrt{\frac{N}{\ell}}, \tag{6.30}$$

where $\varepsilon_{\widetilde{N}} = \widetilde{N}^{-1/2} \to 0$ as $\widetilde{N} \to \infty$.

PROOF   As in [6, (1.8)], we have

$$\langle \psi_{L_k}(\widetilde{T}) - \psi(\widetilde{T}) | \psi_{L_k}(\widetilde{T}) - \psi(\widetilde{T}) \rangle = 2 - \sum_{i=1}^{\ell}[\langle \alpha_{i,k} w_{i,k} | \psi(\widetilde{T}) \rangle \tag{6.31}$$

$$+ \langle \psi(\widetilde{T}) | \alpha_{i,k} w_{i,k} \rangle],$$

and, therefore

$$\sum_{k=1}^{\widetilde{N}} \langle \psi_{L_k}(\widetilde{T}) - \psi(\widetilde{T}) | \psi_{L_k}(\widetilde{T}) - \psi(\widetilde{T}) \rangle = 2\widetilde{N}$$

$$- \sum_{k=1}^{\widetilde{N}} \left[\left\langle \sum_{i=1}^{\ell} \alpha_{i,k} w_{i,k} \middle| \psi(\widetilde{T}) \right\rangle + \left\langle \psi(\widetilde{T}) \middle| \sum_{i=1}^{\ell} \alpha_{i,k} w_{i,k} \right\rangle \right].$$

Let $\mathbb{P}_{L_k}\colon \mathcal{H} \to L_k$ be the orthogonal projection of $\mathcal{H}$ onto L_k. Then

$$\sum_{k=1}^{\widetilde{N}} \left| \left\langle \sum_{i=1}^{\ell} \alpha_{i,k} w_{i,k} \middle| \psi(\widetilde{T}) \right\rangle \right|^2 \leq \sum_{k=1}^{\widetilde{N}} \|\mathbb{P}_{L_k}|\psi(\widetilde{T})\rangle\|^2 \leq 1, \tag{6.32}$$

from which we apply the Cauchy–Schwarz inequality and obtain

$$\sum_{k=1}^{\widetilde{N}} \left| \left\langle \sum_{i=1}^{\ell} \alpha_{i,k} w_{i,k} \middle| \psi(\widetilde{T}) \right\rangle + \left\langle \psi(\widetilde{T}) \middle| \sum_{i=1}^{\ell} \alpha_{i,k} w_{i,k} \right\rangle \right| \leq 2\widetilde{N}^{1/2}. \quad (6.33)$$

Combining (6.31) and (6.33), we have established the left half of the inequality (6.29).

Next, mimicking [6, (19)–(21)], we have

$$\frac{d}{dt}[\langle \psi_{L_k}(t) - \psi(t) | \psi_{L_k}(t) - \psi(t) \rangle] = 2\,\mathrm{Im}\langle \psi_{L_k}(t) | H_{L_k} | \psi(t) \rangle$$

$$\leq 2|\langle \psi_{L_k}(t) | H_{L_k} | \psi(t) \rangle|$$

$$\leq 2\|H_{L_k}|\psi(t)\rangle\| = 2E \left[\sum_{i=1}^{\ell} |\langle w_{i,k} | \psi(t) \rangle|^2 \right]^{1/2}$$

$$= 2E|\mathbb{P}_{L_k}|\psi(t)\rangle|,$$

$$\frac{d}{dt} \sum_{k=1}^{\widetilde{N}} [\langle \psi_{L_k}(t) - \psi(t) | \psi_{L_k}(t) - \psi(t) \rangle] \leq 2E \sum_{k=1}^{\widetilde{N}} |\mathbb{P}_{L_k}|\psi(t)\rangle|,$$

and from $\sum_{k=1}^{\widetilde{N}} |\mathbb{P}_{L_k}|\psi(t)\rangle|^2 = 1$, by an application of the Cauchy–Schwarz inequality again, we obtain

$$\frac{d}{dt} \sum_{k=1}^{\widetilde{N}} [\langle \psi_{L_k}(t) - \psi(t) | \psi_{L_k}(t) - \psi(t) \rangle] \leq 2E\widetilde{N}^{1/2}. \quad (6.34)$$

Integrating (6.34) from 0 to $\widetilde{T}$, noting that $|\psi_{L_k}(0) - \psi(0)\rangle = 0$, we have verified the right half of inequality (6.29). ∎

In general, N/ℓ is not necessarily an integer. Therefore, the disjoint union (6.28) is not always possible. Define $\widetilde{N} = [N/\ell]$, namely, the integral part of the rational number N/ℓ. Then

$$\frac{N}{\ell} = \widetilde{N} + \delta, \quad \text{where} \quad 0 \leq \delta < 1.$$

Then we can rewrite (6.28) as

$$\mathcal{B} = \bigcup_{k=1}^{\widetilde{N}} {}^{\textstyle\cdot} B_k \cup R, \quad (6.35)$$

where $\overset{\cdot}{\underset{k=1}{\overset{\widetilde{N}}{\bigcup}}} B_k$ is an arbitrary collection of disjoint sets of ℓ orthonormal basis elements containing a total of $\widetilde{N}\ell$ of them, and R is the remaining set of orthonormal basis elements with cardinality $\ell\delta(< \ell)$. The proof of Lemma 2.7 extends to this case except for some tedious details of bookkeeping concerning the short set R, which we omit. We therefore have arrived at the following order of time-optimality for the continuous-time generalized Grover algorithm for multiobject search.

THEOREM 6.2

Assume that N/ℓ is large. Then it requires at least

$$\widetilde{T} = \frac{1 - \varepsilon_{N/\ell}}{E}\sqrt{\frac{N}{\ell}}, \quad \varepsilon_{N/\ell} > 0, \quad \varepsilon_{N/\ell} = \mathcal{O}((N/\ell)^{-1/2}),$$

time duration in average for the driven general Hamiltonian system (6.27) to reach the subspace L. □

6.3 Discrete time case: straightforward generalization of Grover's algorithm to multiobject search

In this section we generalize Grover's search algorithm in its original form [7, 8] to the situation where the number of objects satisfying the search criterion is greater than one. We are considering the discrete time case here, which may be regarded as a discrete-time sampled system of the continuous-time case treated in the preceding section. Unlike the continuous-time case, there has been a relatively rich literature studying the generalization of Grover's discrete-time algorithm to multiobject search (see [1]–[4] and the references therein). Our presentation below gives more formalized arguments than those earlier contributions, provides a clearer Hilbert space framework and settles a relevant question contained in [3].

Let the database $\{w_i \mid i = 1, 2, \ldots, N\}$, orthonormal eigenstates $\{|w_i\rangle \mid i = 1, 2, \ldots, N\}$ and the oracle function f be the same as given at the beginning of Section 6.2. The definitions of $\mathcal{H}, L$ and ℓ remain the same.

Define a linear operation in terms of the oracle function f as follows:

$$I_L|w_j\rangle = (-1)^{f(w_j)}|w_j\rangle, \qquad j = 1, 2, \ldots, N. \qquad (6.36)$$

Then since I_L is linear, the extension of I_L to the entire space $\mathcal{H}$ is unique, with an "explicit" representation

$$I_L = \boldsymbol{I} - 2\sum_{j=1}^{\ell}|w_j\rangle\langle w_j|, \qquad (6.37)$$

where $\boldsymbol{I}$ is the identity operator on $\mathcal{H}$. I_L is the operator of *rotation (by π) of the phase* of the subspace L. Note again that the explicitness of (6.37) is misleading because explicit knowledge of $\{|w_j\rangle \mid 1 \leq j \leq \ell\}$ and ℓ in (6.37) is not available. Nevertheless, (6.37) is a well-defined (*unitary*) operator on $\mathcal{H}$ because of (6.36). (Unitarity is a requirement for all operations in a QC.) We now define $|s\rangle$ as in (6.15). Then

$$|s\rangle = \frac{1}{\sqrt{N}}\sum_{i=1}^{N}|w_i\rangle = \frac{1}{\sqrt{N}}\sum_{i=1}^{\ell}|w_i\rangle + \sqrt{\frac{N-\ell}{N}}|r\rangle; \text{ see (6.16) for } |r\rangle.$$

$$(6.38)$$

Now, define another operator, *the inversion about average operation*, just as in Grover [7, 8]:

$$I_s = \boldsymbol{I} - 2|s\rangle\langle s|. \qquad (6.39)$$

Note that I_s in (6.39) is unitary and hence quantum mechanically admissible. I_s is *explicitly known*, constructible with the so-called Walsh–Hadamard transformation.

LEMMA 6.2

Let $\widetilde{L}$ be defined as in (6.8). Then $\{|w_i\rangle, |r\rangle \mid i = 1, 2, \ldots, \ell\}$ forms an orthonormal basis of $\widetilde{L}$. The orthogonal direct sum $\mathcal{H} = \widetilde{L} \oplus \widetilde{L}^{\perp}$ is an orthogonal invariant decomposition for both operators $I_{\widetilde{L}}$ and I_s. Furthermore,

  1. The restriction of I_s on $\widetilde{L}$ admits a unitary matrix representation with respect to the orthonormal basis $\{|w_1\rangle, |w_2\rangle, \ldots, |w_\ell\rangle, |r\rangle\}$:

$$A = [a_{ij}]_{(\ell+1)\times(\ell+1)}, \qquad (6.40)$$

$$a_{ij} = \begin{cases} \delta_{ij} - \dfrac{2}{N}, & 1 \leq i, j \leq \ell, \\ -\dfrac{2\sqrt{N-\ell}}{N}(\delta_{i,\ell+1} + \delta_{j,\ell+1}), & i = \ell+1 \text{ or } j = \ell+1, i \neq j, \\ \dfrac{2\ell}{N} - 1, & i = j = \ell+1. \end{cases}$$

2. *The restriction of I_s of $\widetilde{L}^{\perp}$ is $\mathbb{P}_{\widetilde{L}^{\perp}}$, the orthogonal projection operator on $\widetilde{L}^{\perp}$. Consequently, $I_s|_{\widetilde{L}^{\perp}} = \boldsymbol{I}_{\widetilde{L}^{\perp}}$, where $\boldsymbol{I}_{\widetilde{L}^{\perp}}$ is the identity operator on $\widetilde{L}^{\perp}$.*

PROOF   We have, from (6.38) and (6.39),

$$
\begin{aligned}
I_s &= \boldsymbol{I} - 2\left[\frac{1}{\sqrt{N}}\sum_{i=1}^{\ell}|w_i\rangle + \sqrt{\frac{N-\ell}{N}}|r\rangle\right]\left[\frac{1}{\sqrt{N}}\sum_{j=1}^{\ell}\langle w_j| + \sqrt{\frac{N-\ell}{N}}\langle r|\right] \\
&= \left[\sum_{i=1}^{\ell}|w_i\rangle\langle w_i| + |r\rangle\langle r| + \mathbb{P}_{\widetilde{L}^{\perp}}\right] - \left\{\frac{2}{N}\sum_{i=1}^{\ell}\sum_{j=1}^{\ell}|w_i\rangle\langle w_j|\right. \\
&\quad \left. + \frac{2\sqrt{N-\ell}}{N}\left[\sum_{i=1}^{\ell}(|w_i\rangle\langle r| + |r\rangle\langle w_i|)\right] + 2\left(\frac{N-\ell}{N}\right)|r\rangle\langle r|\right\} \\
&= \sum_{i=1}^{\ell}\sum_{j=1}^{\ell}\left(\delta_{ij} - \frac{2}{N}\right)|w_i\rangle\langle w_j| - \frac{2\sqrt{N-\ell}}{N}\left[\sum_{i=1}^{\ell}(|w_i\rangle\langle r| + |r\rangle\langle w_i|)\right] \\
&\quad + \left(\frac{2\ell}{N} - 1\right)|r\rangle\langle r| + \mathbb{P}_{\widetilde{L}^{\perp}}.
\end{aligned}
\tag{6.41}
$$

The conclusion follows.   ∎

The generalized "Grover's search engine" for multiobject search is now defined as

$$
U = -I_s I_L.
\tag{6.42}
$$

LEMMA 6.3
The orthogonal direct sum $\mathcal{H} = \widetilde{L} \oplus \widetilde{L}^{\perp}$ is an invariant decomposition for the unitary operator U, such that the following holds:

(1) With respect to the orthonormal basis $\{|w_1\rangle, \ldots, |w_{\ell}\rangle, |r\rangle\}$ of $\widetilde{L}, U$ admits a unitary matrix representation

$$
U|_{\widetilde{L}} = [u_{ij}]_{(\ell+1)\times(\ell+1)},
\tag{6.43}
$$

$$
u_{ij} = \begin{cases}
\delta_{ij} - \dfrac{2}{N}, & 1 \leq i, j \leq \ell, \\
\dfrac{2\sqrt{N-\ell}}{N}(\delta_{j,\ell+1} - \delta_{i,\ell+1}), & i = \ell+1 \text{ or } j = \ell+1, i \neq j, \\
1 - \dfrac{2\ell}{N}, & i = j = N+1.
\end{cases}
$$

(2) The restriction of U on $\widetilde{L}^\perp$ is $-\mathbb{P}_{\widetilde{L}^\perp} = -I_{\widetilde{L}^\perp}$.

PROOF   Substituting (6.37) and (6.41) into (6.42) and simplifying, we obtain

$$U = -I_s I_L = \cdots \text{(simplification)}$$

$$= \sum_{i=1}^{\ell}\sum_{j=1}^{\ell} \left(\delta_{ij} - \frac{2}{N} \right) |w_i\rangle\langle w_j| + \frac{2\sqrt{N-\ell}}{N} \sum_{i=1}^{\ell} (|w_i\rangle\langle r| - |r\rangle\langle w_i|)$$

$$+ \left(1 - \frac{2\ell}{N} \right) |r\rangle\langle r| - \mathbb{P}_{\widetilde{L}^\perp}.$$

The proof follows.   ∎

Lemmas 6.2 and 6.3 above effect a reduction of the problem to an invariant subspace $\widetilde{L}$, just as Prop. 6.1 did. However, $\widetilde{L}$ is an $(\ell + 1)$-dimensional subspace where ℓ may also be large. Another reduction of dimensionality is needed to further simplify the operator U.

PROPOSITION 6.3
Define $\mathcal{V}$ as in (6.17). Then $\mathcal{V}$ is an invariant two-dimensional subspace of U such that

(1) $r, s \in \mathcal{V}$;

(2) $U(\mathcal{V}) = \mathcal{V}$.

PROOF   Straightforward verification.   ∎

Let $|\widetilde{w}\rangle$ be defined as in (6.19). Then as in Section 6.2, $\{|\widetilde{w}\rangle, |r\rangle\}$ forms an orthonormal basis of $\mathcal{V}$. We have the second reduction, to dimensionality 2.

THEOREM 6.3
With respect to the orthonormal basis $\{|\widetilde{w}\rangle, |r\rangle\}$ in the invariant subspace $\mathcal{V}, U$ admits the unitary matrix representation

$$U = \begin{bmatrix} \frac{N-2\ell}{N} & \frac{2\sqrt{\ell(N-\ell)}}{N} \\ -\frac{2\sqrt{\ell(N-\ell)}}{N} & \frac{N-2\ell}{N} \end{bmatrix} = \begin{bmatrix} \cos\theta & \sin\theta \\ -\sin\theta & \cos\theta \end{bmatrix}, \qquad (6.44)$$

$$\theta \equiv \sin^{-1}\left(\frac{2\sqrt{\ell(N-\ell)}}{N}\right).$$

PROOF   Use the matrix representation (6.43) and (6.19).   ∎

Since $|s\rangle \in \mathcal{V}$, we can calculate $U^m|s\rangle$ efficiently using (6.44):

$$U^m|s\rangle = U^m\left(\frac{1}{\sqrt{N}}\sum_{i=1}^{\ell}|w_i\rangle + \sqrt{\frac{N-\ell}{N}}|r\rangle\right) \qquad \text{(by (6.38))}$$

$$= U^m\left(\sqrt{\frac{\ell}{N}}|\widetilde{w}\rangle + \sqrt{\frac{N-\ell}{N}}|r\rangle\right)$$

$$= \begin{bmatrix} \cos\theta & -\sin\theta \\ \sin\theta & \cos\theta \end{bmatrix}^m \begin{bmatrix} \sqrt{\frac{\ell}{N}} \\ \sqrt{\frac{N-\ell}{N}} \end{bmatrix}$$

$$= \begin{bmatrix} \cos(m\theta + \alpha) \\ \sin(m\theta + \alpha) \end{bmatrix}, \quad \left(\alpha \equiv \cos^{-1}\sqrt{\frac{\ell}{N}}\right) \qquad (6.45)$$

$$= \cos(m\theta + \alpha) \cdot |\widetilde{w}\rangle + \sin(m\theta + \alpha) \cdot |r\rangle.$$

Thus, the probability of reaching the state $|\widetilde{w}\rangle$ after m iterations is

$$P_m = \cos^2(m\theta + \alpha), \qquad (6.46)$$

cf. (6.23) in the continuous-time case. If $\ell \ll N$, then α is close to $\pi/2$ and, therefore, (6.46) is an increasing function of m initially. This again manifests the notion of amplitude amplification. This probability P_m is maximized if $m\theta + \alpha = \pi$, implying

$$m = \left[\frac{\pi - \alpha}{\theta}\right] = \text{the integral part of } \frac{\pi - \alpha}{\theta}.$$

When ℓ/N is small, we have

$$\theta = \sin^{-1}\left(\frac{2\sqrt{\ell(N-\ell)}}{N}\right)$$

$$= \sin^{-1}\left(2\sqrt{\frac{\ell}{N}}\left[1 - \frac{1}{2}\frac{\ell}{N} - \frac{1}{8}\left(\frac{\ell}{N}\right)^2 \pm \cdots\right]\right)$$

$$= 2\sqrt{\frac{\ell}{N}} + \mathcal{O}((\ell/N)^{3/2}),$$

$$\alpha = \cos^{-1}\sqrt{\frac{\ell}{N}} = \frac{\pi}{2} - \left[\sqrt{\frac{\ell}{N}} + \mathcal{O}((\ell/N^{1/2})^3)\right].$$

Therefore

$$m \approx \frac{\pi - \left\{\frac{\pi}{2} - \left[\sqrt{\frac{\ell}{N}} + \mathcal{O}((\ell/N^{1/2})^3)\right]\right\}}{2\sqrt{\frac{\ell}{N}} + \mathcal{O}((\ell/N)^{3/2})}$$

$$= \frac{\pi}{4}\sqrt{\frac{N}{\ell}}\left[1 + \mathcal{O}\left(\sqrt{\frac{\ell}{N}}\right)\right]. \tag{6.47}$$

COROLLARY 6.4
The generalized Grover's algorithm for multiobject search with operator U given by (6.42) has success probability $P_m = \cos^2(m\theta + \alpha)$ of reaching the state $|\widetilde{w}\rangle \in L$ after m iterations. For ℓ/N small, after $m = \frac{\pi}{4}\sqrt{N/\ell}$ iterations, the probability of reaching $|\widetilde{w}\rangle$ is close to 1. $\qquad\square$

The result (6.47) is consistent with Grover's original algorithm for single object search with $\ell = 1$, which has $m \approx \frac{\pi}{4}\sqrt{N}$.

THEOREM 6.4
(Boyer, Brassard, Høyer and Tapp [3]). *Assume that ℓ/N is small. Then any search algorithm for ℓ objects, in the form of*

$$U_p U_{p-1}\ldots U_1|w_I\rangle,$$

where each $U_j, j = 1, 2,\ldots, p$, is a unitary operator and $|w_I\rangle$ is an arbitrary combination state, takes in average $p = \mathcal{O}(\sqrt{N/\ell})$ iterations in order to reach the subspace L with a positive probability P independent of N and ℓ. Therefore, the generalized Grover algorithm in Cor. 6.4 is of optimal order.

PROOF   This is the major theorem in [3]; see Section 7 and particularly Theorem 8 therein. Note also the work by C. Zalka who considered some measurement effects in [15].   ∎

Unfortunately, if the number ℓ of good items is not known in advance, Corollary 6.4 does not tell us when to stop the iteration. This problem was addressed in [3], and in another way in [4]. In a related context an equation arose that was not fully solved in [3]. We consider it in the final segment of this section. As in [3, Section 3], consider stopping the Grover process after j iterations, and, if a good object is not obtained, starting it over again from the beginning. From Corollary 6.4, the probability of success after j iterations is $\cos^2(j\theta - \alpha)$. By a well-known theorem of probability theory, if the probability of success in one "trial" is p, then the expected number of trials before success is achieved will be $1/p$. (The probability that success is achieved on the kth trial is $p(1-p)^{k-1}$. Therefore, the expected number of trials is

$$\sum_{k=1}^{\infty} kp(1-p)^{k-1} = -p \sum_{k=1}^{\infty} \frac{d}{dp}(1-p)^k = -p \frac{d}{dp}\frac{1-p}{p}, \qquad (6.48)$$

which is $1/p$.) In our case, each trial consists of j Grover iterations, so the expected number of iterations before success is

$$E(j) = j \cdot \sec^2(j\theta + \alpha) .$$

The optimal number of iterations j is obtained by setting the derivative $E'(j)$ equal to zero:

$$0 = E'(j) = \sec^2(j\theta + \alpha) + 2j\theta \sec^2(j\theta - \alpha)\tan(j\theta + \alpha),$$
$$2j\theta = -\cot((j\theta + \alpha)) . \qquad (6.49)$$

(In [3, Section 3], this equation is derived in the form $4\vartheta j = \tan((2j + 1)\vartheta)$, which is seen to be equivalent to (6.49) by noting that $\vartheta = \frac{\theta}{2} = \frac{\pi}{2} + \alpha$. Those authors then note that they have not solved the equation $4\vartheta j = \tan((2j+1)\vartheta)$ but proceed to use an ad hoc equation $z = \tan(z/2)$ with $z = 4\vartheta j$ instead.) Let us now approximate the solution j of (6.49) iteratively as follows. From (6.49),

$$2j\theta \sin(j\theta + \alpha) + \cos(j\theta + \alpha) = 0 ,$$
$$e^{2i(\theta j + \alpha)} = (i2\theta j + 1)/(i2\theta j - 1) , \qquad (6.50)$$

and by taking the logarithm of both sides, we obtain

$$2i(\theta j + \alpha) = 2i\pi n + i \arg\left(\frac{i2\theta j + 1}{i2\theta j - 1}\right) + \ln\left|\frac{i2\theta j + 1}{i2\theta j - 1}\right| , \qquad (6.51)$$

for any integer n. Assume that ℓ/N is small so that j is large, but we are looking for the smallest such positive j. Note that the logarithmic term in (6.51) vanishes, and

$$\arg\left(\frac{i2\theta j+1}{i2\theta j-1}\right) = -2\tan^{-1}\frac{1}{2\theta j}$$

$$= 2\left[\sum_{q=0}^{\infty}\frac{(-1)^{q+1}}{2q+1}\left(\frac{1}{2\theta j}\right)^{2q+1}\right]$$

$$= -\frac{1}{\theta j} + \mathcal{O}((\theta j)^{-3}) \;;$$

by taking $n = 1$ in (6.51), we obtain

$$j = \frac{1}{2i\theta}\left[2i\pi - 2i\alpha - i\cdot\frac{1}{\theta j} + \mathcal{O}((\theta j)^{-3})\right]$$

$$= \frac{1}{\theta}\left[\pi - \alpha - \frac{1}{2\theta j} + \mathcal{O}((\theta j)^{-3})\right] . \qquad (6.52)$$

The first order approximation j_1 for j is obtained by solving

$$j_1 = \frac{1}{\theta}\left(\pi - \alpha - \frac{1}{2\theta j_1}\right) ,$$

$$j_1^2 - \frac{1}{\theta}(\pi - \alpha)j_1 + \frac{1}{2\theta^2} = 0 ,$$

$$j_1 = \frac{1}{2\theta}\left[(\pi - \alpha) + \sqrt{(\pi - \alpha)^2 - 2}\right] . \qquad (6.53)$$

Higher order approximations j_{n+1} for $n = 1, 2, \ldots$, may be obtained by successive iterations

$$j_{n+1} = \frac{1}{\theta}\left[(\pi - \alpha) - \tan^{-1}\frac{1}{2\theta j_n}\right]$$

based on (6.49). This process will yield a convergent solution j to (6.49).

Appendix: Random multi-object search

Given a set of N unsorted objects, among which ℓ of them are the desired objects that we are searching for, how many times on average do we need to search in order to obtain the first desired object?

Because the N objects are unsorted, the above problem is equivalent to a familiar problem in probability theory:

"An urn contains N balls, ℓ of them are black and the rest are white. Each time draw a ball randomly without replacement. Determine the number of times in average needed in order to draw the first black ball." (A.1)

A different version of random multi-object search would correspond to (A.1) but *with replacement* after each drawing. This search method is less efficient than (A.1), but can be treated by a similar approach as given below and, thus, will be omitted.

Even though we believe the solution to (A.1) is available in the literature, we could not locate a precise citation and, therefore, feel the urge to write this Appendix.

We define a random variable

$$
\begin{aligned}
T_b \quad \equiv \quad & \text{number of drawings needed to draw a ball} \\
& \text{randomly and without replacement until} \\
& \text{the first black ball is out.}
\end{aligned}
\tag{A.2}
$$

Our objective is to calculate $E(T_b)$, the expected or the average value of T_b.

Obviously,

$$
T_b \in \{1, 2, \ldots, N - \ell + 1\}.
\tag{A.3}
$$

We use $\binom{n}{j}$ to denote the combinatorial coefficient $n!/[j!(n-j)!]$, and use $P(A)$ to denote the probability of a given event A (measurable with respect to the random variable T_b).

Proposition A.1 *For $j \in \{1, 2, \ldots, N - \ell + 1\}$,*

$$
P(T_b = j) = \frac{\binom{N-\ell}{j-1}}{\binom{N}{j-1}} \cdot \frac{\ell}{N - (j+1)}.
\tag{A.4}
$$

PROOF   By the very definition of T_b in (A.2), we know that

$P(T_b = j) = P$(the first $j - 1$ drawings result in $j - 1$ white balls, but

the j-th drawing results in a black ball).

Therefore (A.4) follows.   ∎

Proposition A.2 *([13, p. 54, (10), (11)]) For any $m, n \in \{0, 1, 2, \ldots\}$ and $m \leq n$,*

$$\sum_{k=0}^{n-m} \binom{m+k}{m} = \sum_{j=m}^{n} \binom{j}{m} = \binom{n+1}{m+1}.$$

PROOF   By Pascal's formula,

$$\binom{j+1}{m+1} = \binom{j}{m+1} + \binom{j}{m}, \tag{A.5}$$

we have

$$\sum_{j=m}^{n} \binom{j}{m} = \sum_{j=m}^{n} \left[\binom{j+1}{m+1} - \binom{j}{m+1} \right]$$

$$= \sum_{j=m}^{n} \binom{j+1}{m+1} - \sum_{j=m-1}^{n-1} \binom{j+1}{m+1}$$

$$= \binom{n+1}{m+1} - \binom{m}{m+1} = \binom{n+1}{m+1}.$$

∎

Theorem A.1

$$E(T_b) = \frac{N+1}{\ell+1}. \tag{A.6}$$

PROOF   From (A.4), we have

$$E(T_b) = \sum_{j=1}^{N-\ell+1} j \cdot \frac{\binom{N-\ell}{j-1} \cdot \ell}{\binom{N}{j-1} \cdot [N-(j-1)]}$$

$$= \sum_{j=1}^{N-\ell+1} j \cdot \ell \cdot \frac{\frac{(N-\ell)!}{(j-1)!(N-\ell-j+1)!}}{\frac{N!}{(j-1)!(N-j+1)!} \cdot (N-j+1)}$$

$$= \frac{\ell(N-\ell)!}{N!} \sum_{j=0}^{N-\ell} (j+1) \frac{(N-j-1)!}{(N-\ell-j)!}$$

$$= \frac{\ell(N-\ell)!}{N!} \sum_{k=0}^{N-\ell} (N-\ell+1-k) \cdot \frac{(k+\ell-1)!}{k!}$$

(where $k = N - \ell - j$)

$$= (N - \ell + 1) \left[\frac{\ell(N - \ell)!}{N!} \sum_{k=0}^{N-\ell} \frac{(k + \ell - 1)!}{k!} \right]$$

$$- \frac{\ell(N - \ell)!}{N!} \sum_{k=1}^{N-\ell} \frac{(k + \ell - 1)!}{(k - 1)!}$$

$$= (N - \ell + 1) \left[\sum_{k=0}^{N-\ell} \frac{\frac{(k+\ell-1)!}{k!(\ell-1)!}}{\frac{N!}{\ell!(N-\ell)!}} \right] - \ell \left[\sum_{k=0}^{N-(\ell+1)} \frac{\frac{(k+\ell)!}{k!\ell!}}{\frac{N!}{\ell!(N-\ell)!}} \right]$$

$$= (N - \ell + 1) \left[\frac{1}{\binom{N}{\ell}} \sum_{k=0}^{N-\ell} \binom{k + \ell - 1}{\ell - 1} \right]$$

$$- \ell \left[\frac{1}{\binom{N}{\ell}} \sum_{k=0}^{N-(\ell+1)} \binom{k + \ell}{\ell} \right]. \tag{A.7}$$

Now, applying Proposition A.2, we obtain

$$\binom{N}{\ell - 1} + \sum_{k=0}^{N-\ell} \binom{k + \ell - 1}{\ell - 1} = \binom{N}{\ell}, \tag{A.8}$$

$$\binom{N}{\ell} + \sum_{k=0}^{N-(\ell+1)} \binom{k + \ell}{\ell} = \binom{N}{\ell + 1} = \frac{N - \ell}{\ell + 1} \binom{N}{\ell}. \tag{A.9}$$

Substituting (A.8) and (A.9) into (A.7), we obtain

$$E(T_b) = \cdots \text{(continuing from (A.7))}$$

$$= (N - \ell + 1) \binom{N}{\ell}^{-1} \binom{N}{\ell - 1} - \ell + (N - \ell + 1)$$

$$- \ell \cdot \frac{N - \ell}{\ell + 1} = \frac{N + 1}{\ell + 1}.$$

∎

Remark A.1 When $\ell = 1$, by (A.6) it takes $(N + 1)/2$ searches on average to obtain the desired single object. In the literature, this is usually cited as $N/2$, which of course differs negligibly for large N.

For $N = 4$ and $\ell = 1$, by (A.6) it takes $(4 + 1)/(1 + 1) = 2.5$ times of search on average to obtain the desired item. But in [5, p. 3408], it is stated that it takes $9/4 = 2.25$ times of search on average. The reason for the discrepancy is a different definition of successful search. The authors of [5] regard the search as completed as soon as the location of the desired item is known, even if that item has not been physically "drawn from the urn." They therefore count the worst case, where the desired item is the last one drawn, as requiring only three steps instead of four. This redefinition could be incorporated into our theorem at the expense of some complication; but it seems to us to be the less natural convention in the scenario of multiple desired objects, only one of which is required to be "produced."

□

Acknowledgments: We thank M.O. Scully for originally expounding [6], B.-G. Englert and M. Hillery for acquainting us with some of the literature of quantum computation, M. M. Kash for a technical discussion and Hwang Lee, D. A. Lidar and J. D. Malley for comments on the manuscript.

References

[1] E. Biham, O. Biham, D. Biron, M. Grassl and D.A. Lidar, Grover's quantum search algorithm for an arbitrary initial amplitude distribution, *Phys. Rev. A* **60** (1999), 2742–2745.

[2] D. Biron, O. Biham, E. Biham, M. Grassl and D.A. Lidar, Generalized Grover search algorithm for arbitrary initial amplitude distribution, in *Quantum Computing and Quantum Communications* Lecture Notes. Comp. Sci., vol. 1509, Springer-Verlag, New York, 1998, pp. 140–147.

[3] M. Boyer, G. Brassard, P. Høyer and A. Tapp, Tight bounds on quantum searching, *Fortsch. Phys.* **46** (1998), 493–506.

[4] G. Brassard, P. Høyer and A. Tapp, Quantum counting, quant-ph/9805082, May 1998.

[5] I.L. Chuang, N. Gershenfeld and M. Kubinec, Experimental implementation of fast quantum searching, *Phys. Rev. Lett.* 80 (1998), 3408–3441.

[6] E. Farhi and S. Gutmann, Analog analogue of a digital quantum computation, *Phys. Rev. A* **57** (1998), 2403–2405.

[7] L.K. Grover, A fast quantum mechanical algorithm for database search, Proc. 28th Annual Symposium on the Theory of Computing, 212–218, ACM Press, New York, 1996.

[8] L.K. Grover, Quantum mechanics helps in searching for a needle in a haystack, *Phys. Rev. Lett.* **78** (1997), 325–328.

[9] L.K. Grover, Quantum computers can search rapidly by using almost any transformation, *Phys. Rev. Lett.* **80** (1998), 4329–4332.

[10] R. Jozsa, Searching in Grover's algorithm, quant-ph/9901021, Jan. 1999.

[11] D. Knuth, *The Art of Computer Programming, Vol. 1, Fundamental Algorithms*, second edition, Addison-Wesley, Reading, MA, 1973.

[12] A. Messiah, *Quantum Mechanics*, Vol. 2, Wiley, New York, 1966.

[13] P. Shor, Algorithms for quantum computation: discrete logarithms and factoring, Proc. 35th IEEE Symposium on the Foundations of Computer Sci., 124–134, 1994.

[14] P. Shor, Polynomial-time algorithms for prime factorization and discrete logarithms on a quantum computer, *SIAM J. Comp.* **26** (1997), 1484–1510.

[15] C. Zalka, Grover's quantum searching algorithm is optimal, *Phys. Rev.* A **60** (1999), 2746–2751. See also `quant-ph/9711070`, Nov. 1997.

Chapter 7

Generalization of Grover's algorithm to multiobject search in quantum computing, Part II: general unitary transformations

Goong Chen* and Shunhua Sun†

Abstract   There are major advantages in a newer version of Grover's quantum algorithm [4] utilizing a general unitary transformation in the search of a single object in a large unsorted database. In this paper, we generalize this algorithm to multiobject search. We show the techniques to achieve the reduction of the problem to one on an invariant subspace of dimension just equal to two.

*Supported in part by Texas A&M University Interdisciplinary Research Initiative IRI 99-22 and by DARPA QuIST grant F49620-01-1-0566.
†Supported in part by a grant from Natural Science Foundation of China.

7.1   Introduction

This paper is a continuation from [1] on quantum computing algorithms for multiobject search.

L.K. Grover's first papers [2, 3] on "quantum search for a needle in a haystack" have stimulated broad interest in the theoretical development of quantum computing algorithms. Let an unsorted database consist of N objects $\{w_j \mid 1 \le j \le N\}$; each object w_j is stored in a quantum computer (QC) memory as an eigenstate $|w_j\rangle$, $j = 1, 2, \ldots, N$, with $\mathcal{B} \equiv \{|w_j\rangle \mid 1 \le j \le N\}$ forming an orthonormal basis of a Hilbert space $\mathcal{H}$. Let $|w\rangle$ be an element of $\mathcal{B}$ which is the (single) object to be searched. Grover's algorithm in [2, 3] is to utilize a unitary operator

$$U \equiv -I_s I_w \tag{7.1}$$

where

$$I_w \equiv \mathbf{I} - 2|w\rangle\langle w|, \qquad (\mathbf{I} \equiv \text{the identity operator on } \mathcal{H}) \tag{7.2}$$

$$I_s \equiv \mathbf{I} - 2|s\rangle\langle s|, \qquad |s\rangle \equiv \frac{1}{\sqrt{N}} \sum_{i=1}^{N} |w_i\rangle, \tag{7.3}$$

to perform the iterations $U^m|s\rangle$, which will lead to the target state $|w\rangle$ with probability close to 1 after approximately $\frac{\pi}{4}\sqrt{N}$ number of iterations. The algorithm is of optimal order.

In a more recent paper [4], Grover showed that the state $|s\rangle$ in (7.3) can be replaced by *any* quantum state $|\gamma\rangle$ with nonvanishing amplitude for each object w_j and, correspondingly, the Walsh–Hadamard operator previously used by him to construct the operator I_s can be replaced by a sufficiently general nontrivial unitary operator. Grover's new "search engine" in [4] is a unitary operator taking the form

$$U = -I_\gamma V^{-1} I_w V \colon \mathcal{H} \to \mathcal{H} \tag{7.4}$$

where V is an *arbitrary* unitary operator. The object w will be attained (with probability close to 1) by iterating $U^m|\gamma\rangle$.

This seems to give the algorithm/software designer large flexibility in conducting quantum computer search and code development. It increases the variety of quantum computational operations that can feasibly be performed by practical software. In particular, it opens the possibility of working with an initial state $|\gamma\rangle$ (in place of $|s\rangle$) that is other than a superposition of exactly $N = 2^n$ (n = number of qubits) alternatives. This suggests a new paradigm in which the whole dataset (not just the key) is encoded in the quantum apparatus. This new point of view may also overcome some of the practical difficulties noted by Zalka [6] in searching a physical database by Grover's method.

In the next section, we study the generalization of (7.4) to multiobject search.

7.2 Multiobject search algorithm using a general unitary transformation

Let $\{|w_i\rangle \mid 1 \le i \le N\}$ be the basis of orthonormal eigenstates representing an unsorted database w_i, $1 \le i \le N$, as noted in Section 7.1. We inherit much of the notation in [1]: let f be an oracle function such that

$$f(w_i) = \begin{cases} 1, 1 \le i \le \ell, \\ 0, \ell+1 \le i \le N, \end{cases}$$

where w_i, $i = 1, 2, \ldots, \ell$, represent the multiobjects under search. We wish to find at least one w_i, for $i = 1, 2, \ldots, \ell$. Let $|\gamma\rangle$ be *any* unit vector in $\mathcal{H}$, and let $L \equiv \text{span}\{|w_i\rangle \mid 1 \le i \le \ell\}$. Define

$$I_\gamma = I - 2|\gamma\rangle\langle\gamma|: \mathcal{H} \to \mathcal{H},$$

and

$$I_L|w_j\rangle = (-1)^{f(w_j)}|w_j\rangle, \qquad j = 1, 2, \ldots, N,$$

and I_L is then uniquely extended linearly to all $\mathcal{H}$ with the representation

$$I_L = I - 2\sum_{i=1}^{\ell}|w_i\rangle\langle w_i|.$$

Both I_γ and I_L are unitary operators. Let V be any unitary operator on $\mathcal{H}$. Now, define

$$U = -I_\gamma V^{-1} I_L V. \tag{7.5}$$

Then U is a unitary operator; it degenerates into Grover's operator U in (7.4) when $\ell = 1$ and further into the old Grover's operator U in (7.1) if $V \equiv I$.

The unit vector $|\gamma\rangle \in \mathcal{H}$ is arbitrary except that we require $V|\gamma\rangle \notin L$. (Obviously, any $|\gamma\rangle$ such that $\langle w_i|\gamma\rangle \neq 0$ for all $i = 1, 2, \ldots, N$, will work, including $|\gamma\rangle \equiv |s\rangle$ in (7.3).) If $V|\gamma\rangle \in L$, then

$$V|\gamma\rangle = \sum_{j=1}^{\ell} g_i|w_i\rangle, \quad g_i \in \mathbb{C}, \quad \sum_{j=1}^{\ell} |g_i|^2 = 1.$$

A measurement of the state $V|\gamma\rangle$ will yield an eigenstate $|w_j\rangle$, for some j: $1 \leq j \leq \ell$, with probability $|g_j|^2$. Thus the search task would have been completed. Thus, let us consider the nontrivial case $V|\gamma\rangle \notin L$. This implies $|\gamma\rangle \notin V^{-1}(L)$ and, hence,

$$\widetilde{L} \equiv \mathrm{span}(\{|\gamma\rangle\} \cup V^{-1}(L)) \tag{7.6}$$

is an $(\ell + 1)$-dimensional subspace of $\mathcal{H}$. It effects a reduction to a lower dimensional invariant subspace for the operator U, according to the following.

LEMMA 7.1
Assume that $\langle\gamma|\gamma\rangle = 1$ and $V|\gamma\rangle \notin L$. Then $U(\widetilde{L}) = \widetilde{L}$.

PROOF   For any j: $1 \leq j \leq \ell$, denote

$$\mu_{\gamma,j} = \langle w_j|V|\gamma\rangle.$$

(1) We have, for j: $1 \leq j \leq \ell$,

$$\begin{aligned}
U(V^{-1}|w_j\rangle) &= -I_\gamma V^{-1}\left(I - 2\sum_{i=1}^{\ell} |w_i\rangle\langle w_i|\right)|w_j\rangle \\
&= -I_\gamma V^{-1}(-|w_j\rangle) \\
&= I_\gamma V^{-1}|w_j\rangle \\
&= (I - 2|\gamma\rangle\langle\gamma|)V^{-1}|w_j\rangle \\
&= V^{-1}|w_j\rangle - 2((\langle\gamma|V^{-1}|w_j\rangle)|\gamma\rangle \\
&= V^{-1}|w_j\rangle - 2\overline{\mu}_{\gamma,j}\gamma \in \widetilde{L}; \tag{7.7}
\end{aligned}$$

(2)

$$U|\gamma\rangle = -I_\gamma V^{-1}\left(I - 2\sum_{i=1}^{\ell} |w_i\rangle\langle w_i|\right)(V|\gamma\rangle)$$

$$= -(I - 2|\gamma\rangle\langle\gamma|) \left[|\gamma\rangle - 2 \sum_{i=1}^{\ell} (\langle w_i|V|\gamma\rangle) V^{-1}|w_i\rangle \right]$$

$$= |\gamma\rangle + 2 \sum_{i=1}^{\ell} \mu_{\gamma,i} V^{-1}|w_i\rangle - 4 \sum_{i=1}^{\ell} \mu_{\gamma,i}\overline{\mu}_{\gamma,i}|\gamma\rangle$$

$$= \left(1 - 4 \sum_{i=1}^{\ell} |\mu_{\gamma,i}|^2 \right) |\gamma\rangle + 2 \sum_{i=1}^{\ell} \mu_{\gamma,i} V^{-1}|w_i\rangle \in \widetilde{L}. \quad (7.8)$$

∎

By Lemma 7.1, the Hilbert space $\mathcal{H}$ admits an orthogonal direct sum decomposition

$$\mathcal{H} = \widetilde{L} \oplus \widetilde{L}^{\perp}$$

such that $\widetilde{L}^{\perp}$ is also an invariant subspace of U. In our subsequent iterations, the actions of U will be restricted to $\widetilde{L}$, as the following Lemma 7.2 has shown. Therefore we can ignore the complementary summand space $\widetilde{L}^{\perp}$.

LEMMA 7.2

Under the same assumptions as Lemma 7.1, we have $U^m|\gamma\rangle \in \widetilde{L}$ for $m \in \mathbb{Z}^+ \equiv \{0, 1, 2, \ldots\}$.

PROOF It follows obviously from (7.6) and Lemma 7.1. ∎

Consider the action of U on $\widetilde{L}$. Even though $|\gamma\rangle, V^{-1}|w_i\rangle, i = 1, \ldots, \ell,$ form a basis of $\widetilde{L}$, these vectors are not mutually orthogonal. We have

$$U \begin{bmatrix} |\gamma\rangle \\ V^{-1}|w_1\rangle \\ V^{-1}|w_2\rangle \\ \vdots \\ V^{-1}|v_\ell\rangle \end{bmatrix} \quad (7.9)$$

$$
= \begin{bmatrix} 1 - 4\sum_{i=1}^{\ell}|\mu_{\gamma,j}|^2 & 2\mu_{\gamma,1} & 2\mu_{\gamma,2} & \cdots & 2\mu_{\gamma,\ell} \\ -2\overline{\mu}_{\gamma,1} & 1 & 0 & \cdots & 0 \\ -2\overline{\mu}_{\gamma,2} & 0 & 1 & & 0 \\ \vdots & \vdots & & \ddots & \vdots \\ -2\overline{\mu}_{\gamma,\ell} & 0 & 0 & \cdots & 1 \end{bmatrix} \begin{bmatrix} |\gamma\rangle \\ V^{-1}|w_1\rangle \\ V^{-1}|w_2\rangle \\ \vdots \\ V^{-1}|w_\ell\rangle \end{bmatrix}, (7.10)
$$

$$
\equiv \mathcal{M} \begin{bmatrix} |\gamma\rangle \\ V^{-1}|w_1\rangle \\ V^{-1}|w_2\rangle \\ \vdots \\ V^{-1}|w_\ell\rangle \end{bmatrix},
$$

according to (7.7) and (7.8). Therefore, with respect to the basis $\{|\gamma\rangle,$ $V^{-1}|w_i\rangle \mid i = 1, \ldots, \ell\}$, the matrix representation of U on $\tilde{L}$ is $\mathcal{M}^T$, the transpose of $\mathcal{M}$. These two $(\ell + 1) \times (\ell + 1)$ matrices $\mathcal{M}$ and $\mathcal{M}^T$ are *nonunitary*, however, because the basis $\{|\gamma\rangle, V^{-1}|w_i\rangle|,\ i = 1, 2, \ldots, \ell\}$ is not orthogonal. This fact is relatively harmless here, as we can further effect a reduction of dimensionality by doing the following. Define a unit vector

$$
|\mu\rangle = 2\sum_{j=1}^{\ell}\mu_{\gamma,j}V^{-1}|w_j\rangle/a, \quad a \equiv \left(4\sum_{j=1}^{\ell}|\mu_{\gamma,j}|^2\right)^{1/2} > 0. \quad (7.11)
$$

THEOREM 7.1

Let $\mathcal{V} \equiv span\{|\gamma\rangle, |\mu\rangle\}$. *Then* $\mathcal{V}$ *is a two-dimensional invariant subspace of* U. *We have*

$$
U \begin{bmatrix} |\gamma\rangle \\ |\mu\rangle \end{bmatrix} = M \begin{bmatrix} |\gamma\rangle \\ |\mu\rangle \end{bmatrix}, \quad M \equiv \begin{bmatrix} 1 - a^2 & a \\ -a & 1 \end{bmatrix}. \quad (7.12)
$$

Consequently, with respect to the basis $\{|\gamma\rangle, |\mu\rangle\}$ *in* $\mathcal{V}$, *the matrix representation of* U *is* M^T.

PROOF   Using (7.7), we have

$$
U|\mu\rangle = 2\sum_{j=1}^{\ell}\mu_{\gamma,j}V^{-1}|w_j\rangle \cdot \frac{1}{a} - 2\sum_{j=1}^{\ell}|\mu_{\gamma,j}|^2 \cdot \frac{1}{a}|\gamma\rangle
$$
$$
= |\mu\rangle - a|\gamma\rangle.
$$

Again, from the definition of $|\mu\rangle$ in (7.11), we see that (7.8) gives

$$U|\gamma\rangle = (1 - a^2)|\gamma\rangle + a|\mu\rangle.$$

Therefore (7.12) follows.   ∎

Theorem 7.1 gives a dramatic reduction of dimensionality to **2**, i.e., the dimension of the invariant subspace $\mathcal{V}$. Again, we note that the matrices M and M^T in (7.12) are *not unitary*.

Any vector $|v\rangle \in \mathcal{V}$ can be represented as

$$|v\rangle = c_1|\gamma\rangle + c_2|\mu\rangle,$$

and so

$$\begin{aligned} U|v\rangle &= U(c_1|\gamma\rangle + c_2|\mu\rangle) \\ &= c_1[(1 - a^2)|\gamma\rangle + a|\mu\rangle] + c_2[-a|\gamma\rangle + |\mu\rangle], \end{aligned}$$

and thus

$$U|v\rangle = M^T \begin{bmatrix} c_1 \\ c_2 \end{bmatrix} = \begin{bmatrix} 1 - a^2 & -a \\ a & 1 \end{bmatrix} \begin{bmatrix} c_1 \\ c_2 \end{bmatrix}, \tag{7.13}$$

where the first component of the vector on the right hand side of (7.13) corresponds to the coefficient of $|\gamma\rangle$ while the second component corresponds to the coefficient of $|\mu\rangle$. Therefore

$$U^m|\gamma\rangle = \begin{bmatrix} 1 - a^2 & -a \\ a & 1 \end{bmatrix}^m \begin{bmatrix} 1 \\ 0 \end{bmatrix}. \tag{7.14}$$

The above can be viewed geometrically ([10]) as follows: $M^T \begin{bmatrix} 1 \\ 0 \end{bmatrix} = \begin{bmatrix} 1 - a^2 \\ a \end{bmatrix}$, for $a > 0$ very small, $a \approx \sin a$, and therefore $\begin{bmatrix} 1 - a^2 \\ a \end{bmatrix}$ is a vector obtained from the unit vector $\begin{bmatrix} 1 \\ 0 \end{bmatrix}$ by rotating it counterclockwise with angle a. It takes approximately

$$m \approx \frac{\pi/2}{a} = \frac{\pi}{2a} = \pi \Big/ 4 \left[\sum_{j=1}^{\ell} |\mu_{\gamma,j}|^2 \right]^{1/2}$$

rotations to closely align the vector $U^m|\gamma\rangle$ with $|\mu\rangle \in V^{-1}L^1$. Thus $V(U^m|\gamma\rangle)$ deviates little from the subspace $L = \text{span}\{|w_i\rangle \mid i = 1, 2, \ldots, \ell\}$. A measurement of $VU^m|\gamma\rangle$ gives one of the eigenstates $|w_j\rangle$, for some j: $1 \le j \le \ell$, with probability nearly equal to 1, and the task of multiobject search is completed with this large probability.

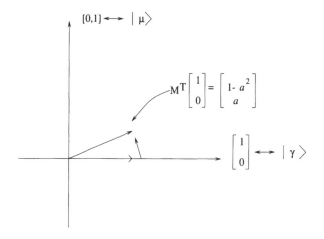

FIGURE 7.1
A geometric view of a single iteration (7.13).

References

[1] G. Chen, S.A. Fulling and J. Chen, Generalization of Grover's algorithm to multiobject search in quantum computing, Part I: Continuous time and discrete time, in Chapter 6 of this volume.

[2] L.K. Grover, A fast quantum mechanical algorithm for database search, Proc. 28th Annual Symposium on the Theory of Computing, 212–218, ACM Press, New York, 1996.

[3] L.K. Grover, Quantum mechanics helps in searching for a needle in a haystack, *Phys. Rev. Lett.* **78** (1997), 325–328.

[4] L.K. Grover, Quantum computers can search rapidly by using almost any transformation, *Phys. Rev. Lett.* **80** (1998), 4329–4332.

[5] R. Jozsa, Searching in Grover's algorithm, `quant-ph/9901021`, Jan. 1999.

[6] C. Zalka, Could Grover's quantum algorithm help in searching an actual database? `quant-ph/9901068`, Jan. 1999.

Quantum Computational Complexity

Chapter 8

Counting complexity and quantum computation

Stephen A. Fenner*

Abstract   We survey several applications to quantum computing of computational complexity theory, especially the complexity of problems related to counting. We show how the connection of quantum computation to counting is very close. We define counting complexity classes, relativization of both classical computation and quantum circuits, and present a number of results from disparate sources, recast into a single consistent formalism. We assume prior knowledge of the mathematical formalism of quantum mechanics, but we present the concepts of computational complexity in an introductory manner.

8.1   Introduction

This chapter is primarily aimed at people (physicists, perhaps) who know more about quantum mechanics than they do about the theory of

*Partially supported by a South Carolina Commission on Higher Education SCRIG Grant R-01-0256.

computation. It explores a close relationship between quantum computing and the complexity (difficulty) of counting the number of solutions
to classical search problems. The latter is well studied, going back to
the 1970s [46]. Although the concept of a quantum computer has also
been around for a long time [8, 21], it was only in the 1990s that quantum computing began to receive intense scrutiny. Our point (and we do
have one) is that the older study of computational complexity, especially
counting complexity, has a lot to say about quantum computers—their
ultimate strengths and limitations.

There are a number of good research papers on the complexity of
quantum computation, relating it to classical complexity. A primary
example is the thorough and detailed treatment of the quantum Turing
machine (QTM) model of Deutsch [15] given by Bernstein and Vazirani
[10]. Their paper proves a number of fundamental results—they give
clear evidence that QTMs are powerful enough to efficiently implement
any "reasonable" quantum algorithm; but more importantly, they construct a universal QTM that can efficiently simulate any other QTM to
good approximation. An efficient universal quantum computer shows
that there is a small, finite handful of primitives that can combine to
implement any reasonable quantum algorithm efficiently—namely, the
quantum operations used by the universal machine. As a result, the
quantum computational model has an especially simple and easy-to-
analyze structure.

An extensive summary of results in quantum complexity can be found
in Gruska [29].

The current chapter collects a number of results from other sources;
some of these are more recent improvements on earlier, better known
results. Although they appear in disparate sources, these results and
the techniques used to prove them are related to each other, and they
deserve to be brought together under a single framework. That is the
main purpose of this chapter. We give detailed proofs of selected results;
often, the technique used in the proof is just as interesting and useful as
the result itself.

We use the quantum circuit model [16] for describing quantum algorithms. Yao [49] showed that quantum circuits were equivalent to QTMs,
each able to simulate the other efficiently to good approximation. By "efficiently" we mean with at most a polynomial slow-down of the machine
or a polynomial blow-up in circuit size. Quantum circuits are conceptually easier than Turing machines for rendering quantum algorithms, and
most researchers prefer them. In Section 8.2, after defining the basic cir-

cuit model for both classical Boolean circuits and quantum circuits, we review the basic concepts of "standard" complexity theory: the classes P and NP, oracles and relativization, reducibility, and completeness. We then build on these concepts to define the function class GapP, which is then used to describe counting problems. Lastly, we prove a close connection between quantum computation and GapP—and hence counting problems.

The rest of the chapter is loosely organized into two parts: the first gives positive results about the power of quantum computation (Section 8.4), and the second gives negative results (Section 8.5). In Section 8.4, we give a proof of the existence (first shown by Green and Pruim [26]) of a black-box problem that is quantum computable with zero error but cannot be computed deterministically, even with free access to an arbitrary NP oracle. The proof serves as a good illustration of oracle construction—a venerable and often-used technique in complexity theory. In Section 8.5, we use the equivalence of GapP and quantum computation to prove a "lowness" property of efficiently quantum computable sets. We show that free access to such a set is useless in solving a counting problem—that is, any problem solvable by counting with access to the set is also solvable by counting without such access. This fact provides strong evidence that quantum computers are not powerful enough to solve arbitrary counting problems.

Quantum complexity is an extremely rich field of study. Due to space limitations, we make no attempt at a complete or unbiased survey. We apologize in advance for leaving out many interesting results, even some that are closely related to the current topic, for the sake of fewer and more detailed proofs.

8.2 Preliminaries

We assume prior knowledge of the mathematical underpinnings of quantum mechanics—Hilbert spaces, operator algebras, Dirac bracket notation, tensor products, and the like. All Hilbert spaces mentioned in this chapter are finite-dimensional. We will also suppose some basic knowledge of the theory of computation, including basic graph theoretic concepts and Turing machines (TMs), and we will briefly review some relevant concepts in complexity theory. More background can be found in several good textbooks on the theory of computation [30, 42] and on complexity theory in particular [4, 5, 24, 35].

We let $\mathbb{N}, \mathbb{Z}, \mathbb{Q}, \mathbb{R}, \mathbb{C}$ stand for the natural numbers $\{0, 1, 2, \ldots\}$, integers, rationals, reals, and complex numbers, respectively. For $z \in \mathbb{C}$, we write $\overline{z}$ for the complex conjugate of z, and we write $|z|$ for $\sqrt{z\overline{z}}$, the absolute value of z.

We let $\|X\|$ stand for the cardinality of finite set X. We fix an alphabet $\Sigma = \{0, 1\}$, and let Σ^* be the set of all finite strings of symbols from Σ. For any string $w \in \Sigma^*$, we let $|w|$ denote the length of w (number of symbols in w). For $n \in \mathbb{N}$, we let Σ^n denote the set of all strings of Σ^* of length n. We often identify classes of finitely describable objects—for example, Boolean values, truth tables, natural numbers, integers, rational numbers, algebraic numbers, graphs, TMs, and strings over various alphabets, as well as pairs, finite lists, and finite sets of these—with strings in Σ^* via reasonable and concise encodings. For example, we identify natural numbers with strings via the usual binary representation, and we identify the Boolean values *true* and *false* with 1 and 0, respectively. We sometimes represent a natural number n in unary notation as 1^n, i.e., a string of n 1s. We will also identify subsets of some (assumed) universal set (such as Σ^n or Σ^*) with their characteristic Boolean functions over the universal set.

All polynomials mentioned will be univariate unless otherwise specified. If a polynomial p represents the running time (number of primitive steps) of an algorithm or the lengths of strings, then we assume it has all integer coefficients, and that $p(n) \geq 0$ for all $n \in \mathbb{N}$. If $c \in \mathbb{R}$, we say $p > c$ to mean that $p(n) > c$ for all $n \in \mathbb{N}$.

A *circuit* is a directed acyclic graph with some arbitrary ordering of the vertices. If C is a circuit, then the vertices with indegree zero are the *initial vertices*, and those with outdegree zero are the *final vertices*. We designate the first few initial vertices of C as the *inputs* to C, the first few final vertices as the *outputs*, and all other vertices are *gates*. Gates that are not initial or final are called *intermediate gates*. The edges are called *wires*. It may seem strange at first to allow initial and final vertices that are not inputs or outputs, but this will be useful when we describe our quantum circuit model. A *Boolean circuit* is a circuit where each intermediate gate is labeled with a Boolean connective $\wedge$, $\vee$, or $\neg$, and all intermediate gates have indegree two except $\neg$-gates, which have indegree one. Noninput initial gates are labeled with the constant 0 (false), and final gates are unlabeled. A labeling of the inputs of a Boolean circuit with Boolean values uniquely determines a labeling of the outputs, computed in the usual manner. So a Boolean circuit with n inputs and m outputs computes a unique function $f : \Sigma^n \to \Sigma^m$.

When we draw a circuit, the initial vertices are on the left and the final vertices on the right. Input and output vertices are represented as bare ends of wires.

8.2.1 Qubits, quantum gates, and quantum circuits

The following definitions are adapted primarily from Nielsen and Chuang [33] and Aharonov, Kitaev, and Nisan [2]. We refer to those sources for more detail and motivation.

For two Hilbert spaces $\mathcal{H}$ and $\mathcal{J}$, we denote the space of linear maps from $\mathcal{H}$ to $\mathcal{J}$ as $\mathcal{L}(\mathcal{H}, \mathcal{J})$, and we abbreviate $\mathcal{L}(\mathcal{H}, \mathcal{H})$ by $\mathcal{L}(\mathcal{H})$, the space of linear operators on $\mathcal{H}$. We will use the density matrix formulation of quantum states; so we regard a quantum state in a Hilbert space $\mathcal{H}$ as a Hermitian operator $\rho \in \mathcal{L}(\mathcal{H})$ that is positive semidefinite ($\rho \geq 0$) and has unit trace ($\mathrm{tr}(\rho) = 1$). A *quantum operation* from $\mathcal{H}$ to $\mathcal{J}$ is a linear map A from $\mathcal{L}(\mathcal{H})$ to $\mathcal{L}(\mathcal{J})$ that is trace-preserving and *completely positive* (A is also called a superoperator). Intuitively, completely positive means that if we embed $\mathcal{H}$ into some larger system, then the standard lifting of A to the larger system preserves positive (semi)definiteness, and thus states get mapped to states. Formally, this means that for any Hilbert space $\mathcal{K}$, the linear map $A \otimes I_\mathcal{K} : \mathcal{L}(\mathcal{H} \otimes \mathcal{K}) \to \mathcal{L}(\mathcal{J} \otimes \mathcal{K})$, where $I_\mathcal{K}$ is the identity map on $\mathcal{L}(\mathcal{K})$, preserves positive definiteness: for any $\rho \in \mathcal{L}(\mathcal{H} \otimes \mathcal{K})$, if $\rho > 0$, then $(A \otimes I_\mathcal{K})(\rho) > 0$. [For $\tau_1 \in \mathcal{L}(\mathcal{H})$ and $\tau_2 \in \mathcal{L}(\mathcal{K})$, $(A \otimes I_\mathcal{K})(\tau_1 \otimes \tau_2) = A(\tau_1) \otimes \tau_2$.]

A *qubit* is any quantum physical system S representable by a two-dimensional Hilbert space $\mathcal{H}_S$, e.g., photon polarization, spin of a spin-$\frac{1}{2}$ particle, etc. We will assume some preferred observable on $\mathcal{H}_S$ with eigenvalues 0 and 1 (the *value* of the qubit), and we will fix an orthonormal basis of $\mathcal{H}_S$ of corresponding eigenvectors $|0_S\rangle$ and $|1_S\rangle$, respectively. This will be called the *standard basis*. We will usually drop the subscripts, in which case the system S is understood implicitly. We let $P_0 = |0\rangle\langle0|$ and $P_1 = |1\rangle\langle1|$ be the orthogonal projection operators onto the subspaces spanned by $|0\rangle$ and $|1\rangle$, respectively.

A combined system $\mathcal{H}_1 \otimes \cdots \otimes \mathcal{H}_n$ of n qubits has dimension 2^n and has a natural product basis of vectors

$$|00\cdots00\rangle = |0_1\rangle|0_2\rangle \cdots |0_{n-1}\rangle|0_n\rangle$$
$$|00\cdots01\rangle = |0_1\rangle|0_2\rangle \cdots |0_{n-1}\rangle|1_n\rangle$$
$$|00\cdots10\rangle = |0_1\rangle|0_2\rangle \cdots |1_{n-1}\rangle|0_n\rangle$$

$$|00\cdots11\rangle = |0_1\rangle|0_2\rangle\cdots|1_{n-1}\rangle|1_n\rangle$$

$$\vdots$$

$$|11\cdots11\rangle = |1_1\rangle|1_2\rangle\cdots|1_{n-1}\rangle|1_n\rangle$$

We will call such a system an n-qubit system, the above basis being its standard basis. We may also write each basis vector above as $|i\rangle$ for $0 \le i < 2^n$, corresponding to the binary representation of i using n bits. For any binary string x of length at most n, we let P_x be the orthogonal projection operator onto the subspace spanned by those basis states $|z\rangle$ where z has x as a prefix—that is, the subspace spanned by $\{|z\rangle : z$ has prefix $x\}$.

For any Hilbert space $\mathcal{H}$, a standard basis $|v_1\rangle, \ldots, |v_n\rangle$ of $\mathcal{H}$ naturally lifts to a standard basis $\{|v_i\rangle\langle v_j| : 1 \le i, j \le n\}$ of $\mathcal{L}(\mathcal{H})$. This means that we can identify quantum operations with matrices over the standard bases involved, and composition of quantum operations corresponds to matrix multiplication.

One can assume that all information in a quantum computation is stored in qubits, so we will restrict each quantum operation to map from one multiqubit system to another (perhaps the same one). A *quantum circuit* C is a circuit whose inputs and outputs are labeled with distinct qubits, whose gates are labeled with quantum operations drawn from some prespecified set, and where the wires leading into each gate are ordered. If the gate has n input wires and m output wires, then its quantum operation will map n-qubit states to m-qubit states. If C has k inputs, then an *input state* of C is some k-qubit state.

We can restrict our quantum gates to three types of quantum operations:

Unitary. Maps an n-qubit state $\rho \in \mathcal{L}(\mathcal{H})$ to the state $U\rho U^\dagger \in \mathcal{L}(\mathcal{H})$, where $U \in \mathcal{L}(\mathcal{H})$ is some unitary operator that we call the *underlying* unitary operator. (The matrix entry of a unitary gate corresponding to the pair of basis vectors $|x\rangle\langle x'|$ and $|y\rangle\langle y'|$ is $\langle x|U|y\rangle\overline{\langle x'|U|y'\rangle}$.) unary operator

Ancilla Introduction. Maps an n-qubit state ρ to the $n+1$-qubit state $\rho \otimes |0\rangle\langle 0|$.[*] For example,

$$\boxed{0} \!\!- |0\rangle\langle 0|$$

---

[*]More accurately, this gate maps the unique zero-qubit "vacuum" state in the Hilbert space $\mathbb{C}$ to the one-qubit state $|0\rangle\langle 0|$.

Partial Trace. Maps an $n + 1$-qubit state $\rho \in \mathcal{L}(\mathcal{H}_1 \otimes \mathcal{H}_2)$ to the n-qubit state $\mathrm{tr}_{\mathcal{H}_2}(\rho) \in \mathcal{H}_1$.* For example,

$$\rho_1 \relbar\joinrel\relbar \rho_1$$
$$\rho_2 \relbar\!\boxed{\mathrm{tr}}$$

(Note that one can effectively trace out several qubits at once simply by tracing out each qubit one at a time; so our "primitive" Partial Trace operation need only act on a single qubit.)

Two important examples of unitary gates are the 1-qubit Hadamard gate

$$|a\rangle\langle b| \relbar\!\boxed{H}\!\relbar \tfrac{1}{2}(|0\rangle + (-1)^a|1\rangle)(\langle 0| + (-1)^b\langle 1|)$$

with underlying $U = \frac{1}{\sqrt{2}} \begin{bmatrix} 1 & 1 \\ 1 & -1 \end{bmatrix}$, and the 2-qubit controlled NOT gate $\wedge_1(\sigma_x)$

$$a \relbar\!\!\bullet\!\!\relbar a$$
$$b \relbar\!\!\oplus\!\!\relbar a \oplus b$$

with underlying $U = \begin{bmatrix} 1 & 0 & 0 & 0 \\ 0 & 1 & 0 & 0 \\ 0 & 0 & 0 & 1 \\ 0 & 0 & 1 & 0 \end{bmatrix}$. The ancilla and partial trace gates act on all the available wires, even though they connect just to wires that are being created or deleted. Other quantum operations can be expressed as compositions of these three types. For instance, the 1-qubit measurement gate, which maps ρ to $P_0\rho P_0 + P_1\rho P_1$, can be implemented by the circuit

$$\rho \relbar\!\!\bullet\!\!\relbar P_0\rho P_0 + P_1\rho P_1$$
$$\boxed{0}\!\relbar\!\oplus\!\relbar\!\boxed{\mathrm{tr}}$$

---

*Similarly to ancilla introduction, this gate is more accurately described as mapping any one-qubit state to the zero-qubit state in $\mathbb{C}$.

Which unitary operations are "reasonable?" This is a somewhat thorny question. The search for a practical implementation of a quantum circuit is still in its infancy, so it is not currently clear what types of gates can be easily built and which cannot. In lieu of practical experience, we must therefore rely on mathematical heuristics. Certainly, we should try to restrict ourselves to some minimal (preferably finite) set of simple operations. It is likely that in practice a quantum gate can only act on a small number of qubits at a time. The matrix entries of a quantum gate should be "simple" numbers in some sense: they should at least be polynomial-time approximable (see below); more preferably, they should be rational numbers with small numerators and denominators, or perhaps algebraic numbers of small degree. Finally, it is much more likely that a gate can be reliably fabricated if it can be simulated in a fault-tolerant manner (there is a large body of literature on fault-tolerant quantum computing; see [36] for a good survey).

Much good work has been done at isolating sets of quantum gates that do well under all these heuristics and that can efficiently simulate much bigger classes of operations to arbitrarily close approximation [12, 38]. We will have more to say on this later. There is an obvious trade-off here, however; restricting the repertoire of gates may require increasing the size and complexity of the circuits and may also lose the exactitude of the computation. An example of this is with the Quantum Fourier Transform (QFT) [39], which is a useful module for a wide variety of quantum algorithms. The QFT on n qubits can be implemented exactly with a reasonably simple circuit containing Hadamard gates and controlled conditional phase shift gates with phase shift $e^{2\pi i/2^n}$, where n grows with the size of the input. There can be no fixed finite palette of gates, however, that can be used to build circuits for computing QFT exactly for all n. This is because all the matrix elements of such a circuit would belong to some fixed finitely generated field extension of $\mathbb{Q}$, but the field generated by the matrix elements of QFT (ranging over all input sizes) is not finitely generated over $\mathbb{Q}$. Thus, a circuit family computing QFT exactly requires an infinite palette of gates, although a finite palette suffices to compute QFT to good approximation (which is enough for all current applications) [14, 39].

8.2.2   Classical complexity

There are several definitions of Turing machine, all equivalent. For concreteness, we use the model (single one-way infinite tape) of Sipser

[42]. One can also forget about Turing machines and think more abstractly about algorithms if one wishes.

8.2.2.1 P and NP

A *language* or *decision problem* is any subset of Σ^*. A language can also naturally be viewed as a predicate, that is, a Boolean- or 0,1-valued function on Σ^*. A Turing machine M *decides* or *computes* a language L if M halts on all inputs (either accepting or rejecting the input), and, for each $w \in \Sigma^*$, $w \in L$ if and only if M accepts input w. Let $t : \Sigma^* \to \mathbb{R}$ be any function. A language L is *computable* in time t if there is a TM M deciding L such that M halts in at most $t(w)$ steps on any input w.

A language is computable in *polynomial time* if it is computable in time $t(w) = p(|w|)$ for some polynomial p. We let P denote the class of all polynomial-time computable languages. P is a *complexity class* that captures the notion of "easy to compute deterministically," at least in a broad theoretical sense. Decision problems in P have fast, deterministic algorithms. The class P is quite *robust* in the sense that its definition does not really depend on which model of computation is used (TMs, random access machines, uniform Boolean circuit families, Pascal programs, etc.); all reasonable computational models give rise to the same class P.

Many decision problems are not known to be in P but are in the more inclusive complexity class NP. A language L is in NP if and only if there is a predicate (language) $R \in P$ and a polynomial p such that, for all $x \in \Sigma^*$,

$$x \in L \iff (\exists y \in \Sigma^r)\,[\,R(x,y)\,],$$

where $r = p(|x|)$. (We tacitly assume a suitable encoding of pairs of strings as single strings.) If a string x is in L, then we call a y satisfying $R(x,y)$ a *certificate* or *witness* to x's membership in L. One thinks of a TM M computing R as a "verifier" that, when handed some alleged proof that $x \in L$, can verify the correctness of the proof in a short amount of time. Thus, P is the class of problems whose instances are easily *decided*, and NP is the class of problems whose solutions can be easily *verified*. A typical NP problem is: given an undirected graph G and integer s, does G have a clique (complete subgraph) of size s? This problem is known as the CLIQUE problem (it is customary to name problems with all capital letters). If G did in fact have a clique of size s, then an easy-to-verify witness would be a description of an actual s-clique in G. Such a clique may be difficult to *find*, but it is easy to verify. The CLIQUE problem is not known to be in P.

If $\mathcal{C}$ is a complexity class, then co$\mathcal{C}$ is the class of complements of languages in $\mathcal{C}$, i.e.,

$$\mathrm{co}\mathcal{C} = \{\Sigma^* - L : L \in \mathcal{C}\} .$$

Clearly P = coP $\subseteq$ NP $\cap$ coNP. It is not known whether the CLIQUE problem is in coNP.

It is a major open question in complexity theory, and in mathematics as a whole, whether or not P = NP. The common intuition is that the two classes are not equal, but a proof of this has yet to be found.

All complexity classes that we mention in this chapter are subclasses of PSPACE, the class of problems solvable by using only a polynomial (in the input size) amount of memory, but unlimited time.

8.2.2.2   Reducibility and completeness

Let A and B be any languages. We say that A is *Turing reducible* (T-reducible) to B ($A \leq_{\mathrm{T}} B$) if there is an algorithm to compute A in polynomial time that can freely ask questions about membership in B. The algorithm is called a *reduction* of A to B.* We now formalize this concept. An *oracle Turing machine* () is a TM M equipped with an additional writable tape (the *query tape*), and three special states $q_?$ (the query state), q_{yes} and q_{no}. Suppose B is any language. A *computation of M with (or relative to) B* is defined similarly to an ordinary computation, except that at any time during the computation M may ask a question of the form, "Is $x \in B$?" for some string $x \in \Sigma^*$, and receive the answer as follows: first, M writes x on its query tape then enters state $q_?$; in the next step, M enters either q_{yes} if $x \in B$ or q_{no} if $x \notin B$. B is called an *oracle*, and we say that M *queries* the oracle, where x is the query. The answer to the query is "recorded" by the resulting state q_{yes} or q_{no}.

We define polynomial time in this model as before, based on the input to the computation, independent of the oracle, and we stipulate that any reduction must run in polynomial time. Turing reducibility captures the notion of relative difficulty. $A \leq_{\mathrm{T}} B$ means that A is easy *relative to* B. Here is another way to say it: A is no more difficult than B, or equivalently, B is at least as difficult as A. It is easy to see that if $A \leq_{\mathrm{T}} B$ and $B \in$ P, then $A \in$ P. Contrapositively, if $A \leq_{\mathrm{T}} B$ and

---

*It is more common to say "polynomial-time reducible" and use the notation $A \leq_{\mathrm{T}}^{p} B$ in this case. Our use is justified because all our reductions will run in polynomial time.

$A \notin$ P, then $B \notin$ P. The class P^B is the class of all languages T-reducible to B.

A restricted form of T-reducibility is called m-reducibility ("many-one" reducibility). We say that A is *m-reducible* to B ($A \leq_m B$) if $A \leq_T B$ by a reduction M which, on any input, makes exactly one query to B, accepts if the answer is yes, and rejects otherwise. Equivalently, $A \leq_m B$ if there is a polynomial-time computable function $f : \Sigma^* \to \Sigma^*$ such that, for every $w \in \Sigma^*$, $w \in A \iff f(w) \in B$.

If $\mathcal{C}$ is a complexity class and L is a language, then we say that L is *$\mathcal{C}$-hard* if $A \leq_m L$ for all $A \in \mathcal{C}$. If in addition $L \in \mathcal{C}$, then L is *$\mathcal{C}$-complete*. The CLIQUE problem, among many other interesting problems, is known to be NP-complete. If any NP-complete problem is shown to be in P, then this would prove that P = NP. We can define hardness and completeness for Turing reducibility as well, in which case we would say that L is *T-hard* or *T-complete* for $\mathcal{C}$.

8.2.2.3 Counting classes

Many interesting complexity classes can be defined in terms of the *number* of witnesses to instances of NP problems. To reify this concept, Valiant [46] defined the class #P consisting of functions $f : \Sigma^* \to \mathbb{N}$ as follows: a function f belongs to #P if and only if there is a P predicate R and a polynomial p such that for all $x \in \Sigma^*$,

$$f(x) = \left\| \left\{ y \in \Sigma^{p(|x|)} : R(x, y) \right\} \right\|.$$

For example, the number of cliques of a given size s in a given graph G is a #P function of (G, s). A variant of #P that is more algebraically useful is the class GapP [18] consisting of functions $f : \Sigma^* \to \mathbb{Z}$. A function f is in GapP if and only if $f = g - h$ for some $g, h \in$ #P. By manipulating predicates in the right way, the following closure properties of GapP are routinely verified:

PROPOSITION 8.1
[[18]]

1. *The identity function $\mathbb{Z} \to \mathbb{Z}$ is in GapP, and if $f : \Sigma^* \to \mathbb{Z}$ is computable in polynomial time and $g \in$ GapP, then $g \circ f \in$ GapP. These two facts imply that $f \in$ GapP also. Furthermore, #P $\subseteq$ GapP.*

2. *If $f \in$ GapP, then $-f \in$ GapP.*

3. *If $g, h \in$ GapP, then $g + h \in$ GapP. More generally, if $f \in$ GapP, and p is a polynomial, then $s \in$ GapP, where*

$$s(x) = \sum_{y:|y|\leq p(|x|)} f(x,y).$$

4. *If $g, h \in$ GapP, then $gh \in$ GapP (gh is the pointwise product of g and h, not the composition). More generally, if $f \in$ GapP and q is a polynomial, then $p \in$ GapP, where*

$$p(x) = \prod_{i=0}^{q(|x|)} f(x,i).$$

In other words, GapP contains all easy-to-compute functions and is closed under negation, uniform exponential sums, and uniform polynomial products. We will use the following useful fact about GapP in the proof of Theorem 8.2.

LEMMA 8.1
[[18]] If $f \in$ GapP, then there is a P predicate R and a polynomial p such that

$$f(x) = \frac{1}{2}\left(\|\{y \in \Sigma^r : R(x,y)\}\| - \|\{y \in \Sigma^r : \neg R(x,y)\}\|\right),$$

for all $x \in \Sigma^*$ with $r = p(|x|)$.

PROOF   Let $S_1, S_2 \in$ P be predicates and q a polynomial such that $f(x) = \|\{y \in \Sigma^s : S_1(x,y)\}\| - \|\{y \in \Sigma^s : S_2(x,y)\}\|$ for all $x \in \Sigma^*$ with $s = q(|x|)$. Let $p = q + 1$, and let R be the predicate

"On input (x, z) with $x, z \in \Sigma^*$:
1. Let $r = p(|x|)$.
2. If $|z| \neq r$ then reject.
3. If $z = 0y$ for some y and $S_1(x,y)$, then accept.
4. If $z = 1y$ for some y and $\neg S_2(x,y)$, then accept.
5. Reject."

A straightforward calculation shows that R and p satisfy the lemma.
∎

Lemma 8.2 below will show us that GapP is very useful in describing quantum computation. Meanwhile, GapP can provide simple characterizations of most counting complexity classes. We'll define only the two most relevant ones here: PP [25] and $C_=P$ [48].

- A language L is in PP if there is an $f \in$ GapP such that, for all $x \in \Sigma^*$,

$$x \in L \iff f(x) > 0.$$

- A language L is in $C_=P$ if there is an $f \in$ GapP such that, for all $x \in \Sigma^*$,

$$x \in L \iff f(x) = 0.$$

A typical problem in PP is: "Given an undirected graph G and natural numbers s and n, does G contain more than n distinct cliques of size s?" A typical problem in $C_=P$ is: "Given an undirected graph G and natural numbers s and n, does G contain exactly n distinct cliques of size s?"

To illustrate the uses of GapP we present a simple proof of the following well-known facts about counting classes:

PROPOSITION 8.2
[[25, 41, 48]] NP $\subseteq$ coC$_=$P $\subseteq$ PP *and* C$_=$P $\subseteq$ PP.

PROOF   Let L be a language in NP with corresponding P predicate R and polynomial p. R and p also naturally define a function $f \in$ #P such that $x \in L \iff f(x) > 0 \iff f(x) = 0$ for all $x \in \Sigma^*$. By Closure Property (1) of Proposition 8.1, f is also in GapP, so $L \in$ coC$_=$P via f.

Now let L be any language in coC$_=$P. There is a $g \in$ GapP such that $x \in L \iff g(x) \neq 0$ for all $x \in \Sigma^*$. By Closure Property (4) of Proposition 8.1, the function $g^2 \in$ GapP, and $x \in L \iff (g(x))^2 > 0$. Thus $L \in$ PP via the function g^2.

The proof that C$_=$P $\subseteq$ PP is similar, except that we use the GapP function $1 - g^2$ instead of g^2.   ∎

For our purposes, the most important property of GapP is that it is closed under uniform multiplication of a polynomial number of matrices, each of which can have exponential size. Roughly speaking, if a small number of large matrices have entries computed by a GapP function, then their product's entries are also computed by a GapP function. Since quantum operations are essentially matrix multiplication, this means we can simulate them with GapP.

We now make this notion precise. Suppose a is a GapP function that takes two parameters $i, j \in \mathbb{N}$ represented in binary, and possibly other parameters $\vec{x}$. Then for numbers $m, n \in \mathbb{N}$ (which may depend on $\vec{x}$) we let $[\, a(\vec{x}) \,]^{m \times n}$ denote the $m \times n$ matrix whose (i, j)th entry is $a(\vec{x}; i, j)$, for $0 \le i < m$ and $0 \le j < n$. A lemma similar to the following was proved by Fortnow and Rogers [22].

LEMMA 8.2
Let $a(\vec{x}, y; i, j)$ be a GapP function and let $s(\vec{x}, y)$ be a polynomial-time computable function. Then there is a GapP function $b(\vec{x}, y; i, j)$ such that for all $\vec{x}$ and for all $r \in \mathbb{N}$,

$$[\, b(\vec{x}, 1^r) \,]^{s_r \times s_0} =$$
$$[\, a(\vec{x}, 1^r) \,]^{s_r \times s_{r-1}} [\, a(\vec{x}, 1^{r-1}) \,]^{s_{r-1} \times s_{r-2}} \cdots [\, a(\vec{x}, 1) \,]^{s_1 \times s_0},$$

where $s_\ell = s(\vec{x}, 1^\ell)$ for $0 \le \ell \le r$.

PROOF   For $0 \le i_r < s_r$ and $0 \le i_0 < s_0$, the (i_r, i_0)th entry on the right hand side of the above equation is

$$\sum_{i_1=0}^{s_1-1} \sum_{i_2=0}^{s_2-1} \cdots \sum_{i_{r-1}=0}^{s_{r-1}-1} \prod_{u=1}^{r} a(\vec{x}, u; i_u, i_{u-1}).$$

This is a uniform exponential size sum of uniform polynomial size products of a GapP function. By the closure properties of GapP given in Proposition 8.1, it is a GapP function of $\vec{x}$, 1^r, i_r, and i_0. ∎

8.2.2.4   Relativization

Many results in complexity theory are "oracle" results. Let A be any language and $\mathcal{C}$ be one of the classes we have defined thus far in terms of P predicates. We can define the class $\mathcal{C}^A$ analogously with our definition of $\mathcal{C}$, except that we now allow the P predicate R free access to A as an oracle; that is, R is now a P^A predicate. This is a natural way of *relativizing* a class $\mathcal{C}$ to an oracle A, just as we relativized P earlier. If a language L is in $\mathcal{C}^A$, then we say that "$L \in \mathcal{C}$ relative to A." Most standard techniques and results in complexity theory carry over when the classes involved are relativized to any oracle. For example, GapP^A has all the closure properties of Proposition 8.1 relativized to A, including $\#\text{P}^A \subseteq \text{GapP}^A$. Thus the proof of Proposition 8.2 also easily relativizes to any oracle A to show that $\text{NP}^A \subseteq \text{coC}_=\text{P}^A \subseteq \text{PP}^A$ and $\text{C}_=\text{P}^A \subseteq \text{PP}^A$. An *oracle result* is a statement about complexity classes that holds relative to *some* oracle. Oracle results have great heuristic value in complexity theory: if some property P about complexity classes holds relative to some oracle, then this provides some evidence that P holds without an oracle (unrelativized); at least it shows that the standard, relativizable techniques of complexity theory will not suffice in refuting P, because otherwise P would be false relative to all oracles. In 1975, Baker, Gill, and Solovay [3] constructed an oracle A such that $\text{P}^A \neq \text{NP}^A$. This result suggests that it will not be easy to show that P = NP. However, in the same paper, they constructed another oracle B such that $\text{P}^B = \text{NP}^B$, which suggests that it will not be easy showing that P $\neq$ NP, either. These two oracles underscore the difficulty of the P versus NP question. Any resolution of this question will need new, possibly radically new, techniques.* Most of the current open problems in complexity theory relativize in both directions, and hence are probably difficult to solve. There are many oracle results relating to quantum complexity classes, some of which we will present after we define relativization in the quantum computation model.

Oracles are also useful for defining new complexity classes. If $\mathcal{C}$ and $\mathcal{D}$ are complexity classes, then we define $\mathcal{C}^{\mathcal{D}} = \bigcup \{\mathcal{C}^A : A \in \mathcal{D}\}$. For example, P^{NP} is the class of all languages that are easily decidable given access to an NP oracle. A typical problem in P^{NP} is: "Given a graph G and $s \in \mathbb{N}$, is s the size of the largest clique in G?" This problem is not known to be either in NP or in coNP, but it can be decided using binary search on s with access to the CLIQUE problem as an oracle.

8.2.2.5 Quantum algorithms and FQP

Classical algorithms can be expressed efficiently with Boolean circuits, where an input string of length n is regarded as the Boolean settings

---

*There are some nonrelativizing techniques. A prime example is the technique of low-degree polynomial interpolation, which has been used to show that all PSPACE languages have interactive proofs [32, 37], even though there is an oracle relative to which some languages in coNP do not have interactive proofs [23]. Despite their early promise, however, and despite their great utility in other areas, these techniques have yet to resolve any open questions about more "traditional" time-bounded complexity classes.

of the input nodes of an n-input circuit. Since an algorithm should run on inputs of any length, we model the algorithm by an infinite family $C_0, C_1, C_2, \ldots$ of Boolean circuits, where C_i has i input nodes and handles all input strings of length i. If the algorithm decides a language, then each C_i should have exactly one output; an output value of 1 signifies acceptance, and 0 signifies rejection. If an algorithm is efficient (polynomial time, say) and we wish to express it efficiently by a family of circuits, then we must have some efficient way of generating each circuit in the family. A family $\{C_i\}_{i \in \mathbb{N}}$ of circuits is *p-uniform* if there is a polynomial time algorithm which, on input 1^i, outputs a full description of C_i. In this case, note that each C_i is not too large, and so can be simulated efficiently. Indeed, we have the following well-known proposition (see, e.g., [47]):

PROPOSITION 8.3

For any language L, $L \in P$ if and only if L is decided by a p-uniform family of Boolean circuits.

Quantum algorithms can be expressed using quantum circuits. Here, an input string x of length n is "fed" into an n-input quantum circuit by setting the quantum state of the input wires (qubits) to $|x\rangle\langle x|$. Our practical considerations in Section 8.2.1 suggest that a quantum operation cannot operate reliably on more than a few qubits at a time; therefore, we set a fixed limit on the indegree and outdegree of any quantum gate in any quantum circuit. A constant of three is convenient, although two actually suffices; and in fact, the only two-qubit gate needed is the controlled NOT gate [6]. For efficient quantum computation, we again insist on a p-uniform family of quantum circuits.

The p-uniformity requirement raises some subtle issues. First, one may ask if p-uniformity is too strict; perhaps some *quantum* process for circuit fabrication is more powerful than a classical one. Second, our requirement implies that each quantum gate must be completely represented by a finite (polynomial size) amount of classical information, whereas the matrix representation of a general quantum operation may contain arbitrary complex numbers, which are not finitely representable.

To address the first issue, we must have a precise concept of a circuit-fabricating quantum process. Such a process should be algorithmic and allow for all possible inputs (of all possible lengths). Here we can turn to the QTM model [15], which was historically the first reasonable model of quantum computation. A single, finitely describable QTM can handle inputs of any and all lengths, so a QTM can be used to fabricate quantum circuits. It was later shown by Yao [49] that QTMs can be simulated efficiently by p-uniform families of quantum circuits (and vice versa). Thus, a family of quantum circuits fabricated by a quantum process as above is no more or less computationally powerful than a p-uniform family of quantum circuits.

The second issue is explained based on physics and some intuition. Since quantum computers do not currently exist, the first physical process for building a quantum computer must be described and implemented classically. So it is reasonable to suppose that any physical procedure for efficiently fabricating a quantum computer, including calibrating the various components, must ultimately rely on an efficient classical algorithmic description. In keeping with these considerations, Bernstein and Vazirani settled upon the requirement that all the transition amplitudes in a QTM should be polynomial-time approximable [10]. [A complex number z is *polynomial-time approximable* if there is a polynomial time algorithm that on input 1^r outputs three integers s, x, y such that $|z' - z| \leq 2^{-r}$, where $z' = 2^{-s}(x + iy)$.] An additional *de facto* requirement is that any QTM M only uses a finite set of transition amplitudes, since these amplitudes are part of the finite description of M's transition function. These two requirements are not overly strict. All currently known and forseeable quantum algorithms easily fit this requirement. Furthermore, any algorithm that violated this requirement would, at the very least, require significant justification, and even then it may not be universally accepted by the research community.

These requirements are also not overly lax; the universal QTM constructed by Bernstein and Vazirani [10], which can efficiently approximate any QTM satisfying these requirements, uses only classical transitions (amplitudes in $\{0, 1\}$) and the one-qubit Hadamard transition (amplitudes in $\left\{-\frac{1}{\sqrt{2}}, \frac{1}{\sqrt{2}}\right\}$).

Yao's simulation of a QTM by a quantum circuit family and vice versa [49] does not disturb the set of amplitudes involved too much, except for arithmetic operations. In particular, all the gates used in a p-uniform quantum circuit family simulating a QTM can have matrix entries drawn from some finite set depending only on the QTM being simulated. These entries will also lie in the field over $\mathbb{Q}$ generated by the transition amplitudes of the QTM. The converse also holds: such a p-uniform quantum circuit family F can be efficiently simulated by a QTM whose transition amplitudes all lie in the field over $\mathbb{Q}$ generated by the matrix entries in the gates of F. We will therefore adopt the reasonable requirement that for every circuit family there is a fixed finite set of amplitudes from which all matrix entries of all gates in the family are drawn.

Bernstein and Vazirani's universal QTM together with Yao's results imply that there is a single finite set of amplitudes that works for all efficient quantum circuit families. In fact, circuits with Toffoli gates (described in Section 8.2.3 below) and Hadamard gates suffice to simulate the universal QTM, and hence they suffice for all efficient quantum computation. A Toffoli gate itself can be simulated (exactly) using

- the one-qubit Hadamard gate H,

- the controlled NOT gate $CNOT$, and

- the one-qubit conditional phase shift, or $\pi/8$ gate with underlying unitary matrix $\sigma_z^{1/4} = \begin{bmatrix} 1 & 0 \\ 0 & e^{i\pi/4} \end{bmatrix}$ [6, 33].

Thus we call these three gates a *univeral set* (of quantum gates) in analogy with classical Boolean circuits. The gates in this set can all be implemented fault-tolerantly [12]. Several other universal sets have been found [6]. (One should add to any such list some gates for nonunitary operations, such as measurement and partial trace.) There are other universal sets where all amplitudes in the underlying unitary matrices are drawn from the set $\left\{-1, -\frac{4}{5}, -\frac{3}{5}, 0, \frac{3}{5}, \frac{4}{5}, 1\right\}$ [43].

For any subset $B \subseteq \mathbb{C}$, we say that a quantum circuit family F is *over* B if all matrix elements of all gates of all circuits in F are elements of B. We will get stronger negative results if we do not insist on a fixed finite universal set of gates for all circuit families, but rather allow different kinds of gates in different circuit families. We will, however, generally restrict B to be the field of algebraic numbers. Algebraic numbers are all polynomial-time approximable (by Newton's method, for example), and such a set of operations is clearly universal. Moreover, all currently known quantum algorithms are expressed directly and easily using algebraic amplitudes.

We now define the input-output behavior of a quantum circuit C with n inputs and m outputs. Each gate g of C represents a quantum operation from a multiqubit Hilbert space $\mathcal{H}$ to a multiqubit space $\mathcal{J}$, taking any state ρ of $\mathcal{H}$ to the state $g(\rho)$ of $\mathcal{J}$. Let $g_1, \ldots, g_s$ be a (topologically sorted) list of the gates of C, so that no output wires of any g_i are inputs to g_j for $j \leq i$. Suppose the input qubits are in state

ρ_0 of space $\mathcal{H}_0$. We first apply g_1 by expressing $\mathcal{H}_0$ as the tensor product $\mathcal{H}_0' \otimes \mathcal{H}_0''$, where $\mathcal{H}_0'$ is the space of qubits entering gate g_1 and $\mathcal{H}_0''$ is the space of qubits bypassing g_1. The gate g_1 maps states of $\mathcal{H}_0'$ to states of some space $\mathcal{H}_1'$, in which case the state of all the qubits after applying g_1 is $\rho_1 = (g_1 \otimes \mathcal{I})(\rho_0)$ of the Hilbert space $\mathcal{H}_1 = \mathcal{H}_1' \otimes \mathcal{H}_0''$.

We then apply g_2 to the state ρ_1 in a similar manner, again keeping track of the qubits bypassing g_2, to obtain a state ρ_2 of a Hilbert space $\mathcal{H}_2$. We then apply gate g_3, and so on. After all the gates have been applied, we obtain a state ρ_s of a space $\mathcal{H}_s$ of at least m qubits, where the first m qubits, say, are the designated outputs. (The state ρ_s is independent of the ordering of the gates as long as they are topologically sorted; for if two gates could be swapped in the ordering, then they act on disjoint sets of qubits and hence the corresponding operations commute.) Finally, we observe the values of the m output qubits (a simultaneous projective measurement), from which we obtain some random variable ranging over the 2^m possible outcomes (classical bit strings of length m).

A function f on Σ^* is a *probabilistic function* if, for all $x \in \Sigma^*$, $f(x)$ is a random variable ranging over Σ^m for some m depending on $|x|$. For any set $B \subseteq \mathbb{C}$, we let FQP_B [2] denote the class of all probabilistic functions computed by p-uniform quantum circuit families over B as described above. If we drop the subscript on FQP, then we mean FQP_A, where A is the field of algebraic numbers.

As we mentioned earlier, a partial trace gate acts on all the current qubits at once but we only attach it to one wire. This means we have more freedom to place the gate in a topological sort of all the gates of the circuit. Our freedom does not lead to any ambiguity, though, because the tr-gate commutes with other tr-gates and with unitary gates. For example, both of the circuits

map an input operator $A \otimes B$ to the output operator $(\mathrm{tr}B)UAU^\dagger$. Similarly, ancilla gates commute with each other and with unitary gates. These facts allow us to place all the ancilla gates to the left (first in the topological sort) and all the tr-gates to the right (last), with all the intermediate gates being unitary. We will say that such a circuit is in *standard form* with *width* (of a circuit equal to the number of qubits present at any point between the ancillæ and tr-gates).

Occasionally, when we "trace out" a qubit (that is, apply a tr-gate to it) in a circuit C, we can guarantee that the state of the qubit is $|0\rangle\langle 0|$. If this is the case, we may label the tr-gate with a 0 instead of the usual tr. If we embed C into a larger circuit, asserting that a qubit is in the zero-state allows it to be reused later as an ancilla, making the larger circuit more efficient. Put another way,

$$\boxed{0} \quad \boxed{0}\!- \quad \text{can always be replaced with} \quad \text{_____}$$

Another good reason to reset ancilla qubits to 0 before tracing them out is that this makes the input–output behavior of C to be unitary. This arises especially when we simulate a classical gate (see Section 8.2.3 below): if the ancilla is in an arbitrary "garbage" state (which may be entangled with the other qubits) when we trace it out, then we will in general get a mixed state result, even though the input was a pure state. This happens when the input state is a nontrivial superposition of basis states. Circuits that avoid this problem by resetting their ancillæ before tracing them out are called clean circuits. If a classical circuit is not clean, there is a straightforward way to clean it up that essentially doubles the number of gates. This is often not the most efficient way, though.

8.2.2.6   The classes BPP and BQP

Decision problems with efficient probabilistic algorithms make up the class BPP (Bounded-error Probabilitic Polynomial time). Decision problems with efficient bounded error quantum algorithms make up the class BQP [10]. AWPP is a counting class defined in [19]. After formally defining these classes, we will give proofs that $P \subseteq BPP \subseteq BQP \subseteq AWPP \subseteq PP$. The first inclusion is obvious and well known. The second inclusion was proved by Bernstein and Vazirani [10], the third by Fortnow and Rogers [22], and the last appeared in [19].

DEFINITION 8.1   *A language L is in* BPP *if there is a* P *predicate R and a polynomial p such that, for all $x \in \Sigma^*$ and for $m = p(|x|)$,*

$$x \in L \Rightarrow \|\{y \in \Sigma^* : |y| = m \wedge R(x,y)\}\| \geq (2/3)2^m,$$
$$x \notin L \Rightarrow \|\{y \in \Sigma^* : |y| = m \wedge R(x,y)\}\| \leq (1/3)2^m.$$

Given x and m as above, if y is chosen uniformly at random among the strings of length m, then $R(x,y)$ computes $L(x)$ correctly with a probability of at least $2/3$. By repeating the computation $R(x, y)$ several times with independently chosen y's and then taking the majority answer, we can make the probability of error exponentially close to zero.

PROPOSITION 8.4

If $L \in$ BPP, then for every polynomial q there is a polynomial p and a P predicate R such that, for all $x \in \Sigma^*$ (setting $r = q(|x|)$ and $m = p(|x|)$),

$$x \in L \Rightarrow \|\{y \in \Sigma^* : |y| = m \wedge R(x, y)\}\| \geq 2^m - 2^{m-r},$$
$$x \notin L \Rightarrow \|\{y \in \Sigma^* : |y| = m \wedge R(x, y)\}\| \leq 2^{m-r}.$$

A well-known problem in BPP that is not known to be in P is the PRIMALITY problem: "Given a natural number x, is x prime?" Primality testing is especially useful for public key cryptography.

The class BQP can be defined in terms of FQP. A *binary probabilistic function* is a probabilistic function whose output random variables range over the set $\{0, 1\}$. A language L is in BQP if and only if there is a binary probabilistic function $f \in$ FQP such that, for all $x \in \Sigma^*$,

$$x \in L \Rightarrow \text{Prob}[\, f(x) = 1 \,] \geq 2/3,$$
$$x \notin L \Rightarrow \text{Prob}[\, f(x) = 1 \,] \leq 1/3.$$

By combining several simultaneous computations of f, we can diminish the error probability as we did with BPP.

LEMMA 8.3

If $L \in$ BQP, then for every polynomial p there is a binary probabilistic $f \in$ FQP such that, for all $x \in \Sigma^*$ (setting $r = p(|x|)$),

$$x \in L \Rightarrow \text{Prob}[\, f(x) = 1 \,] \geq 1 - 2^{-r},$$
$$x \notin L \Rightarrow \text{Prob}[\, f(x) = 1 \,] \leq 2^{-r}.$$

The set of possible *exact* distributions output by FQP functions may be quite sensitive to the choice of allowed gates or amplitudes in the quantum circuit model, but the set of possible *approximable* distributions—and hence the class BQP—is not very sensitive at all. As we mentioned before, Bernstein and Vazirani's universal QTM uses only classical transitions (i.e., those that preserve basis states) and the one-qubit Hadamard transform. The corresponding circuit family thus only needs classical gates (see below) and the one-qubit Hadamard gate.

These gates have all amplitudes in $\mathbb{Q}$—actually in the set $\left\{0, \pm\frac{1}{2}, 1\right\}$. We can therefore simulate any QTM *à la* [10] to good approximation and thus get the same class BQP, even if we restrict our gates to using these amplitudes. Actually there is a bit of a cheat here. Although the Hadamard gate (the superoperator) uses rational matrix elements, its underlying unitary operator has elements $\pm\frac{1}{\sqrt{2}} \notin \mathbb{Q}$. Adleman, DeMarrais and Huang [1] and independently Solovay and Yao [43] showed that rational transition amplitudes in the set $\left\{0, \pm\frac{3}{5}, \pm\frac{4}{5}, \pm 1\right\}$ suffice even for the underlying unitary operators for a universal QTM. All these results show that BQP is a very robust class.

An interesting subclass of BQP is EQP (Exact Quantum Polynomial time [1]), defined just as with BQP except no error is allowed—that is, the allowed acceptance probabilities are 1 and 0. Strangely, EQP is even less sensitive to the allowed amplitudes than BQP. Adleman et al. showed that EQP remains the same even if arbitrary complex amplitudes are allowed, and that this is not the case with BQP.

8.2.3   Classical computations on a quantum circuit

A unitary gate is *classical gate* if its matrix elements are in the set $\{0, 1\}$. Classical gates map basis states onto basis states and so do not introduce quantum superpositions. If a circuit's intermediate gates are all classical, then a classical input (basis state) yields a classical output (basis state). An important result in quantum computation is that any Boolean circuit (and hence any classical computation) can be simulated efficiently by a quantum circuit made up of classical gates. The difficulty is that classical gates in a quantum circuit must obey restrictions that Boolean gates in a classical circuit need not. Boolean gates may be irreversible, losing information from input to output, whereas classical gates in a quantum circuit must be reversible since they correspond to unitary operations. In addition, the output of a Boolean gate may be freely duplicated on several output wires, but this is not possible with a classical gate in a quantum circuit.

Both of these difficulties can be overcome by allowing more ancilla qubits for the simulating quantum circuit. Two types of classical gates—the *Toffoli gate* and the NOT gate—

suffice to build circuits simulating any classical Boolean circuit when provided with a small number of extra ancillæ. The Boolean AND and COPY gates are simulated as follows:

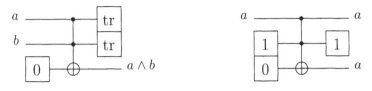

where

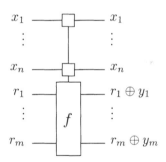

We will allow classical gates in quantum circuits that encapsulate arbitrary efficient classical algorithms, as is justified by the previous considerations. More precisely, for any polynomial-time computable function $f : \Sigma^* \to \Sigma^*$ whose output length depends only on the input length, and for any $n \in \mathbb{N}$, we have the $(n + m)$-qubit gate U_f, which we depict as

$$
\begin{array}{cccc}
x_1 & \!\!-\!\!\square\!\!-\!\! & x_1 \\
\vdots & & \vdots \\
x_n & \!\!-\!\!\square\!\!-\!\! & x_n \\
r_1 & & r_1 \oplus y_1 \\
\vdots & f & \vdots \\
r_m & & r_m \oplus y_m
\end{array}
$$

where $f(x_1 \cdots x_n) = y_1 \cdots y_m$. We call such a gate an *f-gate*. Our decidedly nonstandard depiction of this gate is meant to distinguish the output wires (bottom) from the input wires (top) by connecting the inputs via the little boxes. One can view this gate as shorthand for a quantum subcircuit computing f on strings of length n. Note that f must be computed by a clean circuit to guarantee that U_f is a unitary gate.

If A is an n-ary Boolean function outputting a single bit, then we will call the A-*inversion gate* the unitary gate whose underlying unitary operator I_A is defined by

$$
I_A |x\rangle = (-1)^{A(x)} |x\rangle
$$

for all $x \in \Sigma^n$. Inversion gates are used quite often in quantum algorithmics. The I_A gate is easy to implement with a single A-gate:

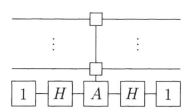

8.2.4   Relativizing quantum computation

To prove oracle results relating to quantum computation, we need to define exactly what it means for a quantum circuit to have access to an oracle. Let $f : \Sigma^* \to \Sigma^*$ be any function (not necessarily computable) whose output length depends only on the input length. There is an f-gate for every $n \in \mathbb{N}$. Note that f-gates cannot be viewed as shorthand for subcircuits, for f may not be computable by any circuit at all. A quantum circuit gains access to f as an oracle by using f-gates. This is completely analogous to the standard way of relativizing Boolean circuits. The relativized class FQP^f can now be defined just as with FQP, except now we allow the quantum circuits to have f-gates and we also allow the polynomial-time circuit-fabrication algorithm access to f as an oracle (the latter allowance is actually unnecessary). A language A can be accessed by a quantum circuit as an oracle since it corresponds to a function with output length 1. In this case, we refer to the oracle gates as A-gates.

An oracle in a quantum algorithm is often referred to as a "black box," and the algorithm is a "black-box algorithm." One can think of a black-box algorithm as computing with some unknown classical gate behind an abstraction barrier. We regard the classical gate—the black box—as another kind of input to the algorithm, and the algorithm extracts information about the black box. Many interesting quantum algorithms are black-box algorithms. We will mention two early black-box algorithms, a variant of one due to Deutsch and Josza [17] and the other due to Simon [40] in Section 8.4.1 below.

8.3 Equivalence of FQP and GapP

In this section we relate FQP closely to GapP. In doing so, we get a deep connection between the complexity of quantum computation and classical counting complexity. The following theorem shows that GapP is at least as powerful a class as FQP. It is adapted from [20], which itself is a straightforward generalization of a result by Fortnow and Rogers [22], which in turn improves an earlier result announced by Valiant (see [10]) essentially showing that BQP $\subseteq$ P^{GapP}.

If we consider some field extension K of $\mathbb{Q}$ with finite basis $\alpha_1, \ldots, \alpha_e$ over $\mathbb{Q}$, then a quantum circuit over K can be finitely described by representing its matrix elements as vectors of rational coefficients of the α_i.

THEOREM 8.1

Fix a field $K \subseteq \mathbb{C}$ with finite dimension over $\mathbb{Q}$, and fix a basis $\alpha_1, \ldots, \alpha_e$ for K over $\mathbb{Q}$ such that $\alpha_1, \ldots, \alpha_d$ (for some $d \leq e$) span $K \cap \mathbb{R}$ (as a vector space over $\mathbb{Q}$). There are GapP functions $h_1, \ldots, h_d$ that behave as follows: let C be any quantum circuit over K with set of matrix coefficients $B \subseteq K$, and suppose C has n inputs and m outputs. Then there is an integer $D_B > 0$ such that for all $x \in \Sigma^n$,

$$\text{Prob}\,[\,C \text{ outputs } y \text{ on input } x\,] = D_B^{-r} \sum_{i=1}^{d} h_i(C, x, y)\alpha_i,$$

where r is the number of unitary gates in C. Moreover, D_B depends only on B and is polynomial-time computable from B (or from C).

In the special case where $K = \mathbb{Q}$ and $\alpha_1 = 1$, there is a GapP function h such that

$$\text{Prob}\,[\,C \text{ outputs } y \text{ on input } x\,] = D_B^{-r} h(C, x, y).$$

COROLLARY 8.1

[[20, 22]] Suppose $f \in$ FQP is computed by a p-uniform quantum circuit family over a finite set B of algebraic numbers. Let K be the field extension of $\mathbb{Q}$ generated by B, and let $\alpha_1, \ldots, \alpha_d \in \mathbb{R}$ span $K \cap \mathbb{R}$. Then there exist an integer $D > 0$, an integer-coefficient polynomial p, and GapP

functions $g_1, \ldots, g_d$ *such that, for all* $x, y \in \Sigma^*$ *(setting* $r = p(|x|)$*),*

$$\text{Prob}\,[\,f(x) = y\,] = D^{-r} \sum_{i=1}^{d} g_i(x, y)\alpha_i.$$

In particular, if $B \subseteq \mathbb{Q}$*, we have* $\text{Prob}\,[\,f(x) = y\,] = D^{-r}g(x, y)$ *for some* GapP *function* g*.*

PROOF   Extend $\alpha_1, \ldots, \alpha_d$ to a basis $\alpha_1, \ldots, \alpha_e$ of K. Apply Theorem 8.1 to get $h_1, \ldots, h_d$ and D_B. Set $D = D_B$. Let $C_0, C_1, C_2, \ldots$ be a p-uniform quantum circuit family computing f, where C_n has n inputs. Let $p(n)$ be a polynomial upper bound on the number r_n of unitary gates in C_n. Then for all $x \in \Sigma^*$ of length n we have

$$\text{Prob}\,[\,f(x) = y\,] = D^{-p(n)} \sum_{i=1}^{d} g_i(x, y)\alpha_i,$$

where $g_i(x, y) = D^{p(n)-r_n} h(C_n, x, y)$ is clearly a GapP function.   ∎

PROOF of Theorem 8.1   We prove the special case where $K = \mathbb{Q}$, and sketch the proof of the general case. A more detailed proof of the general case can be found in [20].

We are given a quantum circuit C with n inputs, m outputs, and set of matrix elements $B = \left\{ \frac{n_1}{d_1}, \ldots, \frac{n_s}{d_s} \right\}$ for integers n_i and d_i. Set $D_B = \text{lcm}(d_1, \ldots, d_s)$. By rearranging gates if necessary, we can assume that C is in standard form with width q and (unitary) gates $g_1, \ldots, g_r$ in some topological order. Each gate g_ℓ corresponds to a quantum operation G_ℓ mapping $\mathcal{L}(\mathcal{H})$ to $\mathcal{L}(\mathcal{H})$ for some fixed q-qubit (2^q-dimensional) Hilbert space $\mathcal{H}$. Thus each G_ℓ is represented by a $Q \times Q$ matrix with entries in B, where $Q = 2^{2q}$. By our choice of D_B, each $D_B G_\ell$ is an integer matrix. It is clear that there is a polynomial-time computable function (and hence a GapP function) $a(C, \ell; i, j)$ which computes the (i, j)th entry of $D_B G_\ell$. By Lemma 8.2, the composition of all the G_ℓ, and thus the computation of the entire circuit, is computed by a function $D_B^{-r} b(C; i, j)$, where $b \in$ GapP because

$$[\,b(C)\,]^{Q \times Q} = [\,a(C, r)\,]^{Q \times Q} \cdots [\,a(C, 1)\,]^{Q \times Q}.$$

(Note that we do not need to supply r as an input to b, because it can be computed from C.)

Let $M_C = D_B^{-r}[\, b(C)\,]^{Q \times Q}$. An input string $x \in \Sigma^n$ is fed to C via the input state $\rho_0 = |x\rangle\langle x| \otimes |0^{q-n}\rangle\langle 0^{q-n}|$ which includes the ancilla qubits. If we write ρ_0 as a column vector V of length Q, then the final state of the circuit (just before the tr-gates are applied and the output qubits are measured) is $\rho = M_C V$ (in column vector form). Using Lemma 8.2 again and reshaping the column vector as a matrix, we get a GapP function $c(C, x; i, j)$ such that ρ is represented by the matrix $D_B^{-r}[\, c(C, x)\,]^{2^q \times 2^q}$.

We can write ρ as the sum

$$\rho = \sum_{y_1, y_2 \in \Sigma^m} |y_1\rangle\langle y_2| \otimes A_{y_1 y_2},$$

where the $|y_1\rangle\langle y_2|$ correspond to the m output qubits and the $A_{y_1 y_2}$ are operators on the space of the remaining $q - m$ qubits. For any $y \in \Sigma^m$, the probability of y being the value of the final output measurement is

$$\mathrm{tr}(P_y \rho) = \mathrm{tr} A_{yy} = D_B^{-r} \sum_{z:|z| = q \text{ and } z \text{ has prefix } y} c(C, x; z, z).$$

The sum on the right-hand side is clearly a GapP function of C, x, and y by Proposition 8.1. Letting $h(C, x, y)$ be this sum proves the special case of the theorem.

We now sketch the proof of the general case. Let K, d, and $\alpha_1 \ldots, \alpha_e$ be given as in the theorem. For all $1 \le i, j, k \le e$, let $c_{jk}^i \in \mathbb{Q}$ be such that $\alpha_j \alpha_k = \sum_{i=1}^e c_{jk}^i \alpha_i$. By rescaling the α_i if necessary, we can assume that all the c_{jk}^i are integers. Now the idea is that we can represent scalars in K such as individual matrix entries uniquely as vectors over $\mathbb{Q}$ of length e. Addition of scalars corresponds to vector addition, and scalar multiplication is computed using the c_{jk}^i. GapP is closed under all of these operations, and so we can generalize Lemma 8.2 to the case where matrix entries are elements of K. We then apply the generalized lemma much as before. Given circuit C with set $B = \{b_1, \ldots, b_s\}$ of matrix elements, we must choose D_B so that we always compute with integer entries in the scalar representations: Let $n_{ij}, d_{ij} \in \mathbb{Z}$ be such that each $b_i = \sum_{j=1}^e \frac{n_{ij}}{d_{ij}} \alpha_j$. We can let D_B be the least common multiple of all the d_{ij}.

Analoguous with the previous argument, we now get GapP functions $c_1(C, x; i, j), \ldots, c_e(C, x; i, j)$ such that the final state

$$\rho = D_B^{-r} \sum_{\ell=1}^e \alpha_\ell [\, c_\ell(C, x)\,]^{2^q \times 2^q}.$$

And for $y \in \Sigma^m$, we get

$$\text{Prob}\,[\,f(x) = y\,] = \sum_{z:|z| = q \text{ and } z \text{ has prefix } y} \rho_{zz},$$

where $\rho_{zz} = D_B^{-r} \sum_{\ell=1}^{e} \alpha_\ell c_\ell(C, x; z, z)$, the (z, z)th entry of the matrix ρ. Since ρ is Hermitian, each ρ_{zz} is real, and so by the linear independence of the α_ℓ we must have $c_\ell(C, x; z, z) = 0$ for $d < \ell \leq e$. Therefore, setting

$$h_\ell(C, x, y) = \sum_{z:|z| = q \text{ and } z \text{ has prefix } y} c_\ell(C, x; z, z),$$

for $1 \leq \ell \leq d$, it is clear that each $h_\ell \in \text{GapP}$, and the equality

$$\text{Prob}\,[\,f(x) = y\,] = D_B^{-r} \sum_{\ell=1}^{d} h_\ell(C, x, y) \alpha_\ell$$

holds as desired.   ∎

COROLLARY 8.2

If $f \in \text{FQP}_B$ where $B = \left\{0, \pm\frac{1}{2}, \frac{1 \pm i}{\sqrt{2}}, \pm 1\right\}$ as in the case for the universal set of gates described in Section 8.2.2.5 above, then there are $h_1, h_2 \in \text{GapP}$ such that

$$\text{Prob}\,[\,f(x) = y\,] = 2^{-r}\left(h_1(x, y) + \sqrt{2}h_2(x, y)\right),$$

where r is a polynomial in $|x|$.

There is a partial converse to Corollary 8.1. The following theorem was shown in [20].

THEOREM 8.2

Let $B = \left\{0, \pm\frac{1}{2}, 1\right\}$. For any $h \in \text{GapP}$ there is an $f \in \text{FQP}_B$ and an integer-coefficient polynomial p such that for all $x \in \Sigma^*$,

$$\text{Prob}\,[\,f(x) = 0^r\,] = 2^{-2(r+1)}(h(x))^2,$$

where $r = p(|x|)$.

PROOF   By Lemma 8.1 there is a predicate $R \in P$ and polynomial p such that for any $x \in \Sigma^*$,

$$h(x) = \frac{1}{2}\left(\|\{y \in \Sigma^r : R(x, y)\}\| - \|\{y \in \Sigma^r : \neg R(x, y)\}\|\right),$$

where $r = p(|x|)$. For $n \in \mathbb{N}$, let C_n be the quantum circuit with n inputs and r ancillæ, where $r = p(n)$, depicted below.

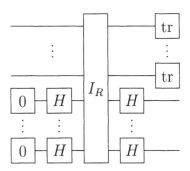

On input $|x\rangle\langle x|$ where $|x| = n$, the circuit C_n transforms the quantum state as follows:

$$|x\rangle\langle x| \overset{0}{\mapsto} |x, 0^r\rangle\langle x, 0^r|$$
$$\overset{H}{\mapsto} 2^{-r} \sum_{y,y' \in \Sigma^r} |x, y\rangle\langle x, y'|$$
$$\overset{I_R}{\mapsto} 2^{-r} \sum_{y,y'} (-1)^{R(x,y)+R(x,y')} |x, y\rangle\langle x, y'|$$
$$\overset{H}{\mapsto} 2^{-2r} \sum_{y,y',z,z'} (-1)^{R(x,y)+R(x,y')+y\cdot z + y'\cdot z'} |x, z\rangle\langle x, z'|$$
$$\overset{\mathrm{tr}}{\mapsto} 2^{-2r} \sum_{y,y',z,z'} (-1)^{R(x,y)+R(x,y')+y\cdot z + y'\cdot z'} |z\rangle\langle z'|.$$

The probability that $f(x)$ equals 0^r is given by the coefficient of $|0^r\rangle\langle 0^r|$ in the final state, which is

$$2^{-2r} \sum_{y,y'} (-1)^{R(x,y)+R(x,y')}$$
$$= 2^{-2r} \left(\sum_y (-1)^{R(x,y)} \right)^2$$
$$= 2^{-2r} (-2h(x))^2$$
$$= 2^{-2(r+1)} (h(x))^2.$$

An easy application of Corollary 8.1 and Theorem 8.2 is a characterization of the class NQP, a quantum analogue of NP, defined by Adleman, DeMarrais, and Huang [1].

DEFINITION 8.2   *[[1]] Let $B \subseteq \mathbb{C}$. A language L is in the class* NQP_B *if there is an* $f \in \mathrm{FQP}_B$ *such that for all* $x \in \Sigma^*$,

$$x \in L \iff \mathrm{Prob}[\, f(x) = 0\,] > 0.$$

THEOREM 8.3 [20]
For any finite set B of algebraic numbers containing $\left\{0, \pm\frac{1}{2}, 1\right\}$, we have $\mathrm{NQP}_B = \mathrm{coC}_=\mathrm{P}$.

PROOF   Let $K \subseteq \mathbb{C}$ be the field generated by B and let $\alpha_1, \ldots, \alpha_d$ be a basis of $K \cap \mathbb{R}$ over $\mathbb{Q}$. Suppose $L \in \mathrm{NQP}_B$ via some $f \in \mathrm{FQP}_B$, and let $D > 0$ and $g_1, \ldots, g_d \in \mathrm{GapP}$ be as in Corollary 8.1. Fix an input x. Since the α_i are linearly independent over $\mathbb{Q}$, we have $\mathrm{Prob}[\, f(x) = 0\,] = 0$ just in the case that $g_1(x, 0) = \cdots = g_d(x, 0) = 0$. It follows that $L \in \mathrm{coC}_=\mathrm{P}$ via the GapP function

$$h(x) = \sum_{i=1}^{d} (g_i(x, 0))^2,$$

that is, $x \in L \iff h(x) \neq 0$.

Conversely, suppose $L \in \mathrm{coC}_=\mathrm{P}$ via some $h \in \mathrm{GapP}$. Let $f \in \mathrm{FQP}_B$ be given by Theorem 8.2. Clearly, $L \in \mathrm{NQP}_B$ witnessed by f.   ∎

8.4   Strengths of the quantum model

In this section we give results that show that the class BQP is large. That is, we place lower bounds on the size of BQP. We first show that BPP $\subseteq$ BQP. This is an old and relatively easy result, first proved by Bernstein and Vazirani [10], but it remains the strongest result of its kind—showing that a previously studied complexity class is contained in BQP.

THEOREM 8.4 [10]

BPP $\subseteq$ BQP$_\mathbb{Q}$.

PROOF   Let L be in BPP with corresponding polynomial p and predicate $R \in$ P. For each $n \in \mathbb{N}$ and $r = p(n)$, consider the circuit with n inputs, $r + 1$ ancillæ, and one output, depicted below.

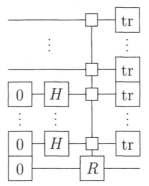

On input $|x\rangle\langle x|$ where $x \in \Sigma^*$, the state after the R-gate is

$$2^{-r} \sum_{y,y' \in \Sigma^r} |x, y, R(x, y)\rangle\langle x, y', R(x, y')|.$$

After the tr-gates are applied, the final state is

$$2^{-r} \sum_{y} |R(x, y)\rangle\langle R(x, y)|.$$

The probability that the output qubit has value 1 is the coefficient of $|1\rangle\langle 1|$ in the final state. This probability is $2^{-r}\|\{y \in \Sigma^r : R(x, y)\}\|$, which is equal to the probability of $R(x, y)$ being true for random y. Thus the above family of circuits computes L as a BQP language. ∎

The other known positive results about BQP show that specific problems are in BQP. These problems come in two types: (1) black-box problems and (2) non-black-box problems (*concrete* problems) that are not known to be in BPP. Primary examples of such concrete problems are INTEGER FACTORIZATION and the DISCRETE LOGARITHM problem [39]. These two problems are easily computable relative to some NP oracle, but have no known efficient randomized solutions.

We will concentrate on black-box problems. These will give us oracle results of the form BQP$^A \not\subseteq \mathcal{C}^A$ for some oracle A, where $\mathcal{C}$ is a complexity class. We will also obtain oracles for EQP as well.

8.4.1   Oracle results

A natural question to ask is: does NP contain BQP, or even EQP? Equivalently, does every problem decidable by an efficient quantum algorithm (with or without error allowed) have easily verifiable solutions? Another natural question is: does BPP = BQP? In other words, can quantum algorithms (with bounded error) be efficiently simulated by classical probabilistic algorithms (with bounded error)? If so, then quantum computation would be a lot less interesting.

We will address the second question first. INTEGER FACTORIZATION is in BQP and it is not *known* to be in BPP, so this gives credibility to the conjecture that BPP $\neq$ BQP, although it is no proof. Credibility of a different sort comes by way of an oracle B such that BPP^B $\neq$ BQP^B, in fact, EQP^B $\not\subseteq$ BPP^B. This oracle can be constructed as an instance of Simon's black-box problem [40], which we now describe. Suppose we are given a black-box function $f : \Sigma^n \to \Sigma^m$ with $n \leq m$, and there is an $s \in \Sigma^n$ such that, for all distinct $x, y \in \Sigma^n$,

$$f(x) = f(y) \iff x \oplus y = s.$$

Simon showed that deciding if $s = 0$ is a BQP problem relative to f. Moreover, the BQP^f algorithm can be used as a subroutine to find s with high probability. Later, Brassard and Høyer [13] showed how to find s with zero error probability, yielding an EQP^f algorithm. The set oracle B just codes the output bits of f. Existence of B signifies that proving BPP = BQP or even EQP $\subseteq$ BPP will at least be very difficult.

We now address the first question. It is clear that if $L \in$ BQP, then $\Sigma^* - L \in$ BQP also. Therefore if BQP $\subseteq$ NP, then BQP $\subseteq$ NP $\cap$ coNP. The same goes for EQP replacing BQP. All concrete problems currently known to be in BQP are also in NP $\cap$ coNP.

Such a containment would be quite difficult to prove, however. There is an oracle A such that EQP^A $\not\subseteq$ NP^A, so any proof that EQP $\subseteq$ NP would need nonrelativizable techniques. A can be obtained via the black-box Balance problem of Deutsch and Jozsa [17]. We will present a stronger oracle result here due to Green and Pruim [26] based on ideas of Boyer, Brassard, Høyer, and Tapp [11]: there is an oracle A such that EQP^A $\not\subseteq$ (P^{NP})^A. This result, as do most oracle results, comes in two pieces: first we describe a particular problem computed by an algorithm (here a quantum algorithm) that uses a single type of black-box gate. Then we concoct a particular black box that plugs into the quantum algorithm so that (1) the quantum algorithm computes the problem with zero probability of error and (2) the problem cannot be solved by any $\mathrm{P^{NP}}$ algorithm, even one with access to the same black box as an oracle.

How do you relativize a class like $\mathrm{P^{NP}}$ to an oracle A, a class that is already defined in terms of oracles? You should at least let the polynomial time TM access an oracle $L \in \mathrm{NP}^A$ (instead of just NP). The "standard" heuristic for relativizing any class is to allow all computations free access to the oracle. In keeping with this, you should also allow the TM direct access to A itself, on a separate oracle tape, say. It turns out that this latter requirement is unnecessary, because information about A can be encoded directly into an appropriately chosen NP^A oracle L, so that the TM can access A indirectly through L. The preceding trick is expressed symbolically by the equation $(\mathrm{P^{NP}})^A = \mathrm{P}^{(\mathrm{NP}^A)}$, with the right-hand side usually being written as just $\mathrm{P^{NP}}^A$. This is the standard definition of $\mathrm{P^{NP}}$ relativized to A.

THEOREM 8.5 [26]

There is an oracle A such that $\mathrm{EQP}^A \not\subseteq \mathrm{P^{NP}}^A$.

PROOF   Green and Pruim use a quantum circuit C_n implicitly described by Boyer et al. [11] to solve a variant of the Deutsch–Josza Balance problem [17] with zero error. Fix a language $A \subseteq \Sigma^*$ as a black box with the promise that, for any length n, A either contains exactly one quarter or exactly three quarters of the strings of length n. The Modified Balance Problem relative to A (MBP^A) is the problem of determining which is the case for each length n, given input 0^n: accepting for one quarter, rejecting for three quarters. The circuit C_n that solves MBP has n inputs and employs two inversion gates I_0 and I_A, where for any $x \in \Sigma^n$,

$$I_0 |x\rangle = \begin{cases} -|x\rangle & \text{if } x = 0, \\ |x\rangle & \text{if } x \neq 0, \end{cases}$$

That is, I_0 is the NOR-inversion gate. Note that I_0 has underlying unitary operator $I - 2|0^n\rangle\langle 0^n|$, where I is the identity operator on n qubits. The ciruit C_n is

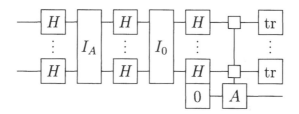

The unitary operator underlying most of C_n is $H^{\otimes n} I_0 H^{\otimes n} I_A H^{\otimes n}$, where $H^{\otimes n}$ is simultaneous application of H on n qubits. (The operator $H^{\otimes n} I_0 H^{\otimes n} I_A$ is used in quantum search algorithms to find an element of A and is sometimes referred to as the Grover iterate [27, 28, 29].)

On input $|0^n\rangle\langle 0^n|$, we compute the state $|\psi\rangle\langle\psi|$ of C_n just before the last A-gate. For any $X \subseteq \Sigma^n$ let $|X\rangle = \sum_{x\in X} |x\rangle$ (unnormalized). For two such sets X and Y, note that $\langle X|Y\rangle = \|X \cap Y\|$. Let $A_n = A \cap \Sigma^n$ with cardinality s. We have $H^{\otimes n}|0^n\rangle = 2^{-n/2}|\Sigma^n\rangle$, and so up to a harmless global phase factor,

$$
\begin{aligned}
|\psi\rangle &= H^{\otimes n} I_0 H^{\otimes n} I_A H^{\otimes n}|0^n\rangle \\
&= 2^{-n/2} H^{\otimes n} (I - 2|0^n\rangle\langle 0^n|) H^{\otimes n} I_A |\Sigma^n\rangle \\
&= 2^{-n/2} (I - 2^{1-n}|\Sigma^n\rangle\langle\Sigma^n|) I_A |\Sigma^n\rangle \\
&= 2^{-n/2} (I - 2^{1-n}|\Sigma^n\rangle\langle\Sigma^n|)(|A_n\rangle - |\Sigma^n - A_n\rangle) \\
&= 2^{-n/2} (|A_n\rangle - |\Sigma^n - A_n\rangle - 2^{1-n}(2s - 2^n)|\Sigma^n\rangle) \\
&= 2^{-n/2} (3|A_n\rangle + |\Sigma^n - A_n\rangle - 2^{2-n} s|\Sigma^n\rangle).
\end{aligned}
$$

If $s = \frac{1}{4}2^n$, then $|\psi\rangle = 2^{1-n/2}|A_n\rangle$; and if $s = \frac{3}{4}2^n$, then $|\psi\rangle = -2^{1-n/2}|\Sigma^n - A_n\rangle$. If we observed the value of the first n qubits at this point, then we are guaranteed to see an element of A_n in the former case and $\Sigma^n - A_n$ in the latter. The last A-gate distinguishes between these two cases with certainty. This completes the description of C_n: if A satisfies the promise on every length, then $\mathrm{MBP}^A \in \mathrm{EQP}^A$ via the family of circuits C_n.

Our remaining job is to construct a particular A satisfying the promise such that $\mathrm{MBP}^A \notin \mathrm{P}^{\mathrm{NP}^A}$ (i.e., MBP^A is not "NP^A-easy"). To do this, we use a time-honored technique from logic and computer science known as diagonalization. A $\mathrm{P}^{\mathrm{NP}^A}$ "machine" involves a P^A predicate $R(x, y)$ and polynomial p, together with a polynomial time oracle TM M computing with oracle L^A, where L^A is the NP^A language determined by R and p. Each such machine has a finite description $\langle R, p, M\rangle$, so we can enumerate these machines as $N_1, N_2, \ldots$ in some way. If MBP^A is NP^A-easy, then there is some i such that N_i correctly computes MBP^A

on each input 0^n. Equivalently, MBP^A is *not* NP-easy if for each i there is a length n_i such that N_i computes the wrong value for MBP^A on input 0^{n_i}. We define A in stages $1, 2, \ldots$. At stage i we "diagonalize against N_i" by explicitly choosing a length n_i and specifying the set $A_{n_i} = A \cap \Sigma^{n_i}$ so as to *force* N_i to give the wrong answer on input 0^{n_i}. All the while, we must ensure that A satisfies the promise at each length; this keeps MBP^A a member of EQP^A.

How do we fool machine N_i on input 0^{n_i}? First we commit just enough of the oracle A to force N_i to give *some* answer. This does not require us to commit too many strings. Then we add just enough additional strings of length n_i to A to satisfy the promise in the sense opposite to N_i's answer. For lengths n other than the n_i, we satisfy the promise trivially by setting A_n to be the lexicographically least 2^{n-2} strings of length n—that is, all strings starting with 00.

For each $i \geq 1$ we let N_i correspond to R_i, p_i, M_i as described above, we let $q_i(n)$ be a strictly monotone polynomial bounding the running time of M_i on all inputs of length $\leq n$, and we let $r_i(n) > 0$ be a strictly monotone polynomial bounding the running time of R_i on all inputs (x, y) where $|x| \leq q(n)$ and $|y| = p(|x|)$. Also as described above, for any oracle X let L_i^X be the NP^X language corresponding to p_i and R_i relative to X. We let $n_0 = 0$, and for $i \geq 1$ we let $n_i \in \mathbb{N}$ be least such that $\frac{1}{4}2^{n_i} \geq q_i(n_i)r_i(n_i)$ and $n_i > r_j(n_{i-1})$ for all $j < i$. Given the way we defined the r_j, we have $n_i > n_{i-1}$ and all queries to our oracle made on behalf of N_j on input 0^{n_j} for $j < i$ have length $< n_i$. (This means that we can freely put strings of length n_i into or out of the oracle without disturbing any of the computations we have already forced.) We set $T = \{n_0, n_1, n_2, \ldots\}$.

At each stage of the construction, we will commit strings to be either in or out of the oracle we are building: for $i \in \mathbb{N}$ we define two sets $A_i \subseteq B_i \subseteq \Sigma^*$ such that $A_0 \subseteq A_1 \subseteq A_2 \subseteq \ldots$ and $B_0 \supseteq B_1 \supseteq B_2 \supseteq \ldots$. A_i is the set of strings we have committed to be in the oracle by the end of Stage i, and B_i is the set of strings we have *not* committed to be *out of* the oracle by the end of Stage i (thus the A_i sets get bigger as i increases, and the B_i sets get smaller). Strings in $B_i - A_i$ are *uncommitted*. All strings of length n_i are committed by the end of Stage i, and all strings with lengths not in T are committed at the beginning of the construction. The final oracle will be $A = \bigcup_i A_i = \bigcap_i B_i$.

Before Stage 1 we let

$$A_0 = \{00y : y \in \Sigma^* \wedge |00y| \notin T\} \quad \text{and} \quad B_0 = A_0 \cup \bigcup_{n \in T} \Sigma^n.$$

Construction of A
Stage $i \geq 1$:
  Set $A_i = A_{i-1}$ and $B_i = B_{i-1}$.
  Simulate M_i on input 0^{n_i}:
    Whenever M_i makes a query x, do:
      If $x \in L_i^C$ for some C with $A_i \subseteq C \subseteq B_i$, then
        Let $y \in \Sigma^{p_i(|x|)}$ be least such that $R_i^C(x, y)$.
        Let Q be the set of queries made while computing $R_i^C(x, y)$.
        /* Commit the queries in Q: */
        Set $A_i = A_i \cup (Q \cap C)$ and $B_i = B_i - (Q - C)$.
        Answer "yes" to M_i's query.
      Else,
        Answer "no" to M_i's query.
  Let $m = \|\Sigma^{n_i} \cap A_i\|$.
  If M_i accepts in our simulation, then:
    Add to A_i the first $\frac{3}{4}2^{n_i} - m$ uncommitted strings in Σ^{n_i}, and
    remove the rest of the length n_i uncommitted strings from B_i.
  Else,
    Add to A_i the first $\frac{1}{4}2^{n_i} - m$ uncommitted strings in Σ^{n_i}, and
    remove the rest of the length n_i uncommitted strings from B_i.
End of Stage i.
Set $A = \bigcup_i A_i$.
End of Construction

We complete the proof by making three observations. First, there are always enough uncommitted strings to add to A_i at the end of Stage i. All strings of length n_i are uncommitted at the start of Stage i because n_i is too big for any of these strings to be committed at earlier stages. Each query set Q in the simulation has cardinality at most $r_i(n_i)$ due to the running time of R_i. M_i can make at most $q_i(n_i)$ queries; so the total number of strings we commit during the simulation is at most $q_i(n_i)r_i(n_i) \leq \frac{1}{4}2^{n_i}$, by the choice of n_i.

Second, N_i cannot "change its mind" about 0^{n_i} after the simulation is finished in Stage i, no matter how we commit strings afterwards. If one of M_i's queries x is in L_i^C for some C such that $A_i \subseteq C \subseteq B_i$, then we commit the queries made by R_i^C as it accepts some appropriate (x, y), so R_i's behavior on (x, y) cannot change, and x will also be in L_i^A. Otherwise, if there is no such C, then $x \notin L_i^A$ since $A_i \subseteq A \subseteq B_i$. Thus we always answer M_i's query according to L_i^A, which makes M_i's accept/reject behavior in the simulation the same as it is relative to A.

Third and finally, A satisfies the promise at all lengths, and M_i (and thus N_i) with oracle A computes MBP^A incorrectly at length n_i. These facts follow from the definitions of A_0 and B_0 and from the number of strings we put into A at the end of Stage i. ∎

8.5 Limitations of the quantum model

Bernstein and Vazirani showed that $\text{BQP} \subseteq \text{P}^{\text{PP}}$ ($= \text{P}^{\#\text{P}} = \text{P}^{\text{GapP}}$), establishing the first connection between quantum complexity and counting complexity. Fortnow and Rogers [22] tightened this connection by showing that $\text{BQP} \subseteq \text{AWPP}$, a counting class we will define shortly. Their results, together with previous results about AWPP [19], give the strongest evidence yet that $\text{NP} \not\subseteq \text{BQP}$, and thus NP-complete problems cannot be decided efficiently with quantum circuits. In particular, there is an oracle G such that $\text{P}^G = \text{BQP}^G \neq \text{NP}^G$. This complements an earlier result of Bennett, Bernstein, Brassard, and Vazirani [9], which states that $\text{NP}^R \not\subseteq \text{BQP}^R$ with probability 1, where the oracle R is chosen at random. (They also show that $\text{NP}^{R'} \cap \text{coNP}^{R'} \not\subseteq \text{BQP}$ with probability 1, where R' is a permutation oracle chosen at random.)

The class AWPP is somewhat analogous to BPP.

DEFINITION 8.3 *[[19, 31]] A language L is in* AWPP *if and only if, for every polynomial q, there is a polynomial-time computable function $p > 0$ and a GapP function f such that*

$$x \in L \Rightarrow (1 - 2^{-r})m \leq f(x) \leq m,$$
$$x \notin L \Rightarrow 0 \leq f(x) \leq 2^{-r}m$$

for all $x \in \Sigma^*$, where $r = q(|x|)$ and $m = p(1^{|x|})$.

It is easy to see that $\text{AWPP} \subseteq \text{PP}$: for $L \in \text{AWPP}$, setting $q(|x|) = 2$ gives an f and $p > 0$ as in the definition, whence the GapP function $g(x) = f(x) - \lfloor p(1^{|x|})/2 \rfloor$ witnesses that $L \in \text{PP}$.

THEOREM 8.6 [22]
$\text{BQP} \subseteq \text{AWPP}$.

PROOF   Let L be a language in BQP, and let q be any polynomial. We will find a polynomial-time computable function $p > 0$ and GapP function f that satisfy Definition 8.3 for L. By Lemma 8.3 there is a binary $g \in \mathrm{FQP}_\mathbb{Q}$ such that

$$x \in L \Rightarrow \mathrm{Prob}[\, g(x) = 1\,] \geq 1 - 2^{-r},$$
$$x \notin L \Rightarrow \mathrm{Prob}[\, g(x) = 1\,] \leq 2^{-r},$$

for any $x \in \Sigma^*$ with $r = q(|x|)$. By Theorem 8.1, there is an integer $D > 0$, polynomial s and GapP function f such that $\mathrm{Prob}[\, g(x) = 1\,] = D^{-s(|x|)} f(x)$ for all $x \in \Sigma^*$, and so

$$x \in L \Rightarrow 1 - 2^{-r} \leq D^{-s} f(x) \leq 1,$$
$$x \notin L \Rightarrow 0 \leq D^{-s} f(x) \leq 2^{-r},$$

where $s = s(|x|)$. The theorem follows immediately by letting $p(1^n) = D^{s(n)}$ for all $n \in \mathbb{N}$.   ∎

COROLLARY 8.3
[[1]] BQP $\subseteq$ PP.

There is a stronger limitation on AWPP: languages in AWPP are *low* for PP—that is, if $A \in$ AWPP, then $\mathrm{PP}^A = \mathrm{PP}$. If some PP-complete language is low for PP, then $\mathrm{PP}^{\mathrm{PP}} = \mathrm{PP}$, which is considered highly unlikely.

PROPOSITION 8.5
[[31]] If $A \in$ AWPP, then $\mathrm{PP}^A = \mathrm{PP}$.

PROOF   Let L be any language in PP^A via some function in GapP^A, which itself corresponds by Lemma 8.1 to a polynomial q and P^A predicate R. That is,

$$x \in L \iff \left\| \left\{ y \in \Sigma^{q(|x|)} : R(x,y) \right\} \right\| - \left\| \left\{ y \in \Sigma^{q(|x|)} : \neg R(x,y) \right\} \right\| > 0.$$

We may assume that $R(x,y)$ depends on (i.e., its computation makes queries to) A only on strings of length $s(|x|)$, where $s > 0$ is some polynomial, and makes exactly $s(|x|)$ such queries. (If this is not the case, then we can replace A with a "padded" version of A:

$$A' = \{ y10^n : y \in A \wedge n \in \mathbb{N} \}.$$

It is obvious that $A' \in \text{AWPP} \iff A \in \text{AWPP}$ and that $\text{PP}^A = \text{PP}^{A'}$. Moreover, there is a $\text{P}^{A'}$ computation of R that satisfies our assumption above.)

The algorithmic nature of R can be decomposed into two pieces—a polynomial-time computable function $a(x, y, b, i)$ and a P predicate $Q(x, y, b)$. Here, $b = b_1 \cdots b_{s(|x|)}$ is a string of bits representing guesses of the answers to R's queries. Given x, y, b, we simulate the computation of $R(x, y)$. When the computation makes its ith query to the oracle, we answer with b_i. In this simulation it is easy to compute what the ith query will be, and we let $a(x, y, b, i)$ be this query. When the simulation ends, we let $Q(x, y, b)$ be the result. Now if b corresponds to all correct query answers (all according to A), then $Q(x, y, b)$ is the same as $R(x, y)$. In other words,

$$Q(x, y, b_1 b_2 \cdots b_{s(|x|)}) = R(x, y)$$

whenever each bit $b_i = A(a(x, y, b, i))$. Clearly, a is computable in polynomial time and $|a(x, y, b, i)| = s(|x|)$.

Let r be a large enough polynomial that we will choose later. The membership of A in AWPP gives us, by Definition 8.3, a polynomial-time computable $p > 0$ and $f \in \text{GapP}$ such that

$$z \in A \Rightarrow (1 - 2^{-r})m \le f(z) \le m, \tag{8.1}$$
$$z \notin A \Rightarrow 0 \le f(z) \le 2^{-r}m, \tag{8.2}$$

for all $z \in \Sigma^*$, where $r = r(|z|)$ and $m = p(1^{|z|}) > 0$. By doubling both p and f, we can assume that m is always even.

Fix an arbitary $x \in \Sigma^*$ and let $n = |x|$, $q = q(n)$, $s = s(n)$, $r = r(s) = r(s(n))$, and $m = p(1^s) = p(1^{s(n)}) > 0$ and even. Let $in_x = \|\{y \in \Sigma^q : R(x, y)\}\|$ and let $out_x = \|\{y \in \Sigma^q : \neg R(x, y)\}\|$. We want to define an $h \in \text{GapP}$ (with no oracle) such that $h(x) > 0 \iff in_x - out_x > 0$ ($\iff x \in L$). Define

$$g(x, y, b) = \prod_{i=1}^{s} [b_i f(a(x, y, b, i)) + (1 - b_i)(m - f(a(x, y, b, i)))]$$

for all $b = b_1 \cdots b_s \in \Sigma^s$ and $y \in \Sigma^q$. It is evident by Proposition 8.1 that $g \in \text{GapP}$. For all $1 \le i \le s$ we have, by (8.1) and (8.2) above,

$$b_i = A(a(x, y, b, i)) \Rightarrow (1 - 2^{-r})m \le \lambda_i \le m,$$
$$b_i \ne A(a(x, y, b, i)) \Rightarrow 0 \le \lambda_i \le 2^{-r}m,$$

where $\lambda_i = b_i f(a(x, y, b, i)) + (1 - b_i)(m - f(a(x, y, b, i)))$. This means that

$$(\forall i)\,[\, b_i = A(a(x, y, i))\,] \Rightarrow (1 - 2^{-r})^s m^s \leq g(x, y, b) \leq m^s, \quad (8.3)$$

$$(\exists i)\,[\, b_i \neq A(a(x, y, i))\,] \Rightarrow 0 \leq g(x, y, b) \leq 2^{-r} m^s. \quad (8.4)$$

The point here is that if all the guesses b_i to query answers are correct, then $g(x, y, b)$ is large (close to m^s), and if not, then $g(x, y, b)$ is close to zero. Now if we define

$$h(x) = \sum_{y \in \Sigma^q} \sum_{b \in \Sigma^s} g(x, y, b)(2Q(x, y, b) - 1) - \frac{m^s}{2},$$

then $h \in \mathrm{GapP}$, which is clear by Proposition 8.1. Roughly speaking, each term in the sum contributes to $h(x)$ a negligible amount if some of the guesses in b are incorrect. If the guesses in b are all correct, then the term contributes approximately m^s, if $Q(x, y, b)$, or approximately $-m^s$, if $\neg Q(x, y, b)$. But if all the guesses in b are correct, then $Q(x, y, b) = R(x, y)$, and so this makes $h(x)$ close enough to $m^s(in_x - out_x)$ that it witnesses that $L \in \mathrm{PP}$. The $m^s/2$ correction is to ensure that $h(x) \leq 0$ when $x \notin L$.

More precisely, for every x, y let $b_{x,y} \in \Sigma^s$ be the unique string of correct oracle answers during the computation of $R(x, y)$. We have

$$h(x) + \frac{m^s}{2} = \sum_y g(x, y, b_{x,y})(2Q(x, y, b_{x,y}) - 1)$$

$$+ \sum_y \sum_{b \neq b_{x,y}} g(x, y, b)(2Q(x, y, b) - 1)$$

$$= \sum_y g(x, y, b_{x,y})(2R(x, y) - 1)$$

$$+ \sum_y \sum_{b \neq b_{x,y}} g(x, y, b)(2Q(x, y, b) - 1).$$

We can bound the second sum on the right-hand side using (8.4) above:

$$\left| \sum_y \sum_{b \neq b_{x,y}} g(x, y, b)(2Q(x, y, b) - 1) \right| < 2^{q+s-r} m^s. \quad (8.5)$$

The first sum splits into two:

$$S = \sum_{y : R(x,y)} g(x, y, b_{x,y}) - \sum_{y : \neg R(x,y)} g(x, y, b_{x,y}).$$

Using (8.3) above, we get

$$(1 - 2^{-r})^s in_x - out_x \leq S/m^s \leq in_x - (1 - 2^{-r})^s out_x. \qquad (8.6)$$

Combining (8.5) and (8.6) gives us

$$(1 - 2^{-r})^s in_x - out_x - 2^{q+s-r} < h(x)/m^s + 1/2$$
$$< in_x - (1 - 2^{-r})^s out_x + 2^{q+s-r}.$$

If $x \in L$, then $out_x + 1 \leq 2^{q-1} \leq in_x - 1$, so

$$h(x)/m^s > (1 - 2^{-r})^s(2^{q-1} + 1) - 2^{q-1} + 1/2 - 2^{q+s-r}. \qquad (8.7)$$

If $x \notin L$, then $in_x \leq 2^{q-1} \leq out_x$, so

$$h(x)/m^s < 2^{q-1}(1 - (1 - 2^{-r})^s) + 2^{q+s-r} - 1/2. \qquad (8.8)$$

Now if we choose the polynomial r to be at least $q + s + 2$, then much tedious calculation reveals that the right-hand side of (8.7) is positive and the right-hand side of (8.8) is negative. Thus $h(x)$ is as desired.

∎

COROLLARY 8.4
All BQP *languages are low for* PP.

The *counting hierarchy* is the chain of classes

$$\mathrm{PP} \subseteq \mathrm{PP}^{\mathrm{PP}} \subseteq \mathrm{PP}^{\mathrm{PP}^{\mathrm{PP}}} \subseteq \cdots.$$

It is widely believed that this hierarchy *does not collapse*, i.e., all classes in the chain are distinct.

COROLLARY 8.5
If BQP = PP, *then* $\mathrm{PP}^{\mathrm{PP}} = \mathrm{PP}$ *and so the counting hierarchy collapses to* PP.

8.5.0.1 Black-box problems

As we mentioned earlier, a significant and growing number of quantum algorithms solve black-box problems. It would be useful for us to understand the inherent limitations of these algorithms. We have some strong, unconditional results that reveal sharp restrictions on the use of quantum circuits to solve black-box problems, unlike the case with concrete problems. Using certain properties of low-degree multivariate polynomials, Beals, Buhrman, Cleve, Mosca, and de Wolf [7] have proven lower bounds on the depths of quantum circuits that decide various important properties of black-box functions.

The key to the results in [7] lies in making two observations: (1) Theorem 8.1 and Lemma 8.1 both relativize *uniformly* to any oracle, as do most results in complexity theory (see Lemma 8.4 below), and (2) a relativized GapP function varies with the oracle in a simple way, according to a low-degree multivariate polynomial. Oracles and black-box problems are conceptually the same, and understanding the implications of these observations will give us a key result in [7] that ties quantum circuits to multivariate polynomials.

Theorem 8.1 and Lemma 8.1 relativize to give us the following lemma, which is the first observation above, and which is implicit in [19]. For simplicity, we will restrict our attention to quantum circuits with rational matrix entries.

LEMMA 8.4

For every oracle X, there is a function $h^X \in \mathrm{GapP}^X$ that behaves as follows: let C be any quantum circuit over $\mathbb{Q}$ with X-gates and set of matrix coefficients $B \subseteq \mathbb{Q}$, and suppose C has n inputs and m outputs. Then for all $z \in \Sigma^n$ and $w \in \Sigma^m$,

$$\mathrm{Prob}\,[\,C \ \text{outputs} \ w \ \text{given input} \ z\,] = D^{-r} h^X(C, z, w),$$

where r is the number of unitary gates in C, and D is the least positive integer such that $DB \subseteq \mathbb{Z}$.

Moreover, h^X is computed uniformly relative to X, i.e., there is a single polynomial p and single polynomial-time oracle TM M that

1. *makes at most $2r$ oracle queries along any path, and*

2. *relative to any oracle X, computes a predicate $R^X \in \mathrm{P}^X$ such that*

$$h^X(x) = \frac{1}{2}\left(\left\|\Sigma^{p(|x|)} \cap R^X(x)\right\| - \left\|\Sigma^{p(|x|)} - R^X(x)\right\|\right),$$

for all $x \in \Sigma^*$, where $R^X(x) = \{y : R^X(x, y)\}$.

(Here we interpret x as encoding the triple (C, z, w).)

PROOF SKETCH The proof is just as with that of Theorem 8.1. The only added feature of C is that it can use X-gates. Each matrix element of an X-gate is either 0 or 1 and is easily computed with two queries to X (recall that X is a superoperator which needs to act on basis vectors of the form $|w\rangle\langle w'|$). It remains to see that, along any single path, M only needs to compute one matrix entry of each X-gate. This follows from a straightforward analysis of how the matrix product is computed in the proof of Theorem 8.1 relativized to X, and we will not go into further detail here. ∎

Given M and p computing h^X as in Lemma 8.4, how does h^X vary with X? To address this question, we associate a real-valued variable v_z for every $z \in \Sigma^*$. We then let any oracle X correspond to a unique setting of each v_z to either 1 (if $z \in X$) or 0 (if $z \notin X$). Now fix an input (x, y) and let the v_z vary over $\{0, 1\}$. The machine M makes oracle queries as it runs on input (x, y), so M's output is some function of all the v_z such that M can possibly query z on input (x, y). This function can be expressed as a multivariate polynomial $q_{(x,y)}$ of low degree over the v_z.

PROPOSITION 8.6

Let M and p compute the function h^X as in Lemma 8.4 above, fix an (x, y) with $|y| = p(|x|)$, and let r be the maximum number of oracle queries made by M on input (x, y) along any computation path. The function $q_{(x,y)}$ described above is a multivariate polynomial over the variables v_z of degree at most r with rational coefficients; hence, the value of $h^X(x)$ is described by a multivariate polynomial s_x over the v_z of degree at most r.

PROOF We use some ideas from the proof of Proposition 8.5. We can assume that M makes exactly r queries on any computation path on input (x, y). Let R^X and h^X be as in Lemma 8.4. Fix some $b = b_1 \cdots b_r \in \Sigma^r$. We simulate M on input (x, y); when M makes its ith query $z_{b,i}$ to the oracle, we answer according to b_i. We say that M *accepts on* b if M accepts in this simulation. Now define the degree r polynomial $m_b = \prod_{i=1}^{r} u_i$, where

$$u_i = \begin{cases} v_{z_{b,i}} & \text{if } b_i = 1, \\ 1 - v_{z_{b,i}} & \text{if } b_i = 0. \end{cases}$$

Let X be any oracle and let the variables v_z be set according to X.

If $X(z_{b,i}) = b_i$ for each i, then M takes the same path and makes the same queries relative to oracle X as it did in our simulation above, and $m_b = 1$; otherwise, one of M's queries $z_{b,i}$ is not answered according to X in the simulation, so $m_b = 0$. If we now set

$$q_{(x,y)} = \sum_{b \in \Sigma^r} m_b a_b,$$

(where $a_b = 1$ if M accepts on b, and $a_b = 0$ otherwise), then $q_{(x,y)}$ is a degree r multivariate polynomial over the v_z that takes the value $R^X(x,y)$ when its variables are set according to X. Thus the multivariate polynomial

$$s_x = \frac{1}{2} \sum_{y \in \Sigma^{p(|x|)}} \left(2q_{(x,y)} - 1\right)$$

has degree r and takes the value $h^X(x)$ when its variables are set according to X. ∎

As our final result, we see that Lemma 8.4 and Proposition 8.6 immediately combine to give us a uniform version of the central lemma in Beals et al. [7]. It shows how the behavior of a quantum circuit C with X-gates varies as the oracle X varies.

PROPOSITION 8.7

Let C be a quantum circuit with n inputs, m outputs, and r many X-gates. For any input $z \in \Sigma^n$, and output $w \in \Sigma^m$, there is a degree $2r$ multivariate polynomial s over the variables v_z such that, for all oracles X,

$$\text{Prob}[\, C \text{ outputs } w \text{ given input } z \,] = s^X,$$

where s^X is the value of s when the variables v_z are set according to X. Moreover, s is the multivariate polynomial s_x defined in the proof of Proposition 8.6 where $x = (C, z, w)$.

8.6   Conclusions

Given the close connection we have seen between quantum complexity and counting complexity, we should continue to expect results in one area to apply to the other, especially the latter to the former, since counting complexity is an older subject. This has clearly been the case so far. For example, the oracle G such that $P^G = BQP^G \neq NP^G$ was discovered in the context of the counting class AWPP in [19] before any of the authors knew about BQP (it was shown there that $P^G = AWPP^G \neq NP^G$). It was only later, when Fortnow and Rogers showed that $BQP \subseteq AWPP$ [22], that the significance of the oracle G to quantum complexity was recognized.

Another example regards the class NQP (Definition 8.2). Showing that $NQP = coC_=P$ [20] immediately imported a wealth of knowledge about $C_=P$ into the quantum realm. For instance, $coC_=P$ contains problems that are at least as hard as any in the *polynomial hierarchy*: $NP \subseteq NP^{NP} \subseteq NP^{NP^{NP}} \subseteq \cdots$ [44, 45], so the same is true of NQP.

Finally, the connection between black-box quantum circuits and low-degree multivariate polynomials [7] allows important results about the decision tree complexity of Boolean functions [34] (already applied to counting classes in [19]) to be applied directly to quantum complexity.

There are many interesting open questions and potentially fruitful lines of further research connecting quantum computation with complexity theory. Can we put BQP inside a class smaller than AWPP— perhaps a counting class? On the other hand, can we put BQP out of a class bigger than P^{NP} with respect to some oracle? How about showing that $BQP^A \not\subseteq NP^{NP^A}$ for some oracle A? What are the analogous relationships between EQP and other complexity classes? Answers to any of these questions will push the field forward considerably.

References

[1] L.M. Adleman, J. DeMarrais, and M.-D.A. Huang. Quantum computability. *SIAM J. Comp.*, 26(5):1524–1540, 1997.

[2] D. Aharonov, A.Yu. Kitaev, and N. Nisan. Quantum circuits with mixed states. *Proceedings of the 30th ACM Symposium on the Theory of Computing*, pages 20–30, 1998, `quant-ph/9806029`.

[3] T. Baker, J. Gill, and R. Solovay. Relativizations of the P = NP question. *SIAM J. Comp.*, 4:431–442, 1975.

[4] J.L. Balcázar, J. Díaz, and J. Gabarró. *Structural Complexity I*, volume 11 of *EATCS Monographs on Theoretical Computer Science*. Springer-Verlag, Berlin, 1988.

[5] J.L. Balcázar, J. Díaz, and J. Gabarró. *Structural Complexity I*, volume 22 of *EATCS Monographs on Theoretical Computer Science*. Springer-Verlag, Berlin, 1990.

[6] A. Barenco, C.H. Bennett, R. Cleve, D.P. DiVincenzo, N. Margolus, P. Shor, T. Sleator, J.A. Smolin, and H. Weinfurter. Elementary gates for quantum computation. *Phys. Rev. A*, 52(5):3457–3467, 1995, quant-ph/9503016.

[7] R. Beals, H. Buhrman, R. Cleve, M. Mosca, and R. de Wolf. Quantum lower bounds by polynomials. *Proceedings of the 39th IEEE Symposium on Foundations of Computer Science*, pages 352–361. IEEE, 1998, quant-ph/9802049.

[8] P.A. Benioff. The computer as a physical system: A microscopic quantum mechanical Hamiltonian model of computers as represented by Turing machines. *J. Statist. Phys.*, 22:563–591, 1980.

[9] C.H. Bennett, E. Bernstein, G. Brassard, and U. Vazirani. Strengths and weaknesses of quantum computation. *SIAM J. Comp.*, 26(5):1510–1523, 1997, quant-ph/9701001.

[10] E. Bernstein and U. Vazirani. Quantum complexity theory. *SIAM J. Comp.*, 26(5):1411–1473, 1997.

[11] M. Boyer, G. Brassard, P. Høyer, and A. Tapp. Tight bounds on quantum searching. *Forsch. Phys.*, 46:493–506, 1998, quant-ph/9605034.

[12] P.O. Boykin, T. Mor, M. Pulver, and V. Roychowdhury. On universal and fault-tolerant quantum computing, 1999, quant-ph/9906054.

[13] G. Brassard and P. Høyer. An exact quantum polynomial-time algorithm for Simon's problem. *Proceedings of the Israeli Symposium on the Theory of Computing and Systems*, pages 12–23, 1997, quant-ph/9704027.

[14] D. Coppersmith. An approximate fourier transform useful in quantum factoring. Research Report RC 19642. IBM, 1994.

[15] D. Deutsch. Quantum theory, the Church-Turing principle and the universal quantum computer. *Proceedings of the Royal Society London A*, 400:97–117, 1985.

[16] D. Deutsch. Quantum computational networks. *Proceedings of the Royal Society London A*, 425:73–90, 1989.

[17] D. Deutsch and R. Jozsa. Rapid solution of problems by quantum computation. *Proceedings of the Royal Society London A*, 439:553–558, 1992.

[18] S. Fenner, L. Fortnow, and S. Kurtz. Gap-definable counting classes. *J. Comp. Syst. Sci.*, 48(1):116–148, 1994.

[19] S. Fenner, L. Fortnow, S. Kurtz, and L. Li. An oracle builder's toolkit. *Proceedings of the 8th IEEE Structure in Complexity Theory Conference*, pages 120–131, 1993.

[20] S. Fenner, F. Green, S. Homer, and R. Pruim. Determining acceptance possibility for a quantum computation is hard for the polynomial hierarchy. *Proceedings of the Royal Society London A*, 455:3953–3966, 1999, quant-ph/9812056.

[21] R.P. Feynman. Simulating physics with computers. *Intl. J. Theoret. Phys.*, 21(6/7):467–488, 1982.

[22] L. Fortnow and J. Rogers. Complexity limitations on quantum computation. *J. Comp. Syst. Sci.*, 59(2):240–252, 1999, cs.CC/9811023.

[23] L. Fortnow and M. Sipser. Are there interactive protocols for co-NP languages? *Inf. Proc. Lett.*, 28:249–251, 1988.

[24] M. Garey and D. Johnson. *Computers and Intractability*. W. H. Freeman and Company, New York, 1979.

[25] J. Gill. Computational complexity of probabilistic complexity classes. *SIAM J. Comp.*, 6:675–695, 1977.

[26] F. Green and R. Pruim. Relativized separation of EQP from P(NP). Manuscript.

[27] L.K. Grover. A fast quantum mechanical algorithm for database search. *Phys. Rev. Lett.*, 78:325, 1997, quant-ph/9605043.

[28] L.K. Grover. A framework for fast quantum mechanical algorithms. *Proceedings of the 30th ACM Symposium on the Theory of Computing*, pages 53–62, 1998, quant-ph/9711043.

[29] J. Gruska. *Quantum Computing*. McGraw-Hill, New York, 1999.

[30] J.E. Hopcroft and J.D. Ullman. *Introduction to Automata Theory, Languages, and Computation*. Addison-Wesley, Reading, MA, 1979.

[31] L. Li. On the counting functions. Technical Report TR-93-12, The University of Chicago, 1993. Ph.D. thesis available at http://www.cs.uchicago.edu/research/publications/techreports/ TR-93-12.

[32] C. Lund, L. Fortnow, H. Karloff, and N. Nisan. Algebraic methods for interactive proof systems. *J. ACM*, 39(4):859–868, 1992.

[33] M.A. Nielsen and I.L. Chuang. *Quantum Computation and Quantum Information*. Cambridge University Press, New York, 2000.

[34] N. Nisan and M. Szegedy. On the degree of boolean functions as real polynomials. *Proceedings of the 24th ACM Symposium on the Theory of Computing*, pages 462–467, New York, 1992. ACM.

[35] C.H. Papadimitriou. *Computational Complexity*. Addison-Wesley, Reading, MA, 1994.

[36] J. Preskill. Fault-tolerant quantum computation, in *Introduction to Quantum Computation and Information*, Lo, Spiller, and Popescu, Eds. World Scientific, River Edge, NJ, 1997.

[37] A. Shamir. IP = PSPACE. *J. ACM*, 39(4):869–877, 1992.

[38] P.W. Shor. Fault-tolerant quantum computation. *Proceedings of the 37th IEEE Symposium on Foundations of Computer Science*, pages 56–65, 1996.

[39] P.W. Shor. Polynomial-time algorithms for prime factorization and discrete logarithms on a quantum computer. *SIAM J. Comp.*, 26:1484–1509, 1997, `quant-ph/9508027`.

[40] D.R. Simon. On the power of quantum computation. *SIAM J. Comp.*, 26(5):1474–1483, 1997.

[41] J. Simon. *On Some Central Problems in Computational Complexity*. Ph.D. thesis, Cornell University, Ithaca, NY, January 1975. Available as Cornell Department of Computer Science Technical Report TR75-224.

[42] M. Sipser. *Introduction to the Theory of Computation*. PWS, Boston, 1997.

[43] R. Solovay and A. Yao. Manuscript, 1996.

[44] S. Toda. PP is as hard as the polynomial-time hierarchy. *SIAM J. Comp.*, 20(5):865–877, 1991.

[45] S. Toda and M. Ogiwara. Counting classes are at least as hard as the polynomial-time hierarchy. *SIAM J. Comp.*, 21(2):316–328, 1992.

[46] L. Valiant. The complexity of computing the permanent. *Theoret. Comp. Sci.*, pages 189–201, 1979.

[47] H. Vollmer. *Introduction to Circuit Complexity*. Springer-Verlag, Berlin, 1999.

[48] K. Wagner. The complexity of combinatorial problems with succinct input representation. *Acta Informatica*, 23:325–356, 1986.

[49] A. Yao. Quantum circuit complexity. *Proceedings of the 34th IEEE Symposium on Foundations of Computer Science*, pages 352–361, 1993.

Quantum Error-Correcting Codes

Chapter 9

Algorithmic aspects of quantum error-correcting codes

Markus Grassl

Abstract   We show how quantum algorithms for encoding and decoding of quantum error-correcting codes (QECC) can be derived from the underlying mathematical structure. For this, we first characterize general quantum error-correcting codes, followed by the discussion of CSS codes and additive quantum error-correcting codes. For the last two families, examples are given that illustrate the constructions.

9.1   Introduction

Quantum error-correcting codes (QECC) are an essential ingredient for the realization of quantum computers. After the first example of a QECC [22], various construction techniques for QECC have been developed (cf., e.g., [4, 5, 7, 11, 15, 24]) and their theory has been developed [17]. The algorithmic aspects, i.e., how efficient quantum algorithms resp. quantum circuits for encoding and decoding can be derived, have got less attention. In this paper, we give an overview how the basic algorithms can be derived.

1-58488-282/4/02/$0.00+$1.50

The paper is organized as follows. In Section 9.2 we present the general characterization of quantum error-correcting codes and give a constructive proof that leads to an in-principle decoding algorithm. The next section deals with QECC that are constructed from certain linear binary codes. It is illustrated how quantum circuits for encoding and decoding can be constructed. Finally, in Section 9.4 we present a generalization of that construction based on the rich algebraic structure of Pauli matrices and their connection to quaternary vectors spaces. Again it is illustrated how the algebraic structure can be exploited to derive quantum circuits for encoding and decoding. The presentation mainly follows [9, Kapitel 3].

9.2   General quantum error-correcting codes

9.2.1   General errors

In this section, we establish the general mathematical frame to characterize quantum error-correcting codes. As the correction of errors is always related to a specific error model, we start with a very general definition of a quantum channel.

DEFINITION 9.1   *A quantum channel* Q *can be modeled as a trace-preserving completely positive linear map from the input space* $\mathcal{H}_{\text{in}}$ *to the output space* $\mathcal{H}_{\text{out}}$. *The map is given by (cf. [18])*

$$\rho \mapsto Q(\rho) := \sum_{i \in \mathcal{I}_Q} A_i \rho A_i^\dagger. \tag{9.1}$$

Here by

$$\{A_i : i \in \mathcal{I}_Q\} \quad \text{with} \sum_{i \in \mathcal{I}_Q} A_i^\dagger A_i = \mathbb{1} \tag{9.2}$$

we denote the error operators of the quantum channel.

In general, the map (9.1) is not invertible. For some quantum channels, however, the restriction of (9.1) to a subspace may have an inverse, i.e., the correction of errors is possible. This yields to the following general definition of a quantum error-correcting code.

DEFINITION 9.2 *Let* Q *be a quantum channel mapping states in* $\mathcal{H}_{\mathrm{in}}$ *to states in* $\mathcal{H}_{\mathrm{out}}$.

A subspace $\mathcal{C} \leq \mathcal{H}_{\mathrm{in}}$ *is a quantum error-correcting code (QECC) for* Q *iff there exists a decoding operation* D *such that for all* $|\psi\rangle \in \mathcal{C}$ $\mathsf{D}(\mathsf{Q}(|\psi\rangle\langle\psi|)) = |\psi\rangle\langle\psi|$.

While this definition of a QECC involves the decoding operation D, the following characterization of a QECC depends on the error operators (9.2) only (cf. [17, Theorem III.2]).

THEOREM 9.1
Let Q *be a quantum channel on* $\mathcal{H}$ *with error operators* $\{A_i : i \in \mathcal{I}_{\mathsf{Q}}\}$.

A subspace $\mathcal{C} \leq \mathcal{H}$ *with orthonormal basis* $\{|\boldsymbol{c}_i\rangle : i \in \mathcal{I}_{\mathcal{C}}\}$ *is a QECC for* Q *iff the following two conditions hold:*

$$\forall k, \ell \in \mathcal{I}_{\mathsf{Q}} \forall i \neq j \in \mathcal{I}_{\mathcal{C}} : \langle \boldsymbol{c}_i | A_k^{\dagger} A_\ell | \boldsymbol{c}_j \rangle = 0 \tag{9.3}$$

$$\forall k, \ell \in \mathcal{I}_{\mathsf{Q}} \forall i, j \in \mathcal{I}_{\mathcal{C}} : \langle \boldsymbol{c}_i | A_k^{\dagger} A_\ell | \boldsymbol{c}_i \rangle = \langle \boldsymbol{c}_j | A_k^{\dagger} A_\ell | \boldsymbol{c}_j \rangle =: \alpha_{k\ell}. \tag{9.4}$$

Equivalently, conditions (9.3) and (9.4) can be formulated in a base independent way as

$$\forall k, \ell \in \mathcal{I}_{\mathsf{Q}} : P_{\mathcal{C}} A_k^{\dagger} A_\ell P_{\mathcal{C}} = \alpha_{k\ell} P_{\mathcal{C}}. \tag{9.5}$$

Here $P_{\mathcal{C}}$ *denotes the projection onto the code* $\mathcal{C}$.

In order to prove this theorem, we need the following

LEMMA 9.1
If conditions (9.3) and (9.4) hold for the error operators $\{A_k : k \in \mathcal{I}_{\mathsf{Q}}\}$, *then they hold also for all linear combinations of the error operators.*

PROOF Consider the operators $A := \sum_{k \in \mathcal{I}_{\mathsf{Q}}} \lambda_k A_k$ and $B := \sum_{\ell \in \mathcal{I}_{\mathsf{Q}}} \mu_\ell A_\ell$. For arbitrary $i \neq j \in \mathcal{I}_{\mathcal{C}}$ we get:

$$\langle \boldsymbol{c}_i | A^{\dagger} B | \boldsymbol{c}_j \rangle = \sum_{k, \ell \in \mathcal{I}_{\mathsf{Q}}} \lambda_k^* \mu_\ell \langle \boldsymbol{c}_i | A_k^{\dagger} A_\ell | \boldsymbol{c}_j \rangle \overset{(9.3)}{=} 0$$

and

$$\langle \boldsymbol{c}_i | A^{\dagger} B | \boldsymbol{c}_i \rangle = \sum_{k, \ell \in \mathcal{I}_{\mathsf{Q}}} \lambda_k^* \mu_\ell \langle \boldsymbol{c}_i | A_k^{\dagger} A_\ell | \boldsymbol{c}_i \rangle$$

$$\overset{(9.4)}{=} \sum_{k,\ell\in\mathcal{I}_Q} \lambda_k^* \mu_\ell \langle c_j | A_k^\dagger A_\ell | c_j \rangle = \langle c_j | A^\dagger B | c_j \rangle.$$

∎

Hence it is sufficient to check conditions (9.3) and (9.4) for a vector space basis $\mathcal{E}$ of the error operators, the so-called error basis (cf. [17]).

PROOF of Theorem 9.1   First we show that conditions (9.3) and (9.4) are sufficient for error correction.

Consider the following arrangement of vectors:

	$A_1\mathcal{C}$	$A_2\mathcal{C}$	$\cdots$	$A_k\mathcal{C}$	$\cdots$
$\mathcal{V}_0$	$A_1\lvert c_0\rangle$	$A_2\lvert c_0\rangle$	$\cdots$	$A_k\lvert c_0\rangle$	$\cdots$
$\mathcal{V}_1$	$A_1\lvert c_1\rangle$	$A_2\lvert c_1\rangle$	$\cdots$	$A_k\lvert c_1\rangle$	$\cdots$
$\vdots$	$\vdots$	$\vdots$	$\ddots$	$\vdots$	
$\mathcal{V}_i$	$A_1\lvert c_i\rangle$	$A_2\lvert c_i\rangle$	$\cdots$	$A_k\lvert c_i\rangle$	$\cdots$
$\vdots$	$\vdots$	$\vdots$		$\vdots$	$\ddots$

$$(9.6)$$

Here $\mathcal{V}_i$ denotes the space spanned by the vectors $\{A_k\lvert c_i\rangle : k\in\mathcal{I}_Q\}$. The space $\mathcal{C}_k := A_k\mathcal{C}$ is generated by the vectors $\{A_k\lvert c_i\rangle : i\in\mathcal{I}_C\}$. Equation (9.3) implies that the spaces $\mathcal{V}_i$ are mutually orthogonal.

For $k=\ell$ equations (9.3) and (9.4) imply that either $\mathcal{C}_k = A_k\mathcal{C}$ is of dimension zero (if $\alpha_{kk}=0$), or that the vectors $\{A_k\lvert c_i\rangle / \sqrt{\alpha_{kk}} : i\in\mathcal{I}_C\}$ form an orthonormal basis of $\mathcal{C}_k$. Then there exist unitary transformations T_k such that for all $k\in\mathcal{I}_Q$ and $i\in\mathcal{I}_C$

$$T_k A_k \lvert c_i\rangle = \sqrt{\alpha_{kk}}\lvert c_i\rangle. \qquad (9.7)$$

Starting from the vectors $\{A_k\lvert c_i\rangle : k\in\mathcal{I}_Q\}$ an orthonormal basis of the space $\mathcal{V}_i$ can be computed using the Gram–Schmidt orthogonalization procedure (cf. [8, 21]):

$$\texttt{for } j\in\mathcal{I}_Q \texttt{ do}$$
$$\lvert b_j^{(i)}\rangle \leftarrow A_j\lvert c_i\rangle$$
$$\lvert b_j^{(i)}\rangle \leftarrow \lvert b_j^{(i)}\rangle - \sum_{\ell<j}\langle b_\ell^{(i)}\lvert b_j^{(i)}\rangle \cdot \lvert b_\ell^{(i)}\rangle$$

```
if |b_j^(i)⟩ ≠ 0 then
```

$$|b_j^{(i)}\rangle \leftarrow 1/\sqrt{\langle b_j^{(i)}|b_j^{(i)}\rangle} \cdot |b_j^{(i)}\rangle$$

```
    end if
  end for
```
(9.8)

The output states are linear combinations of the input states. Hence the (nonzero) elements $|b_j^{(i)}\rangle$ of the resulting basis can be written as $|b_j^{(i)}\rangle = \sum_{k \in \mathcal{I}_Q} \lambda_{jk}^{(i)} A_k |c_i\rangle$ for $j \in \mathcal{J} \subseteq \mathcal{I}_Q$. Furthermore, from (9.8) it follows that the coefficients $\lambda_{jk}^{(i)}$ depend only on the inner products of the vectors $\{A_k|c_i\rangle: k \in \mathcal{I}_Q\}$. Hence, by (9.3) and (9.4), the coefficients $\lambda_{jk}^{(i)}$ are independent of the index i.

For $j \in \mathcal{J}$ we define the operators

$$\tilde{A}_j := \sum_{k \in \mathcal{I}_Q} \lambda_{jk} A_k.$$
(9.9)

Then the vectors $\{|\tilde{v}_j^i\rangle := \tilde{A}_j|c_i\rangle: i \in \mathcal{I}_C, j \in \mathcal{J}\}$ form an orthonormal basis for the space $\bigoplus_{i \in \mathcal{I}_C} \mathcal{V}_i \leq \mathcal{H}$ spanned by all vectors in (9.6). In particular, the vectors $\{\tilde{A}_j c_i: i \in \mathcal{I}_C\}$ are an orthonormal basis of the so-called error spaces $\tilde{C}_j := \tilde{A}_j C$. Similar to (9.7) there exist unitary operators $\tilde{T}_j$ such that for all $j \in \mathcal{J}$ and $i \in \mathcal{I}_C$

$$\tilde{T}_j \tilde{A}_j |c_i\rangle = |c_i\rangle.$$
(9.10)

For the correction of errors, first an observable with eigenspaces $\{\tilde{C}_j := \tilde{A}_j C: j \in \mathcal{J}\}$ is measured. If $\mathcal{W} := \bigoplus_{j \in \mathcal{J}} \tilde{C}_j$ is a proper subspace of the state space $\mathcal{H}$, then let the orthogonal complement $\mathcal{W}^\perp$, i.e., $\mathcal{W} \oplus \mathcal{W}^\perp = \mathcal{H}$, be another eigenspace of the observable. The outcome of the measurement yields information onto which of the error spaces $\tilde{C}_j$ the state has been projected. A projection onto $\mathcal{W}^\perp$ does not occur as $\mathcal{W}^\perp$ does not lie in the image of the quantum channel. Conditioned on the outcome of the measurement, the error can be corrected using the transformation $\tilde{T}_j$.

In order to construct the decoding operator D of Definition 9.2 we define a map with Kraus operators [18]

$$D_j := \tilde{T}_j \sum_{i \in \mathcal{I}_C} \tilde{A}_j |c_i\rangle\langle c_i|\tilde{A}_j^\dagger =: \tilde{T}_j P_{\tilde{C}_j} \quad \text{and} \quad D_0 := P_{\mathcal{W}^\perp}.$$

Here $P_{\tilde{C}_j}$ and $P_{\mathcal{W}^{\perp}}$ denote the projections onto $\tilde{C}_j = A_j\mathcal{C}$ resp. $\mathcal{W}^{\perp}$. The operators D_j, which are products of projections and unitary operations, fulfill the condition

$$D_0^{\dagger}D_0 + \sum_{j\in\mathcal{J}} D_j^{\dagger}D_j = P_{\mathcal{W}^{\perp}} + \sum_{j\in\mathcal{J}} P_{\tilde{C}_j} = \mathbb{1}; \qquad (9.11)$$

hence D is a trace-preserving completely positive linear map. For an arbitrary basis state $|c_{\nu}\rangle$ we get

$D(Q(|c_{\nu}\rangle\langle c_{\nu}|))$

$$= \sum_{\ell\in\mathcal{J}} D_{\ell}\left(\sum_{k\in\mathcal{I}_Q} A_k|c_{\nu}\rangle\langle c_{\nu}|A_k^{\dagger}\right) D_{\ell}^{\dagger}$$

$$= \sum_{i,j\in\mathcal{I}_C}\sum_{k\in\mathcal{I}_Q}\sum_{\ell\in\mathcal{J}} \tilde{T}_{\ell}\tilde{A}_{\ell}|c_i\rangle\langle c_i|\tilde{A}_{\ell}^{\dagger}A_k|c_{\nu}\rangle\langle c_{\nu}|A_k^{\dagger}\tilde{A}_{\ell}|c_j\rangle\langle c_j|\tilde{A}_{\ell}^{\dagger}\tilde{T}_{\ell}^{\dagger}$$

$$\overset{\underset{\mathrm{Lemma\ 9.1}}{(9.3)}}{=} \sum_{k\in\mathcal{I}_Q}\sum_{\ell\in\mathcal{J}} \tilde{T}_{\ell}\tilde{A}_{\ell}|c_{\nu}\rangle\langle c_{\nu}|\tilde{A}_{\ell}^{\dagger}A_k|c_{\nu}\rangle\langle c_{\nu}|A_k^{\dagger}\tilde{A}_{\ell}|c_{\nu}\rangle\langle c_{\nu}|\tilde{A}_{\ell}^{\dagger}\tilde{T}_{\ell}^{\dagger}$$

$$\overset{(9.10)}{=} |c_{\nu}\rangle\langle c_{\nu}| \sum_{k\in\mathcal{I}_Q}\sum_{\ell\in\mathcal{J}} |\langle c_{\nu}|\tilde{A}_{\ell}^{\dagger}A_k|c_{\nu}\rangle|^2$$

$$\overset{(9.11)}{=} |c_{\nu}\rangle\langle c_{\nu}|.$$

(The last equality holds since D is trace-preserving.)

For a detailed proof that conditions (9.3) and (9.4) are necessary we refer to [17, Theorem III.2]. Here we only sketch the main idea. If condition (9.3) is violated then the images of the orthogonal states $|c_i\rangle$ and $|c_j\rangle$ under the error operators A_k resp. A_{ℓ} are nonorthogonal. If condition (9.4) is violated then the angles between the images of $|c_i\rangle$ resp. $|c_j\rangle$ under the error operators A_k and A_{ℓ} differ, resulting in different lengths after projection. In both cases error correction is impossible.

It remains to show that (9.5) holds. For the left hand side we get

$$P_{\mathcal{C}}A_k^{\dagger}A_{\ell}P_{\mathcal{C}} = \left(\sum_{i\in\mathcal{I}_C}|c_i\rangle\langle c_i|\right) A_k^{\dagger}A_{\ell}\left(\sum_{j\in\mathcal{I}_C}|c_j\rangle\langle c_j|\right)$$

$$= \sum_{i,j\in\mathcal{I}_C} |c_i\rangle\left(\langle c_i|A_k^{\dagger}A_{\ell}|c_j\rangle\right)\langle c_j|.$$

Hence equality in (9.5) holds iff both (9.3) and (9.4) hold. ∎

Summarizing we obtain the following decoding algorithm.

ALGORITHM 1

1. *Compute an orthonormal basis of the space $\mathcal{V}_0$ spanned by the vectors $\{A_k|c_0\rangle \colon k \in \mathcal{I}_Q\}$.*

2. *Compute new error-operators $\tilde{A}_j$ according to (9.9).*

3. *Compute unitary operators $\tilde{T}_j$ fulfilling (9.10).*

4. *Use a measurement to (randomly) project onto one of the error-spaces $\tilde{C}_j$.*

5. *Apply the unitary transform $\tilde{T}_j$ to correct the error.*

9.2.2   Local errors

So far we have made no restrictions on the error-operators of the quantum channel. If the channel operates on composed quantum systems or is used several times, one can distinguish how many subsystems are disturbed.

DEFINITION 9.3   *Let Q be a quantum channel on $\mathcal{H}^{\otimes n}$.*
If an error operator A of Q acts trivially on $n - w$ subsystems, i.e., up to a permutation of the subsystems, A is of the form $A = c \cdot \mathbb{1}^{(1)} \otimes \ldots \otimes \mathbb{1}^{(n-w)} \otimes A'$ where $c \in \mathbb{C}$ and $n - w$ is maximal, then the weight of A is $\mathrm{wgt}\, A = w$. By definition, the weight of the zero operator is zero.

Using this definition, codes can be distinguished as follows.

DEFINITION 9.4   *A QECC $C \leq \mathcal{H}^{\otimes n}$ is called t-error-correcting if conditions (9.3) and (9.4) hold for all error operators of weight no greater than t.*

Using Lemma 9.1 it is sufficient to check (9.3) and (9.4) for a vector space basis of the error operators. Such an error basis for the operators on single qubits is given by, for example,

$$P_{00} := |0\rangle\langle 0|, \quad P_{01} := |0\rangle\langle 1|, \quad P_{10} := |1\rangle\langle 0|, \quad \text{and} \quad P_{11} := |1\rangle\langle 1|.$$

Another error basis consists of the Pauli matrices σ_x, σ_y, and σ_z together with the identity $\mathbb{1}$. The Pauli matrices are special cases of the general spin operators s_x, s_y, and s_z introduced by Pauli [20, p. 607f]. The operators s_x, s_y, and s_z obey the relations

$$s_x s_y - s_y s_x = 2i s_z, \quad s_y s_z - s_z s_y = 2i s_x, \quad s_z s_x - s_x s_z = 2i s_y,$$
and $s_x^2 + s_y^2 + s_z^2 = 3$.

For operators on $\mathbb{C}^2$ (spin 1/2) there are the additional relations

$$s_x s_y = -s_y s_x = i s_z, \quad s_y s_z = -s_z s_y = i s_x, \quad s_z s_x = -s_x s_z = i s_y,$$
and $s_x^2 = s_y^2 = s_z^2 = 1$.

$$(9.12)$$

Then a matrix representation is given by

$$\sigma_x := \begin{pmatrix} 0 & 1 \\ 1 & 0 \end{pmatrix}, \quad \sigma_y := \begin{pmatrix} 0 & -i \\ i & 0 \end{pmatrix}, \quad \text{and} \quad \sigma_z := \begin{pmatrix} 1 & 0 \\ 0 & -1 \end{pmatrix}.$$

The following proposition summarizes some of the algebraic properties of the Pauli matrices.

PROPOSITION 9.1

a. The Pauli matrices together with the unit matrix are an orthogonal basis of $\mathbb{C}^{2 \times 2}$ with respect to the inner product $\mathrm{Tr}(A^\dagger B)$.

b. The Pauli matrices generate a group $G := \langle \sigma_x, \sigma_y, \sigma_z \rangle$ of order 16. The center of G is the cyclic group $\mathcal{Z}(G) = \langle i \cdot \mathbb{1} \rangle \leq \mathcal{U}(2)$ of order 4.

c. The factor group $\overline{G} := G/\mathcal{Z}(G)$ is isomorphic to $\mathbb{Z}_2 \times \mathbb{Z}_2$.

d. The Hadamard transform operates on G via conjugation as

$$H \sigma_x H = \sigma_z, \quad H \sigma_z H = \sigma_x, \quad \text{and} \quad H \sigma_y H = -\sigma_y;$$

in particular, σ_x and σ_z are interchanged.

e. The matrix

$$\theta = \frac{1}{2} \begin{pmatrix} i-1 & i+1 \\ i-1 & -(i+1) \end{pmatrix} \in \mathcal{U}(2)$$

operates on G via conjugation as

$$\theta^\dagger \sigma_x \theta = \sigma_z, \quad \theta^\dagger \sigma_y \theta = \sigma_x, \quad \text{and} \quad \theta^\dagger \sigma_z \theta = \sigma_y,$$

i.e., the Pauli matrices are permuted cyclicly.

These error operators can be extended to operators on n qubits.

DEFINITION 9.5   *An error operator on $(\mathbb{C}^2)^{\otimes n}$ that is a tensor product of Pauli matrices and $\mathbb{1}$ is called a Pauli error. The weight of a Pauli error is the number of tensor factors that differ from $\mathbb{1}$.*

The error-correcting properties of a QECC are completely determined by the ability to correct Pauli errors.

THEOREM 9.2

A QECC $C \leq (\mathbb{C}^2)^{\otimes n}$ is t-error-correcting iff all Pauli errors up to weight t can be corrected.

PROOF   This is a direct consequence of the fact that the Pauli errors E where wgt $E \leq t$ are a vector space basis of all error operators A where wgt $A \leq t$.   ∎

Before we transfer the notion of the minimum distance of a classical code to QECC, we consider another special channel, the quantum erasure channel [14]. This channel replaces the input state with some probability by a state $|\odot\rangle$ that is orthogonal to all possible input states. This allows to determine whether an error occurred or not. The error operators of the erasure channel **QEC** on $\mathcal{H}$ with orthonormal basis $\mathcal{B} = \{|i\rangle : i \in \mathcal{I}\}$ are $A_i := |\odot\rangle\langle i|$ for $i \in \mathcal{I}$ and $\mathbb{1}_{\mathcal{H}} = \sum_{i \in \mathcal{I}} |i\rangle\langle i|$. The canonical error operators for the channel **QEC**$^{\otimes n}$ operating on n subsystems are tensor products of A_i and $\mathbb{1}_{\mathcal{H}}$. The number of erasures is the number of tensor factors that differ from $\mathbb{1}_{\mathcal{H}}$.

DEFINITION 9.6   *A QECC $C \leq \mathcal{H}^{\otimes n}$ is t-erasure-correcting iff conditions (9.3) and (9.4) hold for all error operators of the erasure channel on $\mathcal{H}^{\otimes n}$ that have at least $n - t$ factors equal to $\mathbb{1}_{\mathcal{H}}$.*

The relation between the correction of erasures and that of general errors is as follows.

THEOREM 9.3

A QECC $C \leq \mathcal{H}^{\otimes n}$ is t-erasure-correcting iff

$$\forall i \neq j \in \mathcal{I}_C \colon\ \langle c_i|A|c_j \rangle = 0 \tag{9.13}$$

$$\forall i, j \in \mathcal{I}_C \colon\ \langle c_i|A|c_i \rangle = \langle c_j|A|c_j \rangle \tag{9.14}$$

holds for all error operators A on $\mathcal{H}^{\otimes n}$ of weight wgt $A \leq t$.

If a QECC is t-error-correcting then $\min\{2t, n\}$ erasures can be corrected.

PROOF   For erasures of the subsystems k and ℓ we obtain

$$(A_i^{(k)})^\dagger A_j^{(\ell)} = |i\rangle_k \langle \otimes|_k |\otimes\rangle_\ell \langle j|_\ell = \delta_{k\ell} |i\rangle_k \langle j|_k. \tag{9.15}$$

Hence (9.3) and (9.4) can be restricted to erasures on the same subsystems. Equation (9.15) implies (9.3) and (9.4) for operators A with at most t tensor factors of the form $|i\rangle\langle j|$ and at least $n - t$ tensor factors $\mathbb{1}$. These operators span the vector space of all error operators of weight no greater than t.

If a QECC is t-error-correcting then (9.3) and (9.4) hold for all operators up to weight t. If the errors A_k and A_ℓ operate on disjoint subsystems the operator $A_k^\dagger A_\ell$ has weight $\mathrm{wgt}(A_k^\dagger A_\ell) = \mathrm{wgt}(A_k) + \mathrm{wgt}(A_\ell)$. Hence (9.13) and (9.14) hold for all operators up to weight $\min\{2t, n\}$.
∎

DEFINITION 9.7   *The minimum distance of a QECC $C \leq \mathcal{H}^{\otimes n}$ is d iff $d - 1$ or less erasures can be corrected.*

We denote by $C = ((n, K, d))$ a QECC of length n, dimension K, and minimum distance d. This resembles the notation (n, M, d) for general classical codes (cf. [19, p. 38]).

From Theorem 9.3 we obtain

COROLLARY 9.1

A QECC $C = ((n, K, d))$ can correct $\lfloor (d - 1)/2 \rfloor$ errors.

9.3 Binary quantum codes

9.3.1 Construction

Theorem 9.2 and Proposition 9.1 (a) imply that it is sufficient to correct for spin flip errors, i.e., tensor products of $\mathbb{1}$ and σ_x, phase flip errors, i.e., tensor products of $\mathbb{1}$ and σ_z, as well as combinations (products) of them. Spin flip errors correspond to classical bit errors and hence could be corrected using a classical binary code. Phase errors, however, do not have a direct classical counterpart. Proposition 9.1 (d) shows that conjugation by the Hadamard transform interchanges the Pauli matrices σ_x and σ_z, i.e., are turned into phase errors and vice versa.

LEMMA 9.2
Let $C \leq \mathbb{F}_2^n$ be an arbitrary subspace of $\mathbb{F}_2^n$ and let $\boldsymbol{x} \in \mathbb{F}_2^n$ be an arbitrary binary vector. Furthermore, let $C^{\perp}$ denote the space that is orthogonal to C with respect to the inner product $\boldsymbol{x} \cdot \boldsymbol{y} := \sum_{i=1}^{n} x_i y_i$, i.e.,

$$C^{\perp} := \{\boldsymbol{d} \in \mathbb{F}_2^n \mid \forall \boldsymbol{c} \in C: \sum_{i=1}^{n} c_i d_i = 0\}.$$

Then

$$\sum_{\boldsymbol{c} \in C} (-1)^{\boldsymbol{x} \cdot \boldsymbol{c}} = \begin{cases} |C| & \text{for } \boldsymbol{x} \in C^{\perp}, \\ 0 & \text{for } \boldsymbol{x} \notin C^{\perp}. \end{cases} \tag{9.16}$$

PROOF For fixed $\boldsymbol{x} \in C^{\perp}$ and arbitrary $\boldsymbol{c} \in C$, $\boldsymbol{x} \cdot \boldsymbol{c} = 0$, hence $\sum_{\boldsymbol{c} \in C} (-1)^{\boldsymbol{x} \cdot \boldsymbol{c}} = |C|$.

For $\boldsymbol{x} \notin C^{\perp}$ let $D^{\perp} = \langle C^{\perp}, \boldsymbol{x} \rangle$ be the space spanned by $C^{\perp}$ and $\boldsymbol{x}$. Then $D < C$ is a proper subspace of C with codimension one. Hence the space C can be decomposed as $C = D \cup (D + \boldsymbol{c}_0)$ where $\boldsymbol{x} \cdot \boldsymbol{c}_0 = 1$. Finally

$$\sum_{\boldsymbol{c} \in C} (-1)^{\boldsymbol{x} \cdot \boldsymbol{c}} = \sum_{\boldsymbol{d} \in D} (-1)^{\boldsymbol{x} \cdot \boldsymbol{d}} + \sum_{\boldsymbol{d} \in D + \boldsymbol{c}_0} (-1)^{\boldsymbol{x} \cdot \boldsymbol{d}} = \sum_{\boldsymbol{d} \in D} (-1)^{\boldsymbol{x} \cdot \boldsymbol{d}} + (-1)^{\boldsymbol{x} \cdot (\boldsymbol{d} + \boldsymbol{c}_0)}$$

$$= \sum_{\boldsymbol{d} \in D} \left(1 + (-1)^{\boldsymbol{x} \cdot \boldsymbol{c}_0}\right)(-1)^{\boldsymbol{x} \cdot \boldsymbol{d}} = 0.$$

∎

THEOREM 9.4

Let $C \leq \mathbb{F}_2^n$ be a k-dimensional subspace of $\mathbb{F}_2^n$ and let $\boldsymbol{a}, \boldsymbol{b} \in \mathbb{F}_2^n$ be arbitrary binary vectors. Furthermore, let $H_{2^n} := H^{\otimes n}$ denote the Hadamard transform on n qubits, where $H = \frac{1}{\sqrt{2}} \begin{pmatrix} 1 & 1 \\ 1 & -1 \end{pmatrix}$.

Then the Hadamard transform of the state

$$|\psi\rangle := \frac{1}{\sqrt{|C|}} \sum_{c \in C} (-1)^{\boldsymbol{a} \cdot \boldsymbol{c}} |\boldsymbol{c} + \boldsymbol{b}\rangle$$

equals

$$H_{2^n} |\psi\rangle = \frac{(-1)^{\boldsymbol{a} \cdot \boldsymbol{b}}}{\sqrt{|C^\perp|}} \sum_{c \in C^\perp} (-1)^{\boldsymbol{b} \cdot \boldsymbol{c}} |\boldsymbol{c} + \boldsymbol{a}\rangle.$$

PROOF   The Hadamard transform on n qubits can be written as

$$H_{2^n} = \frac{1}{\sqrt{2^n}} \sum_{\boldsymbol{x}, \boldsymbol{y} \in \mathbb{F}_2^n} (-1)^{\boldsymbol{x} \cdot \boldsymbol{y}} |\boldsymbol{x}\rangle \langle \boldsymbol{y}|.$$

Then

$$\begin{aligned}
H_{2^n} |\psi\rangle &= \frac{1}{\sqrt{2^n |C|}} \sum_{\boldsymbol{x}, \boldsymbol{y} \in \mathbb{F}_2^n} (-1)^{\boldsymbol{x} \cdot \boldsymbol{y}} |\boldsymbol{x}\rangle \langle \boldsymbol{y}| \sum_{c \in C} (-1)^{\boldsymbol{a} \cdot \boldsymbol{c}} |\boldsymbol{c} + \boldsymbol{b}\rangle \\
&= \frac{1}{\sqrt{2^n |C|}} \sum_{\boldsymbol{x}, \boldsymbol{y} \in \mathbb{F}_2^n} \sum_{c \in C} (-1)^{\boldsymbol{x} \cdot \boldsymbol{y} + \boldsymbol{a} \cdot \boldsymbol{c}} |\boldsymbol{x}\rangle \langle \boldsymbol{y}|\boldsymbol{c} + \boldsymbol{b}\rangle \\
&= \frac{1}{\sqrt{2^n |C|}} \sum_{\boldsymbol{x} \in \mathbb{F}_2^n} \sum_{c \in C} (-1)^{\boldsymbol{x} \cdot (\boldsymbol{c} + \boldsymbol{b}) + \boldsymbol{a} \cdot \boldsymbol{c}} |\boldsymbol{x}\rangle \\
&= \frac{1}{\sqrt{2^n |C|}} \sum_{\boldsymbol{x} \in \mathbb{F}_2^n} (-1)^{\boldsymbol{b} \cdot \boldsymbol{x}} |\boldsymbol{x}\rangle \sum_{c \in C} (-1)^{(\boldsymbol{x} + \boldsymbol{a}) \cdot \boldsymbol{c}} \\
&\overset{(9.16)}{=} \frac{|C|}{\sqrt{2^n |C|}} \sum_{\boldsymbol{x} \in C^\perp + \boldsymbol{a}} (-1)^{\boldsymbol{b} \cdot \boldsymbol{x}} |\boldsymbol{x}\rangle \\
&= \frac{(-1)^{\boldsymbol{a} \cdot \boldsymbol{b}}}{\sqrt{|C^\perp|}} \sum_{\boldsymbol{d} \in C^\perp} (-1)^{\boldsymbol{b} \cdot \boldsymbol{d}} |\boldsymbol{d} + \boldsymbol{a}\rangle.
\end{aligned}$$

∎

The following construction of QECC was independently developed by Calderbank and Shor [5] and by Steane [23]. Therefore, in the literature the term CSS code is used. (For the theory of the underlying classical linear codes see, e.g., [19].)

THEOREM 9.5
Let $C_1 = [n, k_1, d_1]$ and $C_2 = [n, k_2, d_2]$ be linear binary codes of length n, dimension k_1 resp. k_2, and minimum distance d_1 resp. d_2 with $C_2^{\perp} \le C_1$. Furthermore, let $\mathcal{W} = \{w_1, \ldots, w_K\} \subset \mathbb{F}_2^n$ be a set of representatives of the cosets $C_1/C_2^{\perp}$.
Then the $K = 2^{k_1 - (n - k_2)}$ mutually orthogonal states

$$|\psi_i\rangle = \frac{1}{\sqrt{|C_2^{\perp}|}} \sum_{c \in C_2^{\perp}} |c + w_i\rangle \tag{9.17}$$

are a basis of a QECC $\mathcal{C} = ((n, K, d)) \le (\mathbb{C}_2)^{\otimes n}$. The code can correct at least $\lfloor (d_1 - 1)/2 \rfloor$ spin flip errors and simultaneously at least $\lfloor (d_2 - 1)/2 \rfloor$ phase flip errors. Its minimum distance is $d \ge \min\{d_1, d_2\}$.

Similar to the notation for general linear codes the notation $[\![n, k, d]\!]$ is used for CSS codes of length n, dimension 2^k, and minimum distance d. Steane [24] uses the notation $\{n, K, d_1, d_2\}$ for a QECC of length n and dimension 2^K that can correct up to $\lfloor (d_1 - 1)/2 \rfloor$ spin flip errors and up to $\lfloor (d_2 - 1)/2 \rfloor$ phase flip errors. For $d_1 = d_2 = d$ he uses the notation $\{n, K, d\}$. His notation has the advantage that it distinguishes the two different types of errors. In the literature, however, the former notation $[\![n, k, d]\!]$ dominates.

PROOF of Theorem 9.5 Each state $|\psi_i\rangle$ is a superposition of the basis states $|x\rangle$ with $x \in C_2^{\perp} + w_i$. For $i \ne j$, $\langle \psi_i | \psi_j \rangle = 0$ as the cosets $C_2^{\perp} + w_i$ and $C_2^{\perp} + w_j$ are disjoint.
Each state of the code is a superposition

$$|\psi\rangle = \sum_{i=1}^{K} \alpha_i |\psi_i\rangle = \sum_{i=1}^{K} \alpha_i' \sum_{c \in C_2^{\perp}} |c + w_i\rangle = \sum_{c \in C_1} \beta_c |c\rangle \tag{9.18}$$

where $\alpha_i \in \mathbb{C}$. Here the normalization has been absorbed into the coefficients α_i'. Furthermore we made use of the fact that the vectors $c + w_i$ lie in a coset $C_1/C_2^{\perp}$. In particular, those vectors lie in C_1.

Any combination of spin flip and phase flip errors can be described using two binary vectors $e_x, e_z \in \mathbb{F}_2^n$ as

$$e := (\sigma_x^{e_{x,1}} \sigma_z^{e_{z,1}}) \otimes \ldots \otimes (\sigma_x^{e_{x,n}} \sigma_z^{e_{z,n}}). \tag{9.19}$$

The action of an error (9.19) on the state $|\psi\rangle$ is

$$e|\psi\rangle = \sum_{c\in C_1} \beta_c(-1)^{c\cdot e_z}|c+e_x\rangle. \tag{9.20}$$

Next consider the generator matrices

$$G_1 := \begin{pmatrix} H_2 \\ D_1 \end{pmatrix} \in \mathbb{F}_2^{k_1\times n} \quad \text{and} \quad G_2 := \begin{pmatrix} H_1 \\ D_2 \end{pmatrix} \in \mathbb{F}_2^{k_2\times n}$$

of the codes C_1 resp. C_2, where the matrices H_i are parity parity check matrice for the codes C_i resp. generator matrices for the dual codes $C_i^\perp$. Hence $G_1\cdot H_1^t = 0$ and $G_2\cdot H_2^t = 0$. For the binary code C_1 and an arbitrary binary vector $x\in\mathbb{F}_2^n$, the syndrome $s := x\cdot H_1^t \in \mathbb{F}_2^{n-k_1}$ is an indicator for the error.

The computation of the syndrome can be achieved using $n-k_1$ ancilla qubits which have been initialized in the state $|0\rangle$. The corresponding reversible function is given by $S_1\colon |x\rangle|y\rangle \mapsto |x\rangle|x\cdot H_1^t + y\rangle$. Computing the syndrome yields the state

$$\sum_{c\in C_1} \beta_c(-1)^{c\cdot e_z}|c+e_x\rangle|(c+e_x)\cdot H_1^t\rangle$$
$$= \left(\sum_{c\in C_1} \beta_c(-1)^{c\cdot e_z}|c+e_x\rangle\right) \otimes |e_x\cdot H_1^t\rangle = (e|\psi\rangle) \otimes |e_x\cdot H_1^t\rangle. \tag{9.21}$$

Here it has been used that the syndrome vanishes for all code words and is hence independent of c.

According to Theorem 9.4, Hadamard transform of the basis states $|\psi_i\rangle$ of the code yields

$$H_{2^n}|\psi_i\rangle = \frac{1}{\sqrt{|C_2|}} \sum_{c\in C_2}(-1)^{c\cdot w_i}|c\rangle.$$

For the superposition (9.18) we get

$$H_{2^n}|\psi\rangle = \sum_{i=1}^K \alpha_i H_{2^n}|\psi_i\rangle = \sum_{i=1}^K \alpha_i'' \sum_{c\in C_2}(-1)^{c\cdot w_i}|c\rangle.$$

By Proposition 9.1 (d) the Hadamard transform interchanges the role of e_x and e_z in (9.19). Therefore, Hadamard transform of the erroneous state (9.20) yields

$$H_{2^n}e|\psi\rangle = (H_{2^n}eH_{2^n})\cdot H_{2^n}|\psi\rangle = \sum_{i=1}^K \alpha_i'' \sum_{c\in C_2}(-1)^{c\cdot e_x}(-1)^{c\cdot w_i}|c+e_z\rangle.$$

Similar to (9.21) an error syndrome for the binary code C_2 can be computed using the map $S_2 \colon |\boldsymbol{x}\rangle|\boldsymbol{y}\rangle \mapsto |\boldsymbol{x}\rangle|\boldsymbol{x} \cdot H_2^{\mathbf{t}} + \boldsymbol{y}\rangle$ and $n - k_2$ ancilla qubits. Altogether, we obtain the state

$$\left(H_{2^n}\boldsymbol{e}|\psi\rangle\right) \otimes |\boldsymbol{e}_x \cdot H_1^{\mathbf{t}}\rangle \otimes |\boldsymbol{e}_z \cdot H_2^{\mathbf{t}}\rangle.$$

Finally, another Hadamard transform on the first n qubits results in the state

$$\boldsymbol{e}|\psi\rangle \otimes |\boldsymbol{e}_x \cdot H_1^{\mathbf{t}}\rangle \otimes |\boldsymbol{e}_z \cdot H_2^{\mathbf{t}}\rangle = \left(\sum_{\boldsymbol{c}\in C_1} \beta_{\boldsymbol{c}}(-1)^{\boldsymbol{c}\cdot\boldsymbol{e}_z}|\boldsymbol{c}+\boldsymbol{e}_x\rangle\right) \otimes |\boldsymbol{e}_x \cdot H_1^{\mathbf{t}}\rangle \otimes |\boldsymbol{e}_z \cdot H_2^{\mathbf{t}}\rangle.$$

Measuring the two auxiliary systems gives two (classical) syndrome vectors $\boldsymbol{s}_1 := \boldsymbol{e}_x \cdot H_1^{\mathbf{t}}$ and $\boldsymbol{s}_2 := \boldsymbol{e}_z \cdot H_2^{\mathbf{t}}$. If the number of spin flip errors is less than $(d_1 - 1)/2$, the error vector $\boldsymbol{e}_x$ can be computed from the syndrome $\boldsymbol{s}_1$ using any decoding algorithm for the binary code C_1. Similarly, the error vector $\boldsymbol{e}_z$ can be determined using a decoding algorithm for the code C_2 if the number of phase flip errors is less than $(d_2 - 1)/2$. Knowing the error vectors $\boldsymbol{e}_x$ and $\boldsymbol{e}_z$, the error can be corrected using the unitary transformation (9.19). ∎

Summarizing, we get the following decoding algorithm for CSS codes.

ALGORITHM 2

1. *Using the transformation $S_1 \colon |\boldsymbol{x}\rangle|\boldsymbol{y}\rangle \mapsto |\boldsymbol{x}\rangle|\boldsymbol{x} \cdot H_1^{\mathbf{t}} + \boldsymbol{y}\rangle$ and $n - k_1$ ancilla qubits compute the syndrome $\boldsymbol{s}_1 := \boldsymbol{e}_x \cdot H_1^{\mathbf{t}}$ of the spin flip errors.*

2. *Perform a Hadamard transform on the first n qubits.*

3. *Using the transformation $S_2 \colon |\boldsymbol{x}\rangle|\boldsymbol{y}\rangle \mapsto |\boldsymbol{x}\rangle|\boldsymbol{x} \cdot H_2^{\mathbf{t}} + \boldsymbol{y}\rangle$ and $n - k_2$ ancilla qubits compute the syndrome $\boldsymbol{s}_2 := \boldsymbol{e}_z \cdot H_2^{\mathbf{t}}$ of the phase flip errors.*

4. *Perform a Hadamard transform on the first n qubits.*

5. *Measure the syndromes $\boldsymbol{s}_1$ and $\boldsymbol{s}_2$.*

6. *Using classical decoding algorithms for the codes C_1 and C_2 compute the error vectors $\boldsymbol{e}_x$ and $\boldsymbol{e}_z$ that correspond to the syndromes $\boldsymbol{s}_1$ resp. $\boldsymbol{s}_2$.*

7. *Correct the error using the unitary transformation (9.19).*

It is possible to measure the syndrome s_1 directly after having computed it. Then the spin flip errors can be corrected. After a Hadamard transform, phase flip errors are corrected similarly. This is illustrated in Figure 9.1.

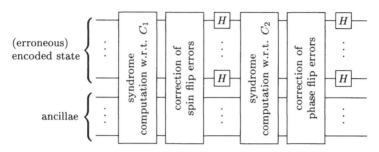

FIGURE 9.1
Decoding scheme for CSS codes based on binary codes $C_2^\perp \le C_1$.

Instead of measuring the syndromes the error can be corrected using unitary transformations that are conditioned on the state of the auxiliary system. In general, one would expect that the resulting transformations have a high complexity as an exponential number of cases has to be distinguished.

Theorem 9.5 reduces the construction of QECC to the search of suitable linear binary codes C_1 and C_2 with $C_2^\perp \le C_1$. While this situation has hardly been considered in the coding theory literature, the special case of weakly self-dual codes has been extensively studied.

COROLLARY 9.2
For a weakly self-dual code $C = [n, k, d]$, i.e., $C \le C^\perp = [n, n-k, d^\perp]$, by Theorem 9.5 (with $C_1 = C_2 = C^\perp$) a CSS-Code $\mathcal{C} = [\![n, n-2k, d']\!]$ with $d' \ge d^\perp$ can be constructed.

Those weakly self-dual CSS codes have the advantage that for both the correction of spin flip errors and the correction of phase flip errors the decoding algorithm for the code $C^\perp$ can be used.

The minimum distances d_1 and d_2 of the codes C_1 resp. C_2 yield only a lower bound on the minimum distance of the CSS codes. In [1] it was shown that not the distance between different codewords, but the distance between the cosets $C_1/C_2^\perp$ determines the minimum distance of the QECC.

THEOREM 9.6

Let C be a CSS code with corresponding binary codes C_1 and C_2 with $C_2^{\perp} \leq C_1$. Furthermore, define $\tilde{d}_1 := \min\{\text{wgt } \boldsymbol{c} : \boldsymbol{c} \in C_1 \setminus C_2^{\perp}\}$ and $\tilde{d}_2 := \min\{\text{wgt } \boldsymbol{c} : \boldsymbol{c} \in C_2 \setminus C_1^{\perp}\}$.

Then the minimum distance $\tilde{d}$ of C is at least $\min\{\tilde{d}_2, \tilde{d}_1\}$.

PROOF The action of an error $\boldsymbol{e}$ in (9.19) with error vectors $\boldsymbol{e}_x$ and $\boldsymbol{e}_z$ on a basis state of the code is

$$\boldsymbol{e}|\psi_i\rangle = \frac{1}{\sqrt{|C_2^{\perp}|}} \sum_{\boldsymbol{c} \in C_2^{\perp}} (-1)^{(\boldsymbol{c}+\boldsymbol{w}_i)\cdot\boldsymbol{e}_z} |\boldsymbol{c} + \boldsymbol{w}_i + \boldsymbol{e}_x\rangle.$$

For $\boldsymbol{e}_x \in C_2^{\perp}$ the state is invariant as only the summands are permuted. For $\boldsymbol{e}_z \in C_1^{\perp}$ the phase factor is constant as $\boldsymbol{c} + \boldsymbol{w}_i \in C_1$, i.e., the state is invariant, too. Therefore, the subsets $C_2^{\perp} \subset C_1$ and $C_1^{\perp} \subset C_2$ can be ignored when computing the minimum distance. ∎

9.3.2 Example: binary Hamming code

The construction of CSS codes and the basic algorithms for encoding and decoding will be illustrated by the following example (cf. [12]).

The binary Hamming code $C = [7, 4, 3]$ contains its dual $C^{\perp} = [7, 3, 4]$. Using Corollary 9.2 a weakly self-dual CSS code $C = [\![7, 1, 3]\!]$ can be constructed. The codes C and $C^{\perp}$ are generated by the rows of the generator matrices G_1 resp. G_2 with

$$G_1 = \begin{pmatrix} 1\,0\,0\,0\,1\,1\,0 \\ 0\,1\,0\,0\,0\,1\,1 \\ 0\,0\,1\,0\,1\,1\,1 \\ 0\,0\,0\,1\,1\,0\,1 \end{pmatrix} \quad \text{and} \quad G_2 = \begin{pmatrix} 1\,0\,0\,1\,0\,1\,1 \\ 0\,1\,0\,1\,1\,1\,0 \\ 0\,0\,1\,0\,1\,1\,1 \end{pmatrix}.$$

The generator matrix G can be transformed such that the subcode $C^{\perp}$ is generated by the last rows and that the rows are in reverse order compared to the usual standard form, i.e.,

$$G = \begin{pmatrix} 0\,0\,0\,1\,1\,0\,1 \\ \hline 0\,0\,1\,0\,1\,1\,1 \\ 0\,1\,0\,1\,1\,1\,0 \\ 1\,0\,0\,1\,0\,1\,1 \end{pmatrix} = \begin{pmatrix} \boldsymbol{g}_4 \\ \boldsymbol{g}_3 \\ \boldsymbol{g}_2 \\ \boldsymbol{g}_1 \end{pmatrix}.$$

The first row $\boldsymbol{w}_1 = \boldsymbol{g}_4$ and the zero vector $\boldsymbol{w}_0 = (0\,0\,0\,0\,0\,0\,0)$ are a system of representatives of the cosets $C/C^{\perp}$. According to (9.17) the basis states of the CSS code are

$$|\psi_0\rangle = \frac{1}{\sqrt{8}} \sum_{i_1,i_2,i_3 \in \{0,1\}} |0g_4 + i_3g_3 + i_2g_2 + i_1g_1\rangle$$

$$\text{and} \quad |\psi_1\rangle = \frac{1}{\sqrt{8}} \sum_{i_1,i_2,i_3 \in \{0,1\}} |1g_4 + i_3g_3 + i_2g_2 + i_1g_1\rangle. \quad (9.22)$$

From these equations the quantum circuit for encoding can be directly derived (cf. Figure 9.2). The circuit computes $|\varphi\rangle = \alpha|0\rangle + \beta|1\rangle \mapsto \alpha|\psi_0\rangle + \beta|\psi_1\rangle = |\psi\rangle$.

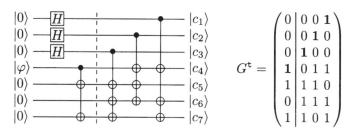

FIGURE 9.2
Encoder for a CSS code derived from G^{t}.

The upper three quantum wires correspond to the summation indices i_1, i_2, and i_3 in (9.22). The Hadamard gates H yield the superposition $(|0\rangle + |1\rangle)/\sqrt{2}$, corresponding to the values 0 and 1 in the summation. The next four columns of CNOT gates correspond to the summands $i_j \, g_j$ for $j = 1,\ldots,4$ in (9.22). If the control qubit j corresponding to i_j is one, the value $g_{j,\ell}$ is added to qubit ℓ, i.e., if $g_{j,\ell} = 1$ then qubit ℓ is inverted iff qubit j is in the state $|1\rangle$. For $g_{j,\ell} = 0$, nothing has to be done. The control qubits of the CNOT gates correspond to the bold ones in the transposed generator matrix G^{t} in Figure 9.2. The target qubits correspond to the remaining ones. On the left-hand side of the bold ones are only zeros. Therefore the state of the corresponding qubit has not yet been changed when the columns are processed from left to right.

For the computation of the error syndrome the mapping $S : |x\rangle|y\rangle \mapsto |x\rangle|x \cdot H^{\text{t}} + y\rangle$ has to be implemented for an arbitrary parity check matrix H of the Hamming code. Here we choose $H = G_2$. The resulting quantum circuit is shown in Figure 9.3.

The first four CNOT gates are used to compute the inner product of the vector $c = (c_1, c_2, \ldots, c_7)$ with the first column of the parity check

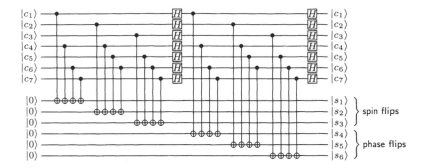

FIGURE 9.3
Syndrome computation for a weakly self-dual CSS code.

matrix H^t. The summation of those c_i with $H_{1j} = 1$ can be realized using CNOT gates. Similarly the following CNOT gates are used to compute the second and third position of the syndrome of the spin flip errors. As the code is weakly self-dual, after a Hadamard transform of the qubits of the code the same sequence of gates is used to compute the syndrome of the phase flip errors. The final Hadamard transform returns the state to the original basis.

9.4 Additive quantum codes

9.4.1 Construction

For the construction of QECC presented in this section we consider the action of the Pauli matrices on the state space $(\mathbb{C}_2)^{\otimes n}$.

DEFINITION 9.8 *The Pauli error group $E_n \leq \mathcal{U}(2^n)$ is generated by the Pauli errors on n qubits, i.e., tensor products of Pauli matrices and identity.*

As it will be demonstrated later, there is a close connection between the Pauli error group and vector spaces over the field $\mathbb{F}_4$. For completeness, we recall some properties of $\mathbb{F}_4$ and the vector space $\mathbb{F}_4^n$.

PROPOSITION 9.2

Let $\mathbb{F}_4 = \{0, 1, \omega, \omega^2\}$ be the field with four elements. Furthermore, let $a \in \mathbb{F}_4$ denote an arbitrary element.

   a. The element ω obeys the relations $\omega^2 + \omega = 1$ and $\omega^3 = 1$.

   b. The conjugation on $\mathbb{F}_4$, i.e., the map $a \mapsto \bar{a} := a^2$, interchanges ω and ω^2, and fixes 0 and 1.

   c. The trace tr is an $\mathbb{F}_2$-linear mapping $\mathrm{tr}\colon \mathbb{F}_4 \to \mathbb{F}_2$ with $\mathrm{tr}(a) = a + \bar{a}$.

   d. The mapping

$$*\colon \mathbb{F}_4^n \times \mathbb{F}_4^n \to \mathbb{F}_2$$
$$(u, v) \quad \mapsto u * v := \sum_{i=1}^{n} u_i \bar{v}_i + \bar{u}_i v_i = \sum_{i=1}^{n} \mathrm{tr}(u_i \bar{v}_i) \qquad (9.23)$$

   defines a symplectic inner product on $\mathbb{F}_4^n$.

The following proposition summarizes the relations between the Pauli error group and vector spaces over $\mathbb{F}_4$ (cf. [3, 10]):

PROPOSITION 9.3

Let E_n denote the Pauli error group on n qubits. Furthermore, let ω be a primitive element of $\mathbb{F}_4$.

   a. The center of E_n is the cyclic group $\mathcal{Z}(E_n) = \langle i \cdot \mathbb{1} \rangle \leq \mathcal{U}(2^n)$ of order 4.

   b. The factor group $\overline{E}_n := E_n / \mathcal{Z}(E_n)$ is isomorphic to $(\mathbb{Z}_2 \times \mathbb{Z}_2)^n \cong \mathbb{F}_2^{2n}$.

   c. Subgroups $S \leq \overline{E}_n$ of the factor group $\overline{E}_n$ are isomorphic to additively closed subsets ($\mathbb{F}_2$ vector spaces) $\gamma^n(\overline{E}_n)$ of the vector space $\mathbb{F}_4^n$. The isomorphism γ^n is the canonical extension of the mapping $\gamma : E_1 \to \mathbb{F}_4$ with

$$\gamma(i^k \mathbb{1}) = 0, \quad \gamma(i^k \sigma_x) = 1, \quad \gamma(i^k \sigma_y) = \omega^2, \quad \text{and} \quad \gamma(i^k \sigma_z) = \omega$$

   (for $k = 0, \ldots, 3$).

   d. The weight of a Pauli error e equals the Hamming weight of the image under γ^n, i.e., $\mathrm{wgt}(e) = \mathrm{wgt}(\gamma^n(e))$.

e. *The conjugation of a subgroup $S \leq E_n$ by the matrix $\Theta := \theta^{\otimes n}$*
   with

$$\theta = \frac{1}{2} \begin{pmatrix} i-1 & i+1 \\ i-1 & -(i+1) \end{pmatrix} \in \mathcal{U}(2)$$

   corresponds to the multiplication of the $\mathbb{F}_2$ vector space $\gamma^n(S)$ with
   the primitive element $\omega \in \mathbb{F}_4$, i.e., for all subgroups $S \leq E_n$

$$\gamma^n(S^\Theta) = \omega \cdot \gamma^n(S).$$

f. *The conjugation of an element $e \in E_n$ by the matrix $(\theta \cdot H)^{\otimes n}$*
   (where $H \in \mathcal{U}(2)$ denotes the unitary Hamadard matrix) corre-
   sponds to the componentwise conjugation of the vector $\gamma^n(e)$.

g. *For all $g, h \in E_n$*

$$gh = (-1)^{\gamma^n(g) * \gamma^n(h)} hg. \tag{9.24}$$

PROOF

a., b. Consequences of Propositions 9.1 (b), 9.1 (c).

c. The relations (9.12) among the Pauli matrices and the relations
   among the elements of $\mathbb{F}_4$ (cf. Proposition 9.6 [a]) imply that γ
   maps products of Pauli matrices to sums of elements of $\mathbb{F}_4$. The
   kernel of γ^n equals the center of E_n. Hence γ^n is injective on $\overline{E}_n$.

d. The weight of a Pauli error is defined as the number of tensor
   factors that differ from $\mathbb{1}$. As γ maps only $\mathbb{1}$ to zero the number
   of nonzero components of $\gamma^n(e)$ equals the weight of e.

e. According to Proposition 9.1 (e) conjugation by θ cyclicly permutes the Pauli matrices. Multiplication with ω operates on the
   images of the Pauli matrices under γ also as a cyclic permutation.
   Hence simultaneously conjugating all tensor factors has the same
   effect as multiplying all components of the image under γ^n with
   ω.

f. Propositions 9.1 (d) and 9.1 (e) imply that conjugation with θH
   fixes σ_x, maps σ_y to σ_z, and maps σ_z to $-\sigma_y$. Hence $\gamma(\sigma_y^{\theta H}) = \gamma(\sigma_z) = (\gamma(\sigma_y))^2$ and $\gamma(\sigma_z^{\theta H}) = \gamma(-\sigma_y) = \gamma(\sigma_y) = (\gamma(\sigma_z))^2$.

g. For $x, y \in E_1$

$$\operatorname{tr}(\gamma(x)\overline{\gamma(y)}) = \gamma(x)\gamma(y)^2 + \gamma(x)^2\gamma(y) = \gamma(x)\gamma(y)\big(\gamma(x) + \gamma(y)\big). \tag{9.25}$$

For $x = y$, $x = 1$, or $y = 1$, (9.25) vanishes. For all other cases, the sum equals one. Hence for any Pauli error g and h the symplectic inner product

$$\gamma^n(g) * \gamma^n(h) = \sum_{i=1}^{n} \operatorname{tr}\left(\gamma(g)_i\overline{\gamma(h)_i}\right) \tag{9.26}$$

equals the number of positions where the tensor factors of g and h are different Pauli matrices. The commutator relations (9.12) imply that g and h commute if the sum (9.26) is even, and $gh = -hg$ else. As any element $e \in E_n$ can be written as $e = i^k g$ with $k = 0, \ldots, 3$ and a Pauli error g, the statement is true for all elements of E_n.

∎

Next we consider sets of $\mathbb{F}_4$ vectors with certain properties.

DEFINITION 9.9   *Let $C \subseteq \mathbb{F}_4^n$ be additively closed.*

By C^* we denote the orthogonal space with respect to the symplectic inner product $*$ defined in (9.23), i.e.,

$$C^* := \{d \in \mathbb{F}_4^n \mid \forall c \in C \colon d * c = 0\}.$$

DEFINITION 9.10   *An additively closed subset $C \subseteq \mathbb{F}_4^n$ —an additive code over $\mathbb{F}_4$—with $C \subseteq C^*$ is called self-orthogonal. A code $C = C^*$ is called self-dual.*

Note that for binary codes we use the the terminology *weakly self-dual* (cf. Section 9.3), while for codes over $\mathbb{F}_4$ we use the terminology *self-orthogonal* (cf. [4]). This reflects the dependence of the orthogonal code $C^\perp$ resp. C^* on the corresponding inner product.

THEOREM 9.7
self-orthogonal.

PROOF   Consequence of Propositions 9.6 (c) and 9.6 (g).   ∎

This theorem allows the identification of abelian subgroups of the Pauli error group and self-orthogonal additive codes over $\mathbb{F}_4$. It is the foundation of the following construction of QECC that has been derived independently by Gottesman from the perspective of abelian subgroups [7] and by Calderbank et al. from the perspective of orthogonal geometry [3]. A comprehensive treatment of those codes can be found in [4].

THEOREM 9.8

Let $S \leq E_n$ be an abelian subgroup of the Pauli error group and let $C :=$ $\gamma^n(S) \subseteq \mathbb{F}_4^n$ be the corresponding self-orthogonal code $C = (n, 2^{n-k}, d)$.

Then any common eigenspace $\mathcal{C} \leq (\mathbb{C}^2)^{\otimes n}$ of S is an additive QECC $\mathcal{C} = [\![n, k, d]\!]$ with $d \geq \min\{\text{wgt}(\mathbf{c}) \colon \mathbf{c} \in C^* \setminus C\}$.

As the group S acts trivially on the code, it is called the stabilizer group of the code $\mathcal{C}$.

PROOF   Without loss of generality we assume $S \cap \mathcal{Z}(E_n) = \{\mathbf{1}\}$. Furthermore, let the code $\mathcal{C}$ be the intersection of the eigenspaces of the elements of S corresponding to the eigenvalue one. Then for all $|\psi\rangle \in \mathcal{C}$ and $s \in S$, $s|\psi\rangle = |\psi\rangle$. Hence for an arbitrary Pauli error e

$$\langle\psi|e|\varphi\rangle = \langle\psi|es|\varphi\rangle = \langle\psi|se|\varphi\rangle \tag{9.27}$$

holds for all $s \in S$ and $|\psi\rangle, |\varphi\rangle \in \mathcal{C}$. Proposition 9.6 (g) implies

$$\langle\psi|es|\varphi\rangle = (-1)^{\gamma^n(e) * \gamma^n(s)} \langle\psi|se|\varphi\rangle.$$

Together with (9.27) we get that $\langle\psi|e|\varphi\rangle$ vanishes if there is an element $s \in S$ with $\gamma^n(e) * \gamma^n(s) = 1$, i.e., if the vector $\gamma^n(e)$ does not lie in the orthogonal code C^*. Hence conditions (9.13) and (9.14) hold for all Pauli errors e with $\gamma^n(e) \neq C^*$. For $\gamma^n(e) \in C$, $e \in S$ and conditions (9.13) and (9.14) hold.

If the conditions are violated for a Pauli error e, then $\gamma(e) \in C^* \setminus C$. Together with Proposition 9.6 (d) we get $\text{wgt}(e) = \text{wgt}(\gamma(e)) \geq d$.

The proof that the dimension of the code equals 2^k is only sketched. For arbitrary $e \in E_n$, $s \in S$, and $|\psi\rangle \in \mathcal{C}$

$$s(e|\psi\rangle) \overset{(9.24)}{=} (-1)^{\gamma^n(e) * \gamma^n(s)} es|\psi\rangle = (-1)^{\gamma^n(e) * \gamma^n(s)} (e|\psi\rangle), \tag{9.28}$$

i.e., $e|\psi\rangle$ is an eigenvector of s with eigenvalue $(-1)^{\gamma^n(e) * \gamma^n(s)}$. Hence E_n operates on the eigenspaces of S. It can be shown that there are 2^{n-k} different eigenspaces and that the eigenspaces have all the same dimension (cf. [4, Theorem 1]). This proves that the dimension of each eigenspace equals 2^k. ∎

DEFINITION 9.11   *An additive quantum code $C = [\![n, k, d]\!]$ is called linear if the associated additive self-orthogonal code $C = [n, (n-k)/2] \le \mathbb{F}_4^n$ is $\mathbb{F}_4$-linear.*

CSS codes are special cases of additive quantum codes.

LEMMA 9.3

   a. *Any CSS code is an additive quantum code.*

   b. *Any weakly self-dual CSS code is a linear quantum code.*

PROOF   Let $C_1 = [n, k_1]$ and $C_2 = [n, k_2]$ with $C_2^\perp \le C_1$ be the binary codes corresponding to the CSS code. The code $C := \omega C_1^\perp + \omega^2 C_2^\perp \subseteq \mathbb{F}_4^n$ is additively closed. Furthermore, $C^* = \omega^2 C_1 + \omega C_2$. From $C_2^\perp \le C_1$ and $C_1^\perp \le C_2$ it follows that C is self-orthogonal. For weakly self-dual CSS codes $C_1 = C_2$. This implies that the additive code C is closed under multiplication with ω, i.e., C is a linear code. ∎

9.4.2   Example: quantum Hamming code

From the algorithmic point of view the questions of encoding and computation of an error syndrome arise. Both aspects will be illustrated for the quantum Hamming code (cf. [2]).

The Hamming code of second order over $\mathbb{F}_4$ is generated by

$$G_4 = \begin{pmatrix} 1 & 0 & 1 & \omega^2 & \omega^2 \\ 0 & 1 & \omega^2 & \omega^2 & 1 \\ \hline 0 & 0 & 1 & \omega & 1 \end{pmatrix}$$

(cf. [19, Ch. 7, §3]). The first two rows of G_4 generate a linear self-orthogonal code $C = [5, 2, 4]$ with $C^* = [5, 3, 3]$. Additively, the codes $C = (5, 2^4, 4)$ and $C^* = (5, 2^6, 3)$ are generated by the rows of the matrices

$$G_2 = \begin{pmatrix} 1 & 0 & 1 & \omega^2 & \omega^2 \\ \omega & 0 & \omega & 1 & 1 \\ 0 & 1 & \omega^2 & \omega^2 & 1 \\ 0 & \omega & 1 & 1 & \omega \end{pmatrix} = \begin{pmatrix} g_1 \\ g_2 \\ g_3 \\ g_4 \end{pmatrix} \qquad (9.29)$$

resp.

$$H = \left(\begin{array}{ccccc} 1 & 0 & 1 & \omega^2 & \omega^2 \\ \omega & 0 & \omega & 1 & 1 \\ 0 & 1 & \omega^2 & \omega^2 & 1 \\ 0 & \omega & 1 & 1 & \omega \\ \hline 0 & 0 & 1 & \omega & 1 \\ 0 & 0 & \omega & \omega^2 & \omega \end{array}\right) = \left(\begin{array}{c} g_1 \\ g_2 \\ g_3 \\ g_4 \\ h_5 \\ h_6 \end{array}\right).$$

For the Pauli errors

$$g_1 := \sigma_x \otimes \mathbb{1} \otimes \sigma_x \otimes \sigma_y \otimes \sigma_y$$
$$g_2 := \sigma_z \otimes \mathbb{1} \otimes \sigma_z \otimes \sigma_x \otimes \sigma_x$$
$$g_3 := \mathbb{1} \otimes \sigma_x \otimes \sigma_y \otimes \sigma_y \otimes \sigma_x$$
$$g_4 := \mathbb{1} \otimes \sigma_z \otimes \sigma_x \otimes \sigma_x \otimes \sigma_z$$

we have $\gamma^4(g_i) = g_i$. Hence those errors generate an abelian group $S = \langle g_1, g_2, g_3, g_4 \rangle$ with $\gamma^4(S) = C$. The resulting QECC C is a common eigenspace of S. The operator

$$P_S := \frac{1}{|S|} \sum_{s \in S} s$$

is a projection onto the space that is stabilized by S. Therefore the image $P_S|\psi\rangle$ of an arbitrary quantum state $|\psi\rangle$ either lies in C or vanishes. The basic idea of the encoding procedure presented in [6] is to replace the projection P_S by a unitary transformation. As S is an abelian group of exponent two, P_S can be rewritten as

$$P_S = \frac{1}{|S|} \prod_{g \in \text{Gen}(S)} (\mathbb{1} + g), \tag{9.30}$$

where $\text{Gen}(S)$ denotes a minimal generating set of S.

First we consider the action of $(\mathbb{1} + \sigma_x)/2$ and $(\mathbb{1} + \sigma_y)/2$. The images of the state $|0\rangle$ are

$$\frac{1}{2}(\mathbb{1} + \sigma_x)|0\rangle = \frac{1}{2}(|0\rangle + |1\rangle) \quad \text{and} \quad \frac{1}{2}(\mathbb{1} + \sigma_y)|0\rangle = \frac{1}{2}(|0\rangle - i|1\rangle).$$

The same superpositions can be obtained operating on the state $|0\rangle$ by the Hadamard transform H resp. the matrix $\begin{pmatrix} 1 & 0 \\ 0 & -i \end{pmatrix} \cdot H$. For the operation on n qubits we get

$$
\left(\mathbb{1} + (\sigma_x \otimes \tilde{g})\right)\left(|0\rangle \otimes |\varphi\rangle\right) = \left(|0\rangle \otimes |\varphi\rangle + |1\rangle \otimes \tilde{g}|\varphi\rangle\right)
$$
$$
= \underbrace{\left(|0\rangle\langle 0| \otimes \mathbb{1} + |1\rangle\langle 1| \otimes \tilde{g}\right)}_{U_1}\left((|0\rangle + |1\rangle) \otimes |\varphi\rangle\right).
$$

(9.31)

Here U_1 is single-controlled quantum gate. The superposition of the first qubit can be achieved by a Hardamard gate. Similarly we get

$$
\left(\mathbb{1} + (\sigma_y \otimes \tilde{g})\right)\left(|0\rangle \otimes |\varphi\rangle\right) = \left(|0\rangle\langle 0| \otimes \mathbb{1} + |1\rangle\langle 1| \otimes \tilde{g}\right)\left((|0\rangle - i|1\rangle) \otimes |\varphi\rangle\right).
$$

(9.32)

This shows that a factor $(\mathbb{1} + g)$ of the projector P_S in (9.30) corresponds to a unitary transformation when operating on a product state with factor $|0\rangle$ and if the corresponding tensor factor of g is either σ_x or σ_y. This can be achieved by transforming the generators of S to standard form. The necessary transformations correspond to transformations on the generator matrix G of the additive code.

DEFINITION 9.12   *The generator matrix G of an additive code $C = (n, 2^{n-k})$ is in standard form iff the following two conditions hold:*

   a. $G_{i,i+k} \in \{1, \omega^2 = 1 + \omega\}$ for $i = 1, \ldots, n - k$.

   b. $G_{i,j} \in \{0, \omega\}$ for $i = 1, \ldots, n - k - 1$ and $j = i + k + 1, \ldots, n$.

In [6] it was shown that—similar to Gaussian elimination—addition of rows, permutation of columns, multiplication of a row with ω, and conjugation of a column (cf. Propositions 9.6 [e] and 9.6 [f]) any generator matrix can be transformed into standard form.

A standard form of the matrix G_2 in (9.29) is given by

$$
G := \begin{pmatrix} 1 & 1 & \omega & 0 & \omega \\ 1+\omega & 0 & 1+\omega & & \omega & \omega \\ 1+\omega & \omega & & \omega & 1+\omega & 0 \\ 1 & & \omega & & 0 & \omega & 1 \end{pmatrix}.
$$

Similar to CSS codes (cf. Figure 9.2) a quantum circuit for encoding an additive QECC can be directly derived from the generator matrix in standard form. (Note that there is an error in the algorithm sketched in [6] as some of the phase gates are put at the wrong position.)

The code in our example encodes one qubit $|\varphi\rangle$ which corresponds to the uppermost quantum wire in Figure 9.4. The generators of the stabilizer group S correspond to Hadamard gates—and if necessary a phase gate as well—and a single-controlled quantum gate according to (9.31) and (9.32). The gates σ_z drawn with dashed lines are redundant as they operate trivially on the state $|0\rangle$. Omitting these gates, the standard form of the generator matrix yields a quantum circuit in upper triangular form.

$$
G^t = \begin{pmatrix}
1 & \omega^2 & w^2 & 1 \\
1 & 0 & \omega & \omega \\
\omega & 1+\omega & \omega & 0 \\
0 & \omega & 1+\omega & \omega \\
\omega & \omega & 0 & 1
\end{pmatrix}.
$$

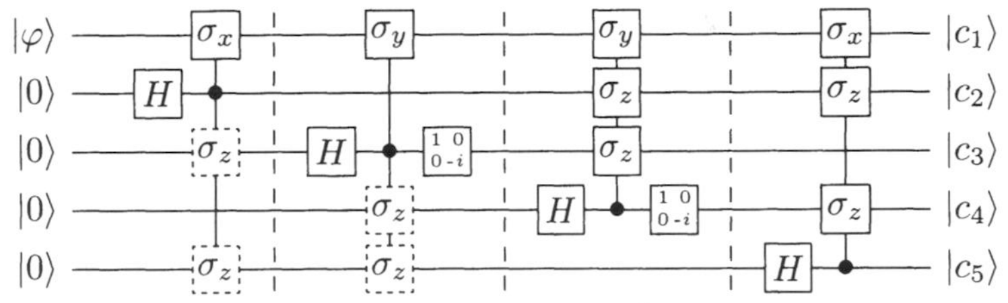

FIGURE 9.4
Encoder for the quantum Hamming code $\mathcal{C} = [\![5,1,3]\!]$ derived from a generator matrix in standard form.

In order to compute an error syndrome note that (9.28) implies that an error e interchanges the two eigenspaces corresponding to the eigenvalues $+1$ and -1 of an element $s \in S$ if $\gamma^n(e) * \gamma^n(s) = 1$. Hence an error syndrome can be obtained by determining for each of the generators of the stabilizer group in which of the eigenspaces the state lies. As the Pauli matrices are hermitian operators this can in principle be done by measurements. Instead of a direct measurement one can use an auxiliary system. By conjugation with H or θ any generator can be transformed into a tensor product of σ_z and $\mathbb{1}$. The eigenstates of σ_z are $|0\rangle$ and $|1\rangle$. Hence the parity of the number of qubits in state $|1\rangle$ determines in which eigenspace the state lies. The complete circuit for computing the syndrome is shown in Figure 9.5. The CNOT gates invert the syndrome qubit iff the state of the—if necessary conjugated—control qubit is $|1\rangle$, i.e., iff the corresponding eigenvalue is -1.

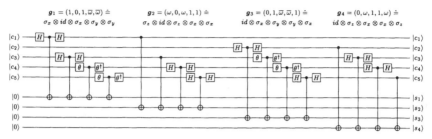

FIGURE 9.5
Computation of the syndrome for the quantum Hamming code
$\mathcal{C} = [\![5,1,3]\!]$ via *nondemolition measurements* of the generators g_i of the
stabilizer group.

The four qubit state $|s\rangle = |s_1\rangle|s_2\rangle|s_3\rangle|s_4\rangle$ corresponds to the representation of the $\mathbb{F}_4$-syndrome of the Hamming code as $\mathbb{F}_2$-vector. Measuring the syndrome, the corresponding error vector can be computed using a decoding algorithm for the Hamming code.

9.5   Conclusions

We have presented the basic algorithms to construct quantum circuits for encoding and decoding two major classes of quantum error-correcting codes. Replacing the Pauli error group by another unitary error basis (cf. [17]) it should be straightforward to transfer the encoding technique presented in Section 9.4.2. Of course, for certain classes of codes, such as cyclic codes, more efficient algorithms can be derived (see [13, 15]).

References

[1] T. BETH AND M. GRASSL, Improved decoding of quantum error correcting codes from classical codes, in *PhysComp96*, T. Toffoli, M. Biafore, and J. Leão, Eds., Boston, Nov. 1996, pp. 28–31.

[2] ——, The quantum Hamming and hexacodes, *Fortschritte der Physik*, 46 (1998), pp. 459–491.

[3] A. R. CALDERBANK, E. M. RAINS, P. W. SHOR, AND N. J. A. SLOANE, Quantum error correction and orthogonal geometry, *Phys. Rev. Lett.*, 78 (1997), pp. 405–408.

[4] A. R. CALDERBANK, E. M. RAINS, P. W. SHOR, AND N. J. A. SLOANE, Quantum error correction via codes over GF(4), *IEEE Transactions on Information Theory*, 44 (1998), pp. 1369–1387.

[5] A. R. CALDERBANK AND P. W. SHOR, Good quantum error-correcting codes exist, *Phys. Rev.*, A54 (1996), pp. 1098–1105.

[6] R. CLEVE AND D. GOTTESMAN, Efficient computations of encodings for quantum error correction, *Phys. Rev.*, A56 (1997), pp. 76–82.

[7] D. GOTTESMAN, A class of quantum error-correcting codes saturating the quantum Hamming bound, *Phys. Rev.*, A54 (1996), pp. 1862–1868.

[8] J. P. GRAM, Ueber die Entwickelung reeller Functionen in Reihen mittels der Methode der kleinsten Quadrate, *Crelles J.*, 94 (1883), pp. 41–73.

[9] M. GRASSL, Fehlerkorrigierende Codes für Quantensysteme: Konstruktionen und Algorithmen, Dissertation, Fakultät für Informatik, Universität Karlsruhe, 2001.

[10] M. GRASSL AND T. BETH, Codierung und Decodierung zyklischer Quantencodes, in *Fachtagung Informations- und Mikrosystemtechnik*, B. Michaelis and H. Holub, Eds., Magdeburg, Mar. 25–27 1998, Otto-von-Guericke-Universität Magdeburg, Fakultät für Elektrotechnik, Institut für Prozeßmeßtechnik und Elektronik (IPE), LOGISCH GmbH, pp. 137–144.

[11] ——, Quantum BCH codes, in *Proceedings X. International Symposium on Theoretical Electrical Engineering*, W. Mathis and T. Schindler, Eds., Magdeburg, Sept. 6–9 1999, Universität Magdeburg, pp. 207–212.

[12] ——, Relations between classical and quantum error-correcting codes, in *Workshop "Physik und Informatik"*, W. Kluge, Ed., DPG-Frühjahrstagung, Heidelberg, 1999, pp. 45–58.

[13] ———, Cyclic quantum error-correcting codes and quantum shift registers, *Proceedings of the Royal Society London*, A456 (2000), pp. 2689–2706.

[14] M. GRASSL, T. BETH, AND T. PELLIZZARI, Codes for the quantum erasure channel, *Phys. Rev.*, A56 (1997), pp. 33–38.

[15] M. GRASSL, W. GEISELMANN, AND T. BETH, Quantum Reed-Solomon codes, in *Proceedings Applied Algebra, Algebraic Algorithms and Error-Correcting Codes (AAECC-13)*, M. Fossorier, H. Imai, S. Lin, and A. Poli, Eds., Lecture Notes in Computer Science, vol. 1719, Honolulu, Hawaii, Nov. 15–19 1999, Springer, pp. 231–244.

[16] E. KNILL AND R. LAFLAMME, Theory of quantum error-correcting codes, Phys. Rev., A55 (1997), pp. 900–11.

[17] A. KLAPPENECKER AND M. RÖTTELER, Clifford codes, in Chapter 10 of this volume.

[18] K. KRAUS, States, Effects, and Operations, *Lect. Notes Phys.*, vol. 190, Springer, Berlin, 1983.

[19] F. J. MACWILLIAMS AND N. J. A. SLOANE, *The Theory of Error-Correcting Codes*, North-Holland, Amsterdam, 1977.

[20] W. PAULI JR., Zur Quantenmechanik des magnetischen Elektrons, *Zeitschrift für Physik*, 43 (1927), pp. 601–623.

[21] E. SCHMIDT, Entwickelung willkuerlicher Functionen nach Systemen vorgeschriebener, Inauguraldissertation, Universität Göttingen, 1905.

[22] P. W. SHOR, Scheme for reducing decoherence in quantum computer memory, *Phys. Rev.*, A52 (1995), pp. R2493–R2496.

[23] A. M. STEANE, Error correcting codes in quantum theory, *Phys. Rev. Lett.*, 77 (1996), pp. 793–797.

[24] ———, Simple quantum error correcting codes, *Phys. Rev.*, A54 (1996), pp. 4741–4751.

Chapter 10

Clifford codes

Andreas Klappenecker and Martin Rötteler

Abstract   Quantum error control codes allow to detect and correct errors that are due to decoherence effects. We review some basic properties of these codes and give some constructions. Our main focus will be on a construction of quantum error control codes that have been introduced by Knill in 1996 with the intention to generalize stabilizer codes. These so-called Clifford codes can be constructed and analyzed with tools from representation theory of finite groups. We show that a large class of Clifford codes are actually stabilizer codes. And we construct the smallest example of a Clifford code that is not a stabilizer code.

10.1   Introduction

A quantum computer takes advantage of entangled states stored in the state of atoms, nuclear spins, photons, or other quantum systems. The interaction of the quantum computer with its environment leads to decoherence errors which alter the state of the memory. The protection of the memory against these errors is a crucial part in the construction of a resilient quantum computer. We describe in this chapter a generalization of stabilizer codes that allows to protect the encoded states.

1-58488-282/4/02/$0.00+$1.50

The first quantum error correcting codes were introduced by Shor [21] and Steane [24] about six years ago. The existence of such codes is a remarkable fact, since it shows that an infinite variety of errors affecting a single quantum bit can be corrected by a finite number of operations. Moreover, the subsequent development of fault-tolerant architectures [22] made it clear that quantum computing can overcome the imprecision problems that defeated the successful implementation of analogue computers.

The theory of quantum error control codes developed rapidly after the initial results by Shor and Steane. Calderbank and Shor showed that good quantum error correcting codes exist [7]. Their construction of a quantum error control code started from a classical binary linear code C containing its dual code $C^\perp$. This construction was independently discovered by Steane [23] and the quantum codes are now known as Calderbank–Shor–Steane codes or shortly CSS codes.

A more general class of quantum error control codes has been introduced by Gottesman [10] and Calderbank, Rains, Shor, and Sloane [5]. These codes are known as stabilizer codes or as additive codes. Most quantum error control codes known to date are constructed as stabilizer codes. The popularity of stabilizer codes stems from the fact that a large body of theory developed for classical error control codes can be translated into the quantum realm, as is explained in the seminal paper [6].

Some practical aspects of quantum codes have been discussed in the literature as well. For instance, Cleve and Gottesman give a construction of encoding circuits for binary stabilizer codes [8]. Grassl and Beth derive encoding and decoding circuits for cyclic codes [12].

The majority of publications on quantum error control codes is confined to the binary case: the encoding of several quantum bits into a larger set of quantum bits. This is somewhat surprising, since the popular implementation models of quantum computing — cavity QED, trapped cold ions, or bulk spin NMR — all allow, at least in principle, to use more than just two level quantum systems. Moreover, the concatenation of codes used in fault-tolerant architectures [1] is most naturally understood in terms of quantum codes with bigger alphabets. We allow arbitrary alphabet sizes for that reason.

10.2 Motivation

A quantum computer stores its information in the state of quantum systems. The computational state space of a quantum system is a finite dimensional complex vector space $\mathbb{C}^d$, sometimes referred to as a qudit. This space is equipped with a standard orthonormal basis which is traditionally expressed in terms of Dirac's ket notation: $|0\rangle, \ldots, |d-1\rangle$.

The combination of several quantum systems yields the state space of the quantum computer

$$\mathcal{H} = \mathbb{C}^{d_1} \otimes \mathbb{C}^{d_2} \otimes \cdots \otimes \mathbb{C}^{d_n}. \tag{10.1}$$

Notice that the quantum systems might have different dimensions d_i.

A quantum error control code is a subspace of $\mathcal{H}$. A well-designed quantum code allows to correct errors affecting only a few quantum systems, that is, a few tensor factors in (10.1). A small example will help to illustrate this feature.

Suppose that we want to protect the state of a single d-level quantum system. This can be done, for instance, by encoding the base states $|k\rangle$ into nine qudits by

$$|k\rangle \mapsto \frac{1}{d^{3/2}} \left(\sum_{j=0}^{d-1} \omega^{kj} |jjj\rangle \right) \otimes \left(\sum_{j=0}^{d-1} \omega^{kj} |jjj\rangle \right) \otimes \left(\sum_{j=0}^{d-1} \omega^{kj} |jjj\rangle \right), \tag{10.2}$$

where $k \in \{0, \ldots, d-1\}$ and $\omega = \exp(2\pi i/d)$. This quantum error control code is a straightforward generalization of Shor's code [21] to the nonbinary case.

The code (10.2) is able to correct an arbitrary error in one of the nine qudits. To see this, we note that the code is given by a concatenation of two codes. The inner code is a repetition code encoding a base state into three replicas

$$|k\rangle \mapsto |k\rangle \otimes |k\rangle \otimes |k\rangle \quad \text{with} \quad k \in \{0, \ldots, d-1\}.$$

This code can correct a shift error $X_\ell |k\rangle = |k + \ell \bmod d\rangle$ applied to a single qudit. The outer code protects against a single phase error $Z_\ell |k\rangle = \omega^{\ell k} |k\rangle$. It is obtained from the repetition code by applying the discrete Fourier transform to each component, that is,

$$|k\rangle \mapsto F|k\rangle \otimes F|k\rangle \otimes F|k\rangle, \quad \text{with} \quad k \in \{0, \ldots, d-1\},$$

where $F|k\rangle = d^{-1/2} \sum_{j=0}^{d-1} \omega^{kj} |j\rangle$.

The concatenated code (10.2) is then able to correct a single shift error, a single phase error, or a combination of both. In fact, the code is even able to correct all linear combinations of these errors, since the error recovery is a linear operation. Therefore, Shor's code (10.2) is able to correct an *arbitrary* error in one of the nine qudits.

Our point of view in the following sections will slightly differ from our approach taken in this motivating example. The error-correcting properties of a quantum error control code do not depend on the particular choice of basis nor on the choice of encoding map. Thus, we prefer a basis free approach in the following sections, since an analogue of (10.2) would be awkward in larger dimensions.

10.3   Quantum error control codes

A *quantum error control code* is a subspace Q of a finite dimensional Hilbert space

$$\mathcal{H} = \mathbb{C}^{d_1} \otimes \mathbb{C}^{d_2} \otimes \cdots \otimes \mathbb{C}^{d_n}. \tag{10.3}$$

We refer to $\mathcal{H}$ as the *ambient space* of Q. The *dimension* of the code Q is its dimension as a complex vector space.

Let E be an error operator acting on a quantum code Q. We say that E is *detectable* by Q if and only if

$$\langle w| E |w\rangle = \langle u| E |u\rangle \tag{10.4}$$

holds for all $u, w \in Q$ with $\|u\| = \|w\|$.

LEMMA 10.1
Let E be a detectable error on a quantum error control code Q. Then

$$\langle w|E|u\rangle = 0 \tag{10.5}$$

holds for all $w, u \in Q$ with $\langle w|u\rangle = 0$.

PROOF   The statement is clearly true if $u = 0$ or $w = 0$. So, without loss of generality, we can assume that u and w are nonzero and normalized to the same length $\|w\| = \|u\|$. A simple calculation shows that $\langle w|E|u \rangle$ can be expressed in terms of the polarization identity

$$\langle w|E|u \rangle = \frac{1}{4}[\langle u + w|E|u + w \rangle - \langle u - w|E|u - w \rangle]$$
$$+ \frac{i}{4}[\langle u + iw|E|u + iw \rangle - \langle u - iw|E|u - iw \rangle].$$

However, $\|u + w\| = \|u - w\|$ and $\|u + iw\| = \|u - iw\|$ and therefore the terms in the brackets are zero by the length condition (10.4). ∎

Denote by P_Q the orthogonal projection from the ambient space $\mathcal{H}$ onto the quantum error control code Q. A simple consequence of the previous result is that an error operator E is detectable by Q if and only if the projection condition

$$P_Q E P_Q = \lambda_E P_Q, \qquad \lambda_E \in \mathbb{C} \tag{10.6}$$

holds.

Suppose that we want to be able to correct for a certain set S of errors acting on Q. We want to be able to reliably distinguish between different encoded states that have been affected by correctable errors. Therefore, it is necessary that orthogonal states $u, w \in Q$ remain orthogonal

$$\forall u, w \in Q: \langle u|w \rangle = 0 \implies \langle u|E_1^\dagger E_2|w \rangle = 0$$

for all possible errors E_1 and E_2 in S. However, this simply means that $E = E_1^\dagger E_2$ must be a detectable error.

LEMMA 10.2
Suppose that an error operator E acting on a quantum error control code Q satisfies

$$\forall u, w \in Q: \langle u|w \rangle = 0 \implies \langle u|E|w \rangle = 0. \tag{10.7}$$

Then E is a detectable error.

PROOF Let B be an orthonormal basis of Q. Suppose that u and w are distinct elements of B, then $\langle u + w|u - w \rangle = 0$, hence

$$0 = \langle u + w|E|u - w \rangle = \langle u|E|u \rangle - \langle w|E|w \rangle.$$

Therefore, $\langle u|E|u \rangle = \langle w|E|w \rangle$ for all $u, w \in B$. It follows that (10.4) holds for arbitrary $u, w \in Q$ with $\|u\| = \|w\|$. ∎

It has been shown by Knill and Laflamme [17] (see also Bennett et al. [4]) that this error correction condition is not only necessary but also sufficient. Thus, to summarize, a set of errors S can be *corrected* by a quantum error control code Q if and only if all errors in the set

$$S^\dagger S = \{E_1^\dagger E_2 \mid E_1, E_2 \in S\}$$

are detectable by Q. An elementary proof of this fact is given in Chapter 9 by M. Grassl in this volume [11].

The detectable errors also lead us to the notion of minimum distance of a quantum error control code. The minimum distance is an essential parameter of a code, since it determines how many localized errors can be corrected by this code.

A *local error operator* is a linear operator E of the form

$$E = M_1 \otimes M_2 \otimes \cdots \otimes M_n,$$

where M_i is a linear operator acting on the tensor component $\mathbb{C}^{d_i}$ in the ambient space (10.3). A local error operator is thus compatible with the tensor product structure of the ambient space $\mathcal{H}$. The *weight* of the local error operator is given by the number of elements M_i that are not scalar multiples of the identity.

The code Q has *minimum distance* at least d if and only if all local errors of weight less than d are detectable by Q. A quantum error control code with minimum distance $d = 2t + 1$ allows to correct decoherence errors affecting up to t qudits.

REMARKS

a. A detailed analysis of general quantum error control codes can be found in Knill and Laflamme [17]. Another early account is given by Bennett et al. [4]. We refer to articles by Knill, Laflamme, and Viola [18] and by Zanardi [25] for more recent discussions of the general theory of quantum error control codes.

b. The notion of detectable errors has been explicitly introduced in [17] in the form (10.6). The equivalent form (10.4) has been used by Rains [19] in his definition of minimum distance of a quantum code. Alternatively, one can define a detectable error by the orthogonality condition (10.7), as is shown by Lemma 10.1 and 10.2. Detectable errors have been studied in detail by Ashikhmin, Barg, Knill, and Litsyn in [2, 3].

10.4 Nice error bases

We introduced the notion of detectable errors in the previous section. The detectability condition (10.4) is linear in the error operators. This suggests the following approach: choose a basis of the linear operators that is particularly convenient for the construction of quantum error control codes. We will be able to characterize the operators in this basis that can be detected by the constructed code. Hence, the code will be able to detect all linear combinations of these error operators. The main benefit is that we have only a discrete number of conditions to check, and the code constructions resemble (and sometimes mimic) constructions of classical codes.

Let us motivate the definition of a nice error basis by way of a familiar example. Consider the set of Pauli matrices σ_x, σ_y, σ_z together with the identity matrix $\mathbf{1}_2$:

$$\mathbf{1}_2 = \begin{pmatrix} 1 & 0 \\ 0 & 1 \end{pmatrix}, \quad \sigma_x = \begin{pmatrix} 0 & 1 \\ 1 & 0 \end{pmatrix}, \quad \sigma_y = \begin{pmatrix} 0 & -i \\ i & 0 \end{pmatrix}, \quad \sigma_z = \begin{pmatrix} 1 & 0 \\ 0 & -1 \end{pmatrix}. \quad (10.8)$$

This set forms an orthonormal basis of the vector space of complex 2×2 matrices $\mathrm{Mat}_2(\mathbb{C})$ with respect to the normalized trace inner product $\langle A|B \rangle = \mathrm{tr}(A^\dagger B)/2$. Thus, we can express a 2×2 matrix A conveniently in the form

$$A = \frac{1}{2} \left(\mathrm{tr}(A)\mathbf{1} + \mathrm{tr}(\sigma_x^\dagger A)\sigma_x + \mathrm{tr}(\sigma_y^\dagger A)\sigma_y + \mathrm{tr}(\sigma_z^\dagger A)\sigma_z \right).$$

Moreover, the multiplication of the matrices (10.8) resembles the composition operation in the Kleinian group of four elements $V_4 = \mathbb{Z}/2\mathbb{Z} \times \mathbb{Z}/2\mathbb{Z}$ if we ignore phase factors. Indeed, if we assign the Pauli matrices to the elements of V_4 in the following way

$$\hat{p}(0,0) = \mathbf{1}_2, \quad \hat{p}(1,0) = \sigma_x, \quad \hat{p}(0,1) = \sigma_z, \quad \hat{p}(1,1) = \sigma_y,$$

then the product of the representing matrices $\hat{p}$ of the elements (a, b) and (c, d) yields — up to a scalar factor — the representing matrix of $(a, b) + (c, d)$. In other words, the matrices (10.8) form a projective representation $\hat{p}$ of the group V_4.

This example motivates the following definition:

DEFINITION 10.1   *Let G be a finite group of square order d^2. The identity of this group is denoted by 1. A nice error basis on $\mathbb{C}^d$ is a set $\mathcal{E} = \{\hat{\rho}(g) \mid g \in G\}$ of unitary matrices such that*

(i) $\hat{\rho}(1)$ is the identity matrix,

(ii) $\mathrm{tr}\hat{\rho}(g) = 0$ for all elements $g \in G$ with $g \neq 1$,

(iii) $\hat{\rho}(g)\hat{\rho}(h) = \omega(g,h)\hat{\rho}(gh)$ for all $g, h \in G$,

where $\omega(g,h)$ is a nonzero complex number depending on $g, h \in G$. We call G the index group *of the nice error basis $\mathcal{E}$.*

The condition (i) and (iii) simply state that $\hat{\rho}$ is a projective representation with factor system ω. The conditions (i) – (iii) imply that $\mathcal{E}$ is an orthonormal basis of the vector space of linear operators acting on $\mathbb{C}^d$ with respect to the normalized trace inner product $\langle A|B\rangle = \mathrm{tr}(A^\dagger B)/d$.

Notice that a nice error basis for $\mathcal{H} = \mathbb{C}^{d_1} \otimes \mathbb{C}^{d_2} \otimes \cdots \otimes \mathbb{C}^{d_n}$ can be obtained by choosing nice error bases for each component and taking tensor products. The corresponding index group is then given by the direct product of the index groups of the components.

We give some more examples of nice error bases before proceeding with the construction of quantum error control codes. The first example is given by the error operators that we have seen in Section 10.2 in the discussion of Shor's code.

Example 10.1
Denote by ω the primitive dth root of unity $\omega = \exp(2\pi i/d)$. Let $X_\ell |k\rangle = |k + \ell \bmod d\rangle$ and $Z_\ell |k\rangle = \omega^{\ell k}|k\rangle$. Then

$$\mathcal{E} = \{\, X_k Z_\ell \mid (k, \ell) \in G \,\}$$

is a nice error basis on $\mathbb{C}^d$ with index group $G = \mathbb{Z}/d\,\mathbb{Z} \times \mathbb{Z}/d\,\mathbb{Z}$. In particular, we obtain in the four-dimensional case $d = 4$ the error matrices

$$X_1 = \begin{pmatrix} . & . & . & 1 \\ 1 & . & . & . \\ . & 1 & . & . \\ . & . & 1 & . \end{pmatrix}, \quad Z_1 = \begin{pmatrix} 1 & . & . & . \\ . & \omega & . & . \\ . & . & \omega^2 & . \\ . & . & . & \omega^3 \end{pmatrix},$$

which generate the nice error basis (i.e., one can obtain all other basis elements by forming the products $X_1^k Z_1^\ell$).   □

In dimension 4, there also exist error bases with *nonabelian* index groups. This is the smallest dimension where this can happen, since all groups of order p^2, with p prime, are abelian. Therefore, there do not exist any nonabelian index groups in dimensions 2 or 3. We will see in the following sections that these nonabelian index groups will allow us to unravel some interesting properties of quantum error control codes.

Example 10.2
In this example, we consider a finite group G generated by three elements a, b, c subject to the relations

$$a^2 = b^2 = [a, b] = 1 \quad \text{and} \quad a^c = b, \quad b^c = a, \quad c^4 = 1,$$

where $[a, b] = a^{-1}b^{-1}ab$ denotes the group-theoretical commutator. This is a group of order 16. It is the extension of a cyclic group $\langle c \rangle$ of order 4 by the direct product $\langle a \rangle \times \langle b \rangle$ of two cyclic groups of order 2. The representing matrices of the generators of G are given by

$$\hat{\rho}(a) = \begin{pmatrix} . & . & -1 & . \\ . & . & . & -1 \\ -1 & . & . & . \\ . & -1 & . & . \end{pmatrix} \quad \hat{\rho}(b) = \begin{pmatrix} . & . & . & -i \\ . & . & i & . \\ . & -i & . & . \\ i & . & . & . \end{pmatrix} \quad \hat{\rho}(c) = \begin{pmatrix} . & 1 & . & . \\ 1 & . & . & . \\ . & . & -i & . \\ . & . & . & i \end{pmatrix},$$

where . is an abbreviation for 0. These representing matrices generate a nice error basis in 4 dimensions. The group G has the property that it is nonabelian, but all proper subgroups of G are abelian. ☐

Example 10.3
Let G be the finite group generated by a, b, c subject to the relations

$$a^4 = b^2 = (ab)^2 = 1 \quad \text{and} \quad c^2 = [a, c] = [b, c] = 1.$$

It is the direct product of the dihedral group $D_8 \cong \langle a, b \rangle$ of order 8 and the cyclic group $C_2 \cong \langle c \rangle$. The group G is the index group of a nice error basis in 4 dimensions, which is generated by

$$\hat{\rho}(a) = \begin{pmatrix} \omega & . & . & . \\ . & \omega^7 & . & . \\ . & . & \omega^5 & . \\ . & . & . & \omega^3 \end{pmatrix} \quad \hat{\rho}(b) = \begin{pmatrix} . & 1 & . & . \\ 1 & . & . & . \\ . & . & . & 1 \\ . & . & 1 & . \end{pmatrix} \quad \hat{\rho}(c) = \begin{pmatrix} . & . & 1 & . \\ . & . & . & 1 \\ 1 & . & . & . \\ . & 1 & . & . \end{pmatrix}$$

where . and ω are abbreviations for 0 and $\exp(2\pi i/8)$, respectively. ☐

REMARKS

a. Rather surprisingly, the definition of a nice error basis severely restricts the possible index groups. It is shown in [14] that an index group of a nice error basis has to be a solvable group. A complete classification of all nice error bases up to dimensions 11 is also derived in that paper.

b. A nice error basis can also be defined as a faithful irreducible unitary projective representation of degree n of a finite group of order n^2.

10.5   Stabilizer codes

The most well-known construction of quantum error control codes is given by the so-called stabilizer construction. Stabilizer codes have been introduced by Gottesman [10] and Calderbank, Rains, Shor, and Sloane [5] for two-level quantum systems. This approach is particularly appealing, since the construction of the quantum codes can be reduced to the construction of classical error control codes over the finite field with four elements [6]. We will discuss a more general setting that allows to combine quantum systems with a different number of levels.

We have introduced the concept of a nice error basis in the last section. Notice that the matrices of a nice error basis do *not* form a group, since they are not closed under multiplication. For instance, the set of matrices (10.8) does not contain the product $\sigma_x\sigma_y$. However, we can obtain a matrix group in a canonical way from a nice error basis by taking the closure under multiplication and inverse operations. We call this group the *error group* associated with a nice error basis.

Unfortunately, the error group of a nice error basis can be infinite, a situation we would like to avoid. We say that two nice error bases $\mathcal{E}_1$, $\mathcal{E}_2$ are equivalent if and only if we can find a unitary matrix U such that

$$\mathcal{E}_1 = \{\hat\rho(g)\,|\,g \in G\} \quad \text{and} \quad \mathcal{E}_2 = \{U\hat\rho(g)U^\dagger\,|\,g \in G\}.$$

It turns out that for each nice error basis there exists an equivalent error basis with finite error group [14]. We call a finite group that is isomorphic to a finite error group an *abstract error group*. Abstract error groups allow us to work with ordinary representations instead of projective representations, which is very convenient. In fact, an abstract error group is a so-called ω-covering group of the index group; hence there corresponds to each projective representation of the index group an ordinary representation of the abstract error group.

Let E be an abstract error group. This group has an irreducible faithful unitary representation ρ of degree $\sqrt{[E : Z(E)]}$. Denote by N a normal subgroup of E. A *stabilizer code* is defined as a joint eigenspace Q of the representing matrices $\{\rho(n) \mid n \in N\}$ of this normal subgroup. In other words, there exist eigenvalues $\chi(n)$ such that

$$\rho(n)v = \chi(n)v \qquad (10.9)$$

for all $v \in Q$, and all $n \in N$. For nontrivial codes Q, the normal subgroup N must be abelian — a condition that we will assume in the following.

Note that the eigenvalues $\chi(n)$ in (10.9) constitute a character of the group N. Indeed, we have $\chi(nm) = \chi(n)\chi(m)$ for all $n, m \in N$, since

$$\chi(nm)v = \rho(nm)v = \rho(n)\rho(m)v = \chi(n)\chi(m)v$$

holds for any vector $v \in Q$.

We can give another characterization of a stabilizer code in terms of an orthogonal projector. The projector P onto the code space Q can be made explicit in the following way:

$$P = \frac{1}{|N|} \sum_{n \in N} \chi(n^{-1})\rho(n). \qquad (10.10)$$

The relation $P^2 = P$ is basically a consequence of the orthogonality relations of characters, and we can immediately see that $P^\dagger = P$. Thus P is an orthogonal projection operation. We claim that the image of P is the stabilizer code Q. Indeed, if v is an element of Q, then $Pv = v$, because the character is defined by the eigenvalues of the representing matrices; hence $Q \subseteq \mathrm{im}(P)$. On the other hand, if $v \in \mathrm{im}(P)$, then we obtain

$$\rho(m)Pv = \frac{1}{|N|} \sum_{n \in N} \chi(n^{-1})\rho(mn)v = \chi(m)\frac{1}{|N|} \sum_{n \in N} \chi(n^{-1})\rho(n)v$$

by the multiplicativity of the character χ. Thus, Q is the image of the orthogonal projector P.

Equation (10.10) is the starting point for a more general construction of quantum error control codes, which will be described in the next section.

10.6   Clifford codes

The projection formula (10.10) suggests an immediate extension: replace the abelian normal subgroup by an arbitrary normal subgroup N. The joint eigenspace of the representing matrices is trivial in the case of nonabelian normal subgroups, but the projection formulae can still have nontrivial images. We call the resulting class of codes "Clifford codes" since the construction relies on tools of representation theory developed by Clifford [9].

DEFINITION 10.2   *Let E be an abstract error group with a faithful irreducible unitary representation ρ of degree $\sqrt{[E : Z(E)]}$. Denote by φ the character of E corresponding to this representation, that is, $\varphi(g) = \mathrm{tr}\rho(g)$ for all $g \in E$. Let N be a normal subgroup of the abstract error group E. Denote by χ an irreducible character of N that is a constituent of the restriction of the character φ to N. Then the* Clifford code *with data (E, ρ, N, χ) is defined as the image of the orthogonal projector*

$$P = \frac{\chi(1)}{|N|} \sum_{n \in N} \chi(n^{-1})\rho(n). \qquad (10.11)$$

Clifford codes have been introduced by Knill in [16]. Some remarks concerning this definition are in order. Denote by φ_N the restriction of the character φ to the normal subgroup N. In general, this restricted character is not irreducible. The character χ is one of the irreducible constituents of φ_N. The latter condition ensures that the projector P has a nonzero image.

We want to characterize the error correcting properties of a Clifford code. It turns out that the detectable errors can be determined from the characters alone. In order to give a concise characterization, we need two definitions. The *inertia subgroup* $T(\chi)$ of the character χ is defined by

$$T(\chi) = \{g \in E \mid \chi(gxg^{-1}) = \chi(x) \text{ for all } x \in N\}.$$

The *quasikernel* of a character ϑ of a group T is by definition given by

$$Z(\vartheta) = \{n \in T \mid |\vartheta(n)| = \vartheta(1)\}.$$

The significance of these definitions can be seen as follows. Errors corresponding to elements of the abstract error group E, which are not contained in the inertia subgroup $T(\chi)$, map the code Q to an orthogonal complement. Hence, these errors can be detected by a suitable measurement. On the other hand, we are also interested in errors that act trivially by scalar multiplication on the code Q, hence do not affect the encoded information. We note that the image of P is not only a vector space but also an irreducible $T(\chi)$-module. Denote by ϑ the irreducible character of the group $T(\chi)$ afforded by this module. Then the quasikernel $Z(\vartheta)$ of this character contains all elements m of the abstract error group E such that the matrix $\rho(m)$ acts by scalar multiplication on Q.

We summarize the error correcting properties of Clifford codes in the following theorem:

THEOREM 10.1
Let Q be a Clifford code with the data (E, ρ, N, χ). Denote by ϑ the irreducible character of $T(\chi)$ described above. The code Q is able to correct a set of errors $S \subseteq E$ if and only if the condition $s_1^{-1} s_2 \notin T(\chi) \setminus Z(\vartheta)$ holds for all $s_1, s_2 \in S$.

A detailed proof of this result can be found in [15].

10.7 Clifford codes that are stabilizer codes

We have seen in the previous section that a Clifford code Q is given by the image of a projection operator

$$Q = \operatorname{im}\left(\frac{\chi(1)}{|N|} \sum_{n \in N} \chi(n^{-1}) \rho(n) \right).$$

In the case of an abelian normal subgroup N, we obtain a stabilizer code. It is a little bit more surprising that a nonabelian normal subgroup N might still lead to a stabilizer code. We show in the next theorem that many abstract error groups cannot produce *any* nonstabilizer code:

THEOREM 10.2
Let E be an abstract error group. If the index group $G = E/Z(E)$ is an abelian group or a Redei group (i.e., a nonabelian group where all proper

subgroups are abelian), then all Clifford codes in E are stabilizer codes.

The remainder of this section is devoted to the proof of this theorem. We say that a normal subgroup N of an abstract error group is *large* if and only if

$$N/(Z(E) \cap N) \cong E/Z(E) \qquad (10.12)$$

holds, that is, if we factor out the central elements $Z(E) \cap N$, then we still get a group isomorphic to the full index group $E/Z(E)$.

PROPOSITION 10.1

Let (E, ρ, N, χ) be the data of a Clifford code Q with ambient space $\mathcal{H}$. If the normal subgroup N is large, then Q coincides with its ambient space $\mathcal{H}$ and the projection operation

$$P = \frac{\chi(1)}{|N|} \sum_{n \in N} \chi(n^{-1}) \rho(n)$$

is the identity map. In particular, the Clifford code Q is a stabilizer code.

PROOF   We want to exploit the largeness property of N to show that the projector P is the identity map. We do this in two steps. Our first step is to show that the normal subgroup N and the center $Z(E)$ of E generate the error group E. First, we observe that

$$N/(Z(E) \cap N) \cong NZ(E)/Z(E)$$

holds by the second isomorphism theorem, cf. [20, p. 56, Theorem 3.40]. Now $NZ(E)$ is a subgroup of E, and

$$|NZ(E)| = \frac{|N| \cdot |Z(E)|}{|N \cap Z(E)|} \stackrel{\text{by (10.12)}}{=} \frac{|E|}{|Z(E)|} |Z(E)| = |E|$$

holds; therefore $E = NZ(E)$.

In the second step we show that $\rho(g)P\rho(g^{-1}) = P$ holds for all $g \in E$; this implies, by Schur's lemma [13, Lemma 1.5], that P is a scalar multiple of the identity. Our previous discussion shows that we can write an arbitrary group element $g \in E$ in the form $g = nz$ with $n \in N$ and

$z \in Z(E)$. Thus we obtain

$$
\begin{aligned}
\rho(nz)P\rho((nz)^{-1}) &= \frac{\chi(1)}{|N|} \sum_{m \in N} \chi(m^{-1})\rho(nzmz^{-1}n^{-1}) \\
&= \frac{\chi(1)}{|N|} \sum_{m \in N} \chi(m^{-1})\rho(nmn^{-1}) \\
&= \frac{\chi(1)}{|N|} \sum_{m \in N} \chi((n^{-1}mn)^{-1})\rho(m) = P.
\end{aligned}
$$

The last equality follows from the fact that the character χ is a class function, hence $\chi((n^{-1}mn)^{-1}) = \chi(m^{-1})$. Therefore, $P = \alpha\mathbf{1}$ for some scalar α. Since we have $P^2 = P$, we either have $\alpha = 0$ or $\alpha = 1$. However, χ is by definition a constituent of the restricted character $\mathrm{tr}\rho_N$, and thus the projector P is a nonzero map, which proves the claim $P = \mathbf{1}$. ∎

PROPOSITION 10.2

Let E be an abstract error group and $\varphi \in \mathrm{Irr}(E)$ a faithful character of degree $\varphi(1)^2 = [E\!:\!Z(E)]$. Denote by N a normal subgroup of E and let $\chi \in \mathrm{Irr}(\varphi|N)$ be an irreducible constituent of the restricted character φ_N. Note that the restriction of χ to the center $Z(N)$ is a multiple of a linear character $\chi_{Z(N)} = \chi(1)\varphi$ with $\varphi \in \mathrm{Irr}(Z(N))$. If $N/(Z(E) \cap N)$ is abelian, then

$$
\frac{\chi(1)}{|N|} \sum_{n \in N} \chi(n^{-1})\rho(n) = \frac{1}{|Z(N)|} \sum_{n \in Z(N)} \varphi(n^{-1})\rho(n).
$$

Thus, in particular, the Clifford code (N, χ) coincides with the stabilizer code $(Z(N), \varphi)$.

PROOF   Denote by $Z = Z(E) \cap N$ the intersection of N with the center of E. We recall that the support of φ is the center $Z(E)$ and that by definition $(\varphi_N, \chi) \neq 0$. These two facts imply that χ coincides — up to a nonzero constant factor — with φ on Z; hence χ_Z is a faithful character. Thus we can invoke the following lemma:

LEMMA 10.3

Let $\chi \in \mathrm{Irr}(N)$, $Z = N \cap Z(E)$, N/Z abelian, and χ_Z a faithful character. Then $\mathrm{supp}(\chi) = Z(N)$.

PROOF   We can assume without loss of generality that the group N is nonabelian. Let $x \in N - Z(N)$. Then there exists an element $h \in N$ such that $xh \neq hx$. We have $[x, h] = z$ for some $z \in Z$ with $z \neq 1$, since N/Z is abelian. Keeping in mind that $[x, h] = z$ is equivalent to $xz = h^{-1}xh$, we get $\chi(x) = \chi(h^{-1}xh) = \chi(xz) = \omega\chi(x)$, with $\omega \neq 1$, since χ is faithful on the center Z; hence, $\chi(x) = 0$. It follows that $\mathrm{supp}(\chi) = Z(N)$, as claimed.   ∎

It is known that the minimal support condition $\mathrm{supp}(\chi) = Z(N)$ is equivalent to the extremal degree condition $\chi(1)^2 = [N : Z(N)]$, cf. Isaacs [13], Corollary 2.30. Moreover, $\chi_{Z(N)}(n) = \chi(1)\varphi(n)$ for a linear character of $Z(N)$. Therefore,

$$\frac{\chi(1)}{|N|} \sum_{n \in N} \chi(n^{-1})\rho(n) = \frac{\chi(1)^2}{|N|} \sum_{n \in Z(N)} \varphi(n^{-1})\rho(n)$$

$$= \frac{1}{|Z(N)|} \sum_{n \in Z(N)} \varphi(n^{-1})\rho(n)$$

which concludes the proof of Proposition 10.2.   ∎

PROOF (of Theorem 10.2)   We have shown that a normal subgroup N of an error group E can produce only stabilizer codes in case $N/(Z(E) \cap N)$ is large or is abelian. Thus, an index group that is a Redei group or an abelian group cannot produce any Clifford code that is not a stabilizer code.   ∎

REMARKS

a. The proof of Proposition 10.1 showed that the character χ of a large normal subgroup is extendible to a character of E. We can derive similar results whenever the character χ extends to E. This leads to an even larger class of error groups admitting only stabilizer codes.

b. A different proof of the statement of Theorem 10.2 for the case of error groups with abelian index groups has been given in [15]. It is an interesting open problem to characterize all abstract error groups that admit only stabilizer codes.

10.8 A remarkable error group

Let G be the finite group generated by three elements a, b, c subject to the relations

$$a^2 = b^2 = [a, b] = 1 \quad \text{and} \quad a^c = b, \quad b^c = a, \quad c^4 = 1.$$

This is the index group that we have introduced in Example 10.2.

An abstract error group E is obtained by a central extension of the index group G by a cyclic group of order 2. More explicitly, E is presented by four generators a, b, c, d that are subject to the relations

$$a^2 = b^2 = [a, b] = 1, \qquad d^2 = [a, d] = [b, d] = [c, d] = 1,$$

and $c^4 = 1$, $a^c = b$, $b^c = ad$.

The group E is nilpotent of class 3 and of order 32. A faithful irreducible representation of E is given by

$$\rho(a) = \begin{pmatrix} . & . & -1 & . \\ . & . & . & -1 \\ -1 & . & . & . \\ . & -1 & . & . \end{pmatrix} \quad \rho(b) = \begin{pmatrix} . & . & . & -i \\ . & . & i & . \\ . & -i & . & . \\ i & . & . & . \end{pmatrix} \quad \rho(c) = \begin{pmatrix} . & 1 & . & . \\ 1 & . & . & . \\ . & . & . & -i \\ . & . & . & i \end{pmatrix}$$

and the generator d of the center of E is represented by $\rho(d) = -\mathbf{1}$.

What is so remarkable about this error group? It has a nonabelian index group and yet all its Clifford codes are stabilizer codes. This follows from the fact that all nontrivial normal subgroups of G are abelian.

10.9 A weird error group

Let G be the finite group generated by a, b, c subject to the relations

$$a^4 = b^2 = (ab)^2 = 1 \quad \text{and} \quad c^2 = [a, c] = [b, c] = 1.$$

This is the index group that we have introduced in Example 10.3.

Let E be the group generated by a, b, c, d subject to the relations

$$a^4 d = b^2 = (ab)^2 = 1, \qquad c^2 = [a, c]d = [b, c] = 1,$$

and

$$d^2 = [a, d] = [b, d] = [c, d] = 1.$$

This is a group of order 32. The construction ensured that the center of E is generated by d and that the factor group $E/Z(E)$ is isomorphic to G. A faithful irreducible representation ρ of the group E is given by

$$\rho(a) = \begin{pmatrix} \omega & . & . & . \\ . & \omega^7 & . & . \\ . & . & \omega^5 & . \\ . & . & . & \omega^3 \end{pmatrix} \rho(b) = \begin{pmatrix} . & 1 & . & . \\ 1 & . & . & . \\ . & . & . & 1 \\ . & . & 1 & . \end{pmatrix} \rho(c) = \begin{pmatrix} . & . & 1 & . \\ . & . & . & 1 \\ 1 & . & . & . \\ . & 1 & . & . \end{pmatrix}$$

where . and ω are abbreviations for 0 and $\exp(2\pi i/8)$, respectively. Notice that $\rho(d) = -\mathbf{1}$ is a consequence of the relation $a^4 = d$.

Denote by N the normal subgroup in E generated by ab and ac, a dihedral group of order 16. Let χ be an irreducible character of N of degree 2 with $\chi(d) = -2$. There exist two such characters and both are constituents of the restriction of the character $\varphi(x) = \operatorname{tr} \rho(x)$ to the normal subgroup N. One choice yields the orthogonal projection matrix

$$P = \frac{\chi(1)}{|N|} \sum_{n \in N} \chi(n^{-1})\rho(n) = \frac{1}{2} \begin{pmatrix} 1 & . & i & . \\ . & 1 & . & -i \\ -i & . & 1 & . \\ . & i & . & 1 \end{pmatrix}.$$

The image of this projector yields a 2-dimensional Clifford code $Q = \operatorname{im}(P)$. The stabilizer of this code Q is by definition the set

$$S = \{g \in E \mid \exists s_g \in \mathbf{C} \text{ such that } \rho(g)v = s_g v \text{ for all } v \in \operatorname{im}(P)\}.$$

It is not difficult to check that S is given by the center $\langle d \rangle$ of E. The joint eigenspace (containing Q) of S is the full four-dimensional space $\mathbb{C}^4$, which shows that Q is not a stabilizer code. In fact, the code Q is the smallest example of a Clifford code that is not a stabilizer code.

10.10   Conclusions

Clifford codes are highly structured and have many interesting properties. We have demonstrated here for the first time that the concept of Clifford codes goes truly beyond the stabilizer code concept. We discussed the concept of nice error bases and abstract error groups, following the seminal work of Knill. This allowed us to obtain a more flexible definition of stabilizer codes. We have shed some light on the relation between Clifford codes and the class of stabilizer codes, the hitherto most popular code construction. There are many interesting open problems concerning the constructive aspects of the theory developed in this chapter.

Acknowledgments. This chapter has been completed during the European workshop on Quantum Computer Theory, Villa Gualino, Torino, June 2001. We thank Mario Rasetti and Paolo Zanardi of the Institute for Scientific Interchange Foundation for their kind hospitality. A.K. thanks the Santa Fe Institute for support through their Fellow-at-Large program. M.R. has been supported by the European Community under contract IST-1999-10596 (Q-ACTA).

References

[1] D. Aharonov and M. Ben-Or. Fault-tolerant quantum computation with constant error. In *Proc. of the 29th Annual ACM Symposium on Theory of Computation (STOC)*, pages 176–188, New York, 1997. ACM.

[2] A.E. Ashikhmin, A.M. Barg, E. Knill, and S.N. Litsyn. Quantum error detection I: Statement of the problem. *IEEE Trans. on Information Theory*, 46(3):778–788, 2000.

[3] A.E. Ashikhmin, A.M. Barg, E. Knill, and S.N. Litsyn. Quantum error detection II: Bounds. *IEEE Trans. on Information Theory*, 46(3):789–800, 2000.

[4] C.H. Bennett, D.P. DiVincenzo, J.A. Smolin, and W.K. Wootters. Mixed state entanglement and quantum error correction. *Phys. Rev. A*, 54:3824–3851, 1996.

[5] A.R. Calderbank, E.M. Rains, P.W. Shor, and N.J.A. Sloane. Quantum error correction and orthogonal geometry. *Phys. Rev. Lett.*, 78:405–408, 1997.

[6] A.R. Calderbank, E.M. Rains, P.W. Shor, and N.J.A. Sloane. Quantum error correction via codes over GF(4). *IEEE Trans. Inform. Theory*, 44:1369–1387, 1998.

[7] A.R. Calderbank and P. Shor. Good quantum error-correcting codes exist. *Phys. Rev. A*, 54:1098–1105, 1996.

[8] R. Cleve and D. Gottesman. Efficient computations of encodings for quantum error correction. *Phys. Rev. A*, 56(1):76–82, 1997.

[9] A.H. Clifford. Representations induced in an invariant subgroup. *Ann. Math.*, 38(2):533–550, 1937.

[10] D. Gottesman. Class of quantum error-correcting codes saturating the quantum Hamming bound. *Phys. Rev. A*, 54:1862–1868, 1996.

[11] M. Grassl. Algorithmic aspects of quantum error-correcting codes. In Chapter 9 of this volume.

[12] M. Grassl and Th. Beth. Cyclic quantum error-correcting codes and quantum shift registers. *Proc. Royal Soc. London Series A*, 456(2003):2689–2706, 2000.

[13] I.M. Isaacs. *Character Theory of Finite Groups*. Academic Press, New York, 1976.

[14] A. Klappenecker and M. Rötteler. Beyond stabilizer codes I: Nice error bases. Eprint `quant-ph/0010082`, 2000.

[15] A. Klappenecker and M. Rötteler. Beyond stabilizer codes II: Clifford codes. Eprint `quant-ph/0010076`, 2000.

[16] E. Knill. Group representations, error bases and quantum codes. Los Alamos National Laboratory Report LAUR-96-2807, 1996.

[17] E. Knill and R. Laflamme. A theory of quantum error–correcting codes. *Phys. Rev. A*, 55(2):900–911, 1997.

[18] E. Knill, R. Laflamme, and L. Viola. Theory of quantum error correction for general noise. *Phys. Rev. Lett.*, 84(11):2525–2528, 2000.

[19] E.M. Rains. Nonbinary quantum codes. *IEEE Trans. Inform. Theory*, 45:1827–1832, 1999.

[20] J.S. Rose. *A Course on Group Theory*. Dover, New York, 1994.

[21] P. Shor. Scheme for reducing decoherence in quantum memory. *Phys. Rev. A*, 52:2493–2496, 1995.

[22] P. Shor. Fault-tolerant quantum computation. In *Proceedings of the 37th Symposium on the Foundations of Computer Science*, Los Alamitos, 1996. IEEE Computer Society Press.

[23] A.M. Steane. Multiple-particle interference and quantum error correction. *Proc. Roy. Soc. London A*, 452:2551–2577, 1996.

[24] A.M. Steane. Simple quantum error correcting codes. *Phys. Rev. Lett.*, 77:793–797, 1996.

[25] P. Zanardi. Stabilizing quantum information. *Phys. Rev. A*, 63(1):012301, 2001.

Quantum Computing Algebraic and Geometric Structures

Chapter 11

Invariant polynomial functions on k qudits

Jean-Luc Brylinski* and Ranee Brylinski

Abstract   We study the polynomial functions on tensor states in $(\mathbb{C}^n)^{\otimes k}$ which are invariant under $SU(n)^k$. We describe the space of invariant polynomials in terms of symmetric group representations. For k even, the smallest degree for invariant polynomials is n and in degree n we find a natural generalization of the determinant. For n, d fixed, we describe the asymptotic behavior of the dimension of the space of degree d invariants as $k \to \infty$. We study in detail the space of homogeneous degree 4 invariant polynomial functions on $(\mathbb{C}^2)^{\otimes k}$.

11.1   Introduction

In quantum mechanics, a combination of states in Hilbert spaces $H_1,.., H_k$ leads to a state in the tensor product Hilbert space $H_1 \otimes \cdots \otimes H_k$. Such a state will be called here a tensor state. In this paper we take $H_1 = \cdots = H_k = \mathbb{C}^n$ where $n > 1$. Then a tensor state is a joint

*Research supported in part by NSF Grant No. DMS-9803593.

1-58488-282/4/02/$0.00+$1.50

state of k qudits. It would be very interesting to classify tensor states in $(\mathbb{C}^n)^{\otimes k}$ up to the action of the product $U(n)^k$ of unitary groups of local symmetries. A natural approach to this is to study the algebra of invariant polynomials. This approach was developed by Rains [9], by Grassl, Rötteler and Beth [4, 5], by Linden and Popescu [6] and by Coffman, Kundu and Wootters [2]. These authors study the ring of invariant polynomials in the components of a tensor state in $(\mathbb{C}^n)^{\otimes k}$ and in their complex-conjugates. For k qudits, explicit descriptions of invariants are given in [4, 5, 6] and in [2]. In the case of 3 qudits, a complete description of the algebraic structure of the invariants is derived in [8].

In this paper the symmetry group we consider is the product $G = SU(n)^k$ of special unitary groups; one thinks of G as the special group of local symmetries. We study the G-invariant polynomial functions Q on the tensor states in $(\mathbb{C}^n)^{\otimes k}$ (we discuss in Section 11.2 how this is relevant to the description of the G-orbits). We consider polynomials in the entries of a tensor state, in other words, holomorphic polynomials.

Let $\mathcal{R}_{n,k,d}$ be the space of homogeneous degree d polynomial functions on tensor states in $(\mathbb{C}^n)^{\otimes k}$. Let $\mathcal{R}_{n,k,d}^G$ be the space of G-invariants in $\mathcal{R}_{n,k,d}$. See Section 11.2 for more discussion. We reduce the problem of computing $\mathcal{R}_{n,k,d}^G$ to a problem in the invariant theory of the symmetric group $\mathfrak{S}_d$ (Proposition 11.1). In particular, $\mathcal{R}_{n,k,d}^G$ is non zero only if d is a multiple of n. So the "first" case is $d = n$; we examine this in Section 11.3. We find that if k is odd then $\mathcal{R}_{n,k,n}^G = 0$ while if k is even then $\mathcal{R}_{n,k,n}^G$ is 1-dimensional. In the latter case we write down (Section 11.3) explicitly the corresponding invariant polynomial $P_{n,k}$ in $\mathcal{R}_{n,k,n}$; we find $P_{n,k}$ is a natural generalization of the determinant of a square matrix.

For fixed n, d the direct sum $\oplus_k \mathcal{R}_{n,k,d}$ is an associative algebra. We study the asymptotic behavior of $\dim \mathcal{R}_{n,k,d}^G$ as $k \to \infty$ in Section 11.4. In Section 11.5, we specialize to the case of k-qudits, i.e., $n = 2$. We compute the dimension of the space $\mathcal{R}_{2,k,4}^G$ of degree 4 invariants as well as the dimension of the space of invariants in $\mathcal{R}_{2,k,4}^G$ under the natural action of $\mathfrak{S}_k$. We show that $\oplus_k \mathcal{R}_{2,k,4}^{\mathfrak{S}_k \ltimes G}$ is a polynomial algebra on 2 generators. For $k \leq 5$ we describe the representation of $\mathfrak{S}_k$ on $\mathcal{R}_{2,k,4}^G$. For $k = 4$ we find some interesting relations with the results on classification of tensor states in $(\mathbb{C}^2)^{\otimes 4}$ given in [3].

We thank Markus Grassl for his useful comments on the first version of this paper.

Part of this work was carried out while both authors were Professeurs Invités at the CPT and IML of the Université de la Méditerranée in

Marseille, France. They are grateful to the Université de la Méditerranée for its hospitality.

11.2 Polynomial invariants of tensor states

We will consider $(\mathbb{C}^n)^{\otimes k}$ as a space of contravariant tensor states u. Then (once we fix a basis of $\mathbb{C}^n$) u is given by n^k components $u^{p_1 p_2 \cdots p_k}$. We consider the algebra $\mathcal{R}_{n,k}$ of polynomial functions on $(\mathbb{C}^n)^{\otimes k}$. So $\mathcal{R}_{n,k}$ is the polynomial algebra $\mathbb{C}[x_{p_1 p_2 \cdots p_k}]$ in the n^k coordinate functions $x_{p_1 p_2 \cdots p_k}$. We have a natural algebra grading $\mathcal{R}_{n,k} = \oplus_{d=0}^{\infty} \mathcal{R}_{n,k,d}$ where $\mathcal{R}_{n,k,d}$ is the space of homogeneous degree d polynomial functions.

A function in $\mathcal{R}_{n,k,d}$ amounts to a symmetric degree d covariant tensor Q in $(\mathbb{C}^n)^{\otimes k}$. So Q has n^{dk} components $Q_{i_{11} \cdots i_{dk}}$ where we think of the indices i_{ab} as being arranged in a rectangular array of d rows and k columns and $Q_{i_{11} \cdots i_{dk}}$ is invariant under permutations of the rows of the array. Then Q defines the function

$$u \mapsto Q_{i_{11} \cdots i_{dk}} u^{i_{11} i_{12} \cdots i_{1k}} u^{i_{21} i_{22} \cdots i_{2k}} \cdots u^{i_{d1} i_{d2} \cdots i_{dk}} \qquad (11.1)$$

where we used the usual Einstein summation convention. In this way, $\mathcal{R}_{n,k}$ identifies with $S^d((\mathbb{C}^n)^{\otimes k})$.

Now the group $G = SU(n)^k$ acts on our tensor states u and tensors Q as follows. Let the matrix g_{ij} live in the m-th copy of $SU(n)$ and let g^{ij} be the inverse matrix. Then g_{ij} transforms $u^{p_1 p_2 \cdots p_k}$ into $g_{p_m q_m} u^{q_1 q_2 \cdots q_k}$ and $Q_{i_{11} \cdots i_{dk}}$ into $Q_{j_{11} \cdots j_{dk}} g^{j_{1m} i_{1m}} g^{j_{2m} i_{2m}} \cdots g^{j_{dm} i_{dm}}$. The identification of $\mathcal{R}_{n,k,d}$ with $S^d((\mathbb{C}^n)^{\otimes k})$ is G-equivariant.

We are interested in the algebra $\mathcal{R}_{n,k}^G = \oplus_{d=0}^{\infty} \mathcal{R}_{n,k,d}^G$ of G-invariants. We view this as a first step towards studying the orbits of G on $(\mathbb{C}^n)^{\otimes k}$. One can first study the orbits of the complex group $G_{\mathbb{C}} = SL(n, \mathbb{C})^k$ and then decompose the $G_{\mathbb{C}}$-orbits under the G-action. Note that a polynomial is G-invariant if and only if it is $G_{\mathbb{C}}$-invariant. The closed $G_{\mathbb{C}}$ orbits play a special role: they are the most degenerate orbits. Given any orbit Y, its closure contains a unique closed orbit Z; then points in Y degenerate to points in Z. The $G_{\mathbb{C}}$-invariant functions separate the closed orbits; they take the same values on Y and on Z. The set of closed orbits of $G_{\mathbb{C}}$ in $(\mathbb{C}^n)^{\otimes k}$ has the structure of an affine complex algebraic variety with $\mathcal{R}_{n,k}^G$ as its algebra of regular functions. Thus a complete description of $\mathcal{R}^G_{n,k}$ would lead to a precise knowledge of the closed $G_{\mathbb{C}}$-orbits.

Our approach is thus somewhat different from that of [2, 4, 5, 6, 7, 8, 9] who study the invariant functions on $(\mathbb{C}^n)^{\otimes k}$ which are polynomials in the $x_{p_1 \cdots p_k}$ and in their complex conjugates; these can also be described as the invariant polynomial functions on $(\mathbb{C}^n)^{\otimes k} \oplus \overline{(\mathbb{C}^n)^{\otimes k}}$.

At this point it is useful to examine the case $k = 2$. We can identify $(\mathbb{C}^n)^{\otimes 2}$ with the space $M_n(\mathbb{C})$ of square matrices and then $G = SU(n)^2$ acts on $M_n(\mathbb{C})$ by $(g, h) \cdot u = guh^{-1}$. So $\mathcal{R}^G_{n,k,d}$ is the space of homogeneous degree d polynomial functions Q of an n by n matrix u which are bi-$SL(n, \mathbb{C})$-invariant, i.e., $Q(guh^{-1}) = Q(u)$ for $g, h \in SL(n, \mathbb{C})$. Then Q is, up to scaling, the rth power of the determinant D for some r. Hence $d = rn$. It follows that $\mathcal{R}^G_{n,2}$ is the polynomial algebra $\mathbb{C}[D]$. Thus the space of closed orbits for $SL(n, \mathbb{C})^2$ identifies with $\mathbb{C}$, where λ corresponds to the unique closed orbit Z_λ inside the set X_λ of matrices of determinant λ. For $\lambda \neq 0$, $Z_\lambda = X_\lambda$ while for $\lambda = 0$, Z_0 reduces to the zero matrix.

We view $S^d((\mathbb{C}^n)^{\otimes k})$ as the space of invariants for the symmetric group $\mathfrak{S}_d$ acting on $((\mathbb{C}^n)^{\otimes k})^{\otimes d}$. So

$$\mathcal{R}^G_{n,k,d} = (((\mathbb{C}^n)^{\otimes k})^{\otimes d})^{G \times \mathfrak{S}_d} = (((\mathbb{C}^n)^{\otimes d})^{\otimes k})^{G \times \mathfrak{S}_d} \qquad (11.2)$$

Recall the Schur decomposition $(\mathbb{C}^n)^{\otimes d} = \oplus_\alpha S^\alpha(\mathbb{C}^n) \otimes E_\alpha$ where α ranges over partitions of d with at most n rows, $S^\alpha(\mathbb{C}^n)$ is the irreducible covariant representation of $SU(n)$ given by the Schur functor S^α, and E_α is the corresponding irreducible representation of $\mathfrak{S}_d$. We use the convention that E_α is the trivial representation if $\alpha = [d]$, while E_α is the sign representation if $\alpha = [1^d]$. Thus we have

$$((\mathbb{C}^n)^{\otimes d})^{\otimes k} = \sum_{|\alpha_1| = \cdots = |\alpha_k| = d} S^{\alpha_1}(\mathbb{C}^n) \otimes \cdots \otimes S^{\alpha_k}(\mathbb{C}^n) \otimes E_{\alpha_1} \otimes \cdots \otimes E_{\alpha_k}$$

$$(11.3)$$

Now taking the invariants under $G \times \mathfrak{S}_d$ we get

$$\mathcal{R}^G_{n,k,d} = \sum_{|\alpha_1| = \cdots = |\alpha_k| = d} S^{\alpha_1}(\mathbb{C}^n)^{SU(n)} \otimes \cdots \otimes$$

$$S^{\alpha_k}(\mathbb{C}^n)^{SU(n)} \otimes (E_{\alpha_1} \otimes \cdots \otimes E_{\alpha_k})^{\mathfrak{S}_d} \qquad (11.4)$$

The representation $S^{\alpha_j}(\mathbb{C}^n)$, since it is irreducible, has no $SU(n)$-invariants except if $S^{\alpha_j}(\mathbb{C}^n) = \mathbb{C}$ is trivial. This happens if and only if α_j is a rectangular partition with all columns of length n. This proves:

PROPOSITION 11.1
If n does not divide d, then $\mathcal{R}^G_{n,k,d} = 0$. If $d = nr$, then $\mathcal{R}^G_{n,k,d}$ is isomorphic to $(E_\pi^{\otimes k})^{\mathfrak{S}_d}$ where $\pi = [r^n]$.

The permutation action of $\mathfrak{S}_k$ on $(\mathbb{C}^n)^{\otimes k}$ induces an action of $\mathfrak{S}_k$ on $\mathcal{R}^G_{n,k,d}$.

COROLLARY 11.1
The isomorphism of Proposition 11.1 intertwines the $\mathfrak{S}_k$-action on $\mathcal{R}^G_{n,k,d}$ with the action of $\mathfrak{S}_k$ on $(E_\pi^{\otimes k})^{\mathfrak{S}_d}$ given by permuting the k factors E_π.

11.3 The generalized determinant function

Given n and k, we want to find the smallest positive value of d such that $\mathcal{R}^G_{n,k,d} \neq 0$. By Proposition 11.1, the first candidate is $d = n$.

COROLLARY 11.2
$\mathcal{R}^G_{n,k,n} \neq 0$ iff k is even. In that case, $\mathcal{R}^G_{n,k,n}$ is one-dimensional and consists of the multiples of the function $P_{n,k}$ given by

$$P_{n,k}(u) = \sum_{\sigma_2,\cdots,\sigma_k \in \mathfrak{S}_n} \epsilon(\sigma_2) \cdots \epsilon(\sigma_k) \prod_{h=1}^n u^{h h_{\sigma_2} \cdots h_{\sigma_k}} \qquad (11.5)$$

where $h_{\sigma_j} = \sigma_j(h)$.

PROOF   By Proposition 11.1, we need to compute $(E_\pi^{\otimes k})^{\mathfrak{S}_d}$. For $d = n$, $\pi = [1^n]$ and so E_π is the sign representation of $\mathfrak{S}_n$. Then $(E_\pi^{\otimes k})$ is one-dimensional and carries the trivial representation if k is even, or the sign representation if k is odd.

Now for k even, we can easily compute a non zero function $P = P_{n,k}$ in $\mathcal{R}_{n,k,n}$. For $S^\pi(\mathbb{C}^n)$ is the top exterior power $\wedge^n \mathbb{C}^n$. Thus P is a non zero element of the one-dimensional subspace $(\wedge^n \mathbb{C}^n)^{\otimes k}$ of $((\mathbb{C}^n)^{\otimes n})^{\otimes k}$. The tensor components of P are then given by $P_{i_{11} \cdots i_{nk}} = \frac{1}{n!}\epsilon(\sigma_1) \cdots \epsilon(\sigma_k)$ if for each j, the column $i_{1j}, \cdots, i_{nj}$ is a permutation σ_j of $1, \cdots, n$ and 0 otherwise. Then we get

$$P_{n,k}(u) = \frac{1}{n!} \sum_{\sigma_1,\cdots,\sigma_k \in \mathfrak{S}_n} \epsilon(\sigma_1) \cdots \epsilon(\sigma_k) \prod_{h=1}^n u^{h_{\sigma_1} \cdots h_{\sigma_k}} \qquad (11.6)$$

where $h_{\sigma_i} = \sigma_i(h)$. The expression is very redundant, as each term appears $n!$ times. We remedy this by restricting the first permutation σ_1 to be 1. This gives (11.5).   ∎

$P_{n,k}$ is a *generalized determinant*; $P_{n,k}$ is invariant under the $\mathfrak{S}_k$-action. For $k = 2$, (11.5) reduces to the usual formula for the matrix determinant.

Recall that the rank s of a tensor state u in $(\mathbb{C}^n)^{\otimes k}$ is the smallest integer s such that u can be written as $u = v_1 + v_2 + \cdots + v_s$, where the v_i are decomposable tensor states $v_i = w_{i1} \otimes w_{i2} \otimes \cdots \otimes w_{ik}$. There is a relation between the rank and the vanishing of $P_{n,k}$ as follows:

COROLLARY 11.3
If the tensor state u in $(\mathbb{C}^n)^{\otimes k}$ has rank less than n, then $P_{n,k}(u) = 0$.

It is easy to find a tensor state u of rank n such that $P_{n,k}(u)$ is non zero. For instance, $P_{n,k}(u) = 1$ if u has all components zero except $u^{1 \cdots 1} = \cdots = u^{n \cdots n} = 1$. For $k = 2$, $P_{n,k}(u) = 0$ implies u has rank less than n. For bigger (even) k, this is false, if n is large enough. This happens essentially because the rank of u can be very large (at least $\frac{n^k}{kn-k+1}$). Thus $P_{n,k}$ gives only partial information about the rank.

11.4   Asymptotics as $k \to \infty$

Suppose we fix n and d where $d = rn$. Then there is a G-invariant associative graded algebra structure $P \circ Q$ on the direct sum $\oplus_k \mathcal{R}_{n,k,d}^G$. Indeed, the product of tensors induces a $(G \times \mathfrak{S}_d)$-invariant map $V^{\otimes k} \otimes V^{\otimes l} \to V^{\otimes(k+l)}$ where $V = (\mathbb{C}^n)^{\otimes d}$. The induced multiplication on the spaces of $(G \times \mathfrak{S}_d)$-invariants gives the product on $\oplus_k \mathcal{R}_{n,k,d}^G$, where we use the identification in (11.2). This multiplication corresponds, under the isomorphism of Proposition 11.1, to the product map $E_\pi^{\otimes k} \otimes E_\pi^{\otimes l} \to E_\pi^{\otimes(k+l)}$. This structure is very useful. For instance, if $d = n$, then $P_{n,k} \circ P_{n,l} = \frac{1}{n!} P_{n,k+l}$. Thus the determinant $P_{n,2}$ determines $P_{n,2m}$ in that the m-fold product $P_{n,2} \circ \cdots \circ P_{n,2}$ is equal to $(n!)^{-m+1} P_{n,2m}$.

We will study the size of the algebra $\oplus_k \mathcal{R}_{n,k,d}^G$ by finding an asymptotic formula for the dimension of $\mathcal{R}_{n,k,d}^G$. We do this for $r \geq 2$. Indeed for $r = 1$ we already know $\dim \mathcal{R}^G_{n,k,n}$ is 1 if k is even or 0 if k is odd; we call this the *static* case. The asymptotics involve the number

$$p = \dim E_\pi = d! \ \prod_{m=0}^{n-1} \frac{m!}{(m+r)!} \tag{11.7}$$

where $\pi = [r^n]$ as in Proposition 11.1. Our formula for p is immediate from the hook formula for the dimension of an irreducible symmetric group representation.

PROPOSITION 11.2

Assume $d = rn$ *with* $r \geq 2$. *Then* $\dim \mathcal{R}^G_{n,k,d} \sim c\dfrac{p^k}{d!}$ *as* $k \to \infty$, *where* $c = 1$ *with one exception:* $c = 4$ *if* $n = 2, d = 4$.

PROOF   Let $s = \dim \mathcal{R}^G_{n,k,d} = \dim (E_\pi^{\otimes k})^{\mathfrak{S}_d}$. Then $s = \frac{1}{d!} \sum_{\sigma \in \mathfrak{S}_d}$ $\chi(\sigma)^k$ where $\chi : \mathfrak{S}_d \to \mathbb{Z}$ is the character of E_π. If σ acts trivially on E_π, then $\chi(\sigma) = p$. If σ acts non trivially, we claim $|\chi(\sigma)| < p$. To show this, it suffices to show that σ has at least two distinct eigenvalues on E_π; this is because $\chi(\sigma)$ is the sum of the p eigenvalues of σ. Now the set $\mathfrak{T}_d$ of $\sigma \in \mathfrak{S}_d$ which act on E_π by a scalar is a normal subgroup of $\mathfrak{S}_d$. So if $d \geq 5$, then $\mathfrak{T}_d$ is $\{1\}$, the alternating group $\mathfrak{A}_d$ or $\mathfrak{S}_d$. We can easily rule out the latter two possibilities, so $\mathfrak{T}_d = \{1\}$, which proves our claim. If $d \leq 4$, then (since $r > 1$ and $n > 1$), we have $d = 4$, $n = 2$ and $\pi = [2, 2]$. Our claim is clear here since $\mathfrak{S}_4$ acts on E_π through the reflection representation of $\mathfrak{S}_3$ on $\mathbb{C}^2$. Therefore we have $s = c\frac{p^k}{d!} + o(p^k)$ as $k \to \infty$ where c is cardinality of the kernel of $\mathfrak{S}_d \to Aut\,E_\pi$. Our work in the previous paragraph computes c. ∎

Proposition 11.2 implies that the algebra $\oplus_k \mathcal{R}^G_{n,k,d}$ is far from commutative, as it has roughly $1/N$ times the size of the tensor algebra $\oplus_k (\mathbb{C}^p)^{\otimes k}$. We note however that the $\mathfrak{S}_k$-invariants in $\oplus_k \mathcal{R}^G_{n,k,d}$ form a commutative subalgebra, isomorphic to $S(E_\pi)^{\mathfrak{S}_d}$.

11.5   Quartic invariants of k qudits

The case $n = 2$ is of particular interest, as here the qudits are qudits, and this is the case being most discussed in quantum computation. Here we can give some precise non asymptotic results for the first non static case, namely $\mathcal{R}_{2,k,4}^G$. We put $E = E_\pi = E_{[2,2]}$. The proof of Proposition 11.2 easily gives

COROLLARY 11.4
We have $\dim \mathcal{R}_{2,k,4}^G = \frac{1}{3}(2^{k-1} + (-1)^k)$.

The first few values of $\dim \mathcal{R}_{2,k,4}^G$, starting at $k = 1$, are 0, 1, 1, 3, 5, 11, 21, 43. For $k = 2$ and $k = 3$ the unique (up to scalar) invariants are, respectively, the squared determinant $P_{2,2}^2$ and the Cayley hyper-determinant $H_{2,3}$ (see [3]). We note that the hyperdeterminant is very closely related to the residual tangle of three entangled qudits discussd in [2].

It would be useful to study $\mathcal{R}_{2,k,4}^G$ as a representation of $\mathfrak{S}_k$, where $\mathfrak{S}_k$ acts by permuting the k qudits. The $\mathfrak{S}_k$-invariants in $\mathcal{R}_{2,k,4}^G$ are the $(\mathfrak{S}_k \ltimes G)$-invariants in $\mathcal{R}_{2,k,4}$. These $(\mathfrak{S}_k \ltimes G)$-invariant polynomials are very significant as they separate the closed orbits of the extended symmetry group $\mathfrak{S}_k \ltimes SL(2, \mathbb{C})^k$ acting on $(\mathbb{C}^2)^{\otimes k}$. We can compute the dimension of the $\mathfrak{S}_k$-invariants as follows:

PROPOSITION 11.3
The dimension of the space of $\mathfrak{S}_k \ltimes G$-*invariants in* $\mathcal{R}_{2,k,4}$ *is* $M_k = \left[\frac{k}{6}\right] + r_k$ *where* $r_k = 0$ *if* $k \equiv 1 \bmod 6$, *or* $r_k = 1$ *otherwise. Furthermore the algebra* $\oplus_k \mathcal{R}_{2,k,4}^{\mathfrak{S}_k \ltimes G}$ *is the polynomial algebra* $\mathbb{C}[P_{2,2}^2, H_{2,3}]$.

PROOF   We have isomorphisms $\mathcal{R}_{2,k,4}^{\mathfrak{S}_k \ltimes G} \simeq (E^{\otimes k})^{\mathfrak{S}_k \times \mathfrak{S}_3} \simeq S^k(E)^{\mathfrak{S}_3}$ since the representation of $\mathfrak{S}_4$ on E factors through $\mathfrak{S}_3$. Thus the algebra $\oplus_k \mathcal{R}_{2,k,4}^{\mathfrak{S}_k \ltimes G}$ identifies with $S(E)^{\mathfrak{S}_3}$. Now $S(E)^{\mathfrak{S}_3}$ is the algebra of $\mathfrak{S}_3$-invariant polynomial functions on traceless 3×3 diagonal matrices, and so is a polynomial algebra on the functions $A \mapsto Tr(A^2)$ and $A \mapsto Tr(A^3)$. These invariants correspond (up to scaling) to $P_{2,2}^2$ and $H_{2,3}$. The formula for the dimension follows easily.   ∎

For instance, we have: $M_1 = 0$, $M_k = 1$ for $2 \leq k \leq 5$, and $M_6 = 2$. We remark that by replacing $S(E)^{\mathfrak{S}_3}$ by $\wedge(E)^{\mathfrak{S}_3}$, it is easy to prove that the sign representation of $\mathfrak{S}_k$ does not occur in $(E^{\otimes k})^{\mathfrak{S}_4}$ for any $k \geq 2$.

We can determine the $\mathfrak{S}_k$-representation on $\mathcal{R}_{2,k,4}^G$ for small k by explicit trace computations. For $k = 2$ and $k = 3$ we have the trivial

1-dimensional representation. For $k = 4$, we find $\mathcal{R}_{2,4,4}^G$ is the direct sum $E_{[4]} \oplus E_{[2,2]}$. The trivial representation $E_{[4]}$ of $\mathfrak{S}_4$ is spanned by $P_{2,4}^2$, while the 2-dimensional representation $E = E_{[2,2]}$ is spanned by the determinants $\Delta(ijkl)$ introduced in [3]. Here $(ijkl)$ is a permutation of (1234). Given a tensor state $u \in (\mathbb{C}^2)^{\otimes 4}$, we can view it as an element v of $\mathbb{C}^4 \otimes \mathbb{C}^4$, where the first (resp. second) $\mathbb{C}^4$ is the tensor product of the i-th and j-th copies of $\mathbb{C}^2$ (resp. of the k-th and l-th copies). Then $\Delta(ijkl)(u)$ is the determinant of v. As shown in [3], the $\Delta(ijkl)$ span the representation E of $\mathfrak{S}_4$. The significance of the $\Delta(ijkl)$ is that their vanishing describes the closure of the set of tensor states in $(\mathbb{C}^2)^{\otimes 4}$ of rank ≤ 3. For $k = 5$ the representation $\mathcal{R}_{2,5,4}^G$ of $\mathfrak{S}_5$ is $E_{[5]} \oplus E_{[2,1,1,1]}$.

References

[1] J-L. Brylinski, Algebraic measures of entanglement, In Chapter 1 of this book.

[2] V. Coffman, J. Kundu and W. K. Wootters, Distributed entanglement, preprint `quant-ph/9907047`.

[3] I.M. Gelfand, M. Kapranov and A. Zelevinsky, *Discriminants, Resultants and Multidimensional Determinants*, Birkhäuser, Boston (1991).

[4] M. Grassl, M. Rötteler and T. Beth, Computing local invariants of quantum-bit systems, *Phys. Rev.* A, 58 no. 3 (1998), 1833–1839; also on the Arxiv as `quant-ph/9712040`.

[5] M. Grassl, M. Rötteler and T. Beth, Description of multiparticle entanglement through polynomial invariants, Talk of M. Grassl at the Isaac Newton Institute for Mathematical Sciences in July 1999, available on the web as `http://iaks-www.ira.uka.de/home/grassl/publications.html`.

[6] N. Linden and S. Popescu, On multi-particle entanglement, *Forts. der Physik*, 46 (1998), no. 4–5, 567–578, also on the Arxiv as `quant-ph/9711016`.

[7] D. Meyer and N. Wallach, Global entanglement in multiparticle systems, preprint `quant-ph/0108104`.

[8] D. Meyer and N. Wallach, Invariants for multiple qudits: the case of 3 qudits. In Chapter 3 of this book.

[9] E. Rains, Polynomial invariants of quantum codes, *IEEE Trans. on Information Th.*, 46 no. 1 (2000), 54–59.

Chapter 12

Z_2-systolic freedom and quantum codes

Michael H. Freedman, David A. Meyer, and Feng Luo

Abstract   A closely coupled pair of conjectures/questions—one in differential geometry (by M. Gromov), the other in quantum information theory—are both answered in the negative. The answer derives from a certain metrical flexibility of manifolds and a corresponding improvement to the theoretical efficiency of existing local quantum codes. We exhibit this effect by constructing a family of metrics on $S^2 \times S^1$, and other three and four dimensional manifolds. Quantitatively, the explicit "freedom" exhibited is too weak (a $\log^{1/2}$ factor in the natural scaling) to yield practical codes but we cannot rule out the possibility of other families of geometries with more dramatic freedom.

12.0   Preliminaries and statement of results

We define the *p-systole* of a closed Riemannian manifold M to be:

$$\inf_{\alpha \neq 0} p\text{-area}(\alpha)$$

1-58488-282/4/02/$0.00+$1.50
© 2002 by Chapman & Hall/CRC

where α is a smooth oriented p-cycle whose class $[\alpha] \neq 0 \in H_p(M; \mathbb{Z})$. Similarly, we define:

$$\mathbb{Z}_2 - p\text{-systole}(M) = \inf_{\alpha \neq 0} p\text{-area}(\alpha)$$

where α is a smooth unoriented cycle whose class $[\alpha] \neq 0 \in H_p(M; \mathbb{Z}_2)$. In dimension 2, for surfaces different from the 2-sphere, it is known* that

$$(1\text{-systole})^2 \leq \frac{\pi}{2}\,(2\text{-systole}) = \frac{\pi}{2}\text{area}$$

(equality holds for the round projective plane). Moreover, the same inequality holds for $\mathbb{Z}_2$-systoles.

In dimension 3 and greater, oriented and unoriented systoles are quite distinct. The oriented theory has been very well developed (for example see [3, 4] and [14], and "systolic freedom" has been established in many cases.

DEFINITION 12.1 M^d *is (p, q)-free, $p + q = d$, if:*

$$\inf_g \frac{(d\text{-systole})}{(p\text{-systole})(q\text{-systole})} = 0$$

where g varies over Riemannian metrics on M. For example, it has been proven [3]) and ([16] that: (i) every $M^d, d \geq 3$ with $b_1(M) \geq 1$ is $(1, d-1)$-free, and (ii) every simply connected smooth 4-manifold is $(2, 2)$-free.

In contrast, it has been conjectured by Gromov (see [16]) that in all dimensions, when $\mathbb{Z}_2$-systoles are considered, rigidity as in Loewner's theorem rather than freedom should prevail. We construct a counterexample:

THEOREM 12.1
There exists a family of metrics on $S^2 \times S^1$ exhibiting $\mathbb{Z}_2$-$(1, 2)$-systolic freedom.

---

*M. Katz points out that this follows by combining Loewner's (unpublished) result for T^2, Berger's for $\mathbb{R}P^2$ [1], and Gromov's for other closed surfaces [14].

THEOREM 12.2

There exists a family of metrics on $S^2 \times S^2$ exhibiting $\mathbb{Z}_2$-$(2,2)$-systolic freedom.

REMARK 12.1   When examined quantitatively, we will see that our family only reaches freedom by a highly iterated log factor when measured in the natural scaling, whereas the standard examples in the $\mathbb{Z}$ case, due to Gromov, are much more robust, exhibiting a definite power or even an exponential [19]. We find a bit more freedom, scaling like $(\log)^{1/2}$ in "weak" families with variable topology (see Section 12.2). We do not know if there are families or even weak families with greater than logarithmic freedom in the $\mathbb{Z}_2$-setting. For application to practical local quantum codes, this is a critical question.   ∎

Quantum codes were invented in 1995 [21] as the beginning of a solution to the problem of protecting quantum information from errors induced by unwanted interactions of a quantum computer with its environment. Functionally, they are similar to classical codes which protect bit strings (elements of $\mathbb{Z}_2^k$) by encoding them as longer bit strings $\left((b_1, \ldots, b_n) \in \mathbb{Z}_2^n, n > k \right)$ in such a way that noise induced errors which flip some of the bits ($b_i \mapsto b_i + 1$) can be corrected. A *quantum bit* (*qubit*) is an element in two dimensional Hilbert space $\mathbb{C}^2 = \mathbb{C}^{\mathbb{Z}_2}$. An n-qubit code for k-qubits is a 2^k dimensional subspace of $(\mathbb{C}^2)^{\otimes n} = \mathbb{C}^{\mathbb{Z}_2^n}$. For a complete discussion of quantum error correcting codes see Gottesman's thesis [13]; here we will simply posit that a quantum code should allow recovery from a set of errors defined by Hermitian operators of the form $\sigma_{i_1} \otimes \cdots \otimes \sigma_{i_n}$, where $\sigma_{i_j} \in \{\mathrm{id}, \sigma_x, \sigma_y, \sigma_z\}$ and

$$\sigma_x = \begin{pmatrix} 0 & 1 \\ 1 & 0 \end{pmatrix}, \ \sigma_y = \begin{pmatrix} 0 & -i \\ i & 0 \end{pmatrix}, \ \sigma_z = \begin{pmatrix} 1 & 0 \\ 0 & -1 \end{pmatrix}$$

are the *Pauli matrices* acting on $\mathbb{C}^2$ in a fixed "computational" basis written $(|0\rangle, |1\rangle)$.

Classical linear codes encode k-bit strings into n-bit strings by a linear map $\mathbb{Z}_2^k \longrightarrow \mathbb{Z}_2^n$. The image of this map is the kernel of a linear map $P : \mathbb{Z}_2^n \to \mathbb{Z}_2^{n-k}$ with maximal rank. P is the *parity check matrix* for the code; each row of P is a *parity check* whose $\mathbb{Z}_2$-inner product with each valid *codeword* vanishes. The parity checks form an abelian group under addition, with rank $n - k$. An *error* is an element $e \in \mathbb{Z}_2^n$ with $e_i = 1$ at the bits which are flipped by the noise, so that a codeword $b \in \mathbb{Z}_2^n$ becomes $b + e$. Since $P(b + e) = Pe$, if Pe is different for each of some set of possible errors, each error can be identified by the results of the parity checks and then corrected—independently of the codeword.

The quantum analogue of a set of parity checks that define a classical code is a set of *stabilizer operators*. These are mutually commuting Hermitian operators, which form an abelian group under multiplication. They have the same form as the error operators; so each has eigenvalues ± 1. Just as classical codewords vanish under the action of parity checks, elements of a quantum code are fixed by each stabilizer.

We consider a special class of stabilizer codes: the CSS (Calderbank–Shor [8] and Steane [23]) codes. The group of stabilizers for a CSS code has a set of generators each of which involves only σ_x or only σ_z operators as its nonidentity tensor factors. As σ_x acts on a qubit to interchange the computational basis vectors, it can be described as a bit flip and there is a naturally corresponding classical parity check. Thus to a set of s_x bit flip (σ_x) stabilizers there corresponds a parity check matrix $P_x : \mathbb{Z}_2^n \longrightarrow \mathbb{Z}_2^{s_x}$ with entries set to 1 or 0 corresponding to the presence or absence of a σ_x factor. We can also construct a parity check matrix $P_z : \widehat{\mathbb{Z}}_2^n \to \mathbb{Z}_2^{s_z}$ corresponding to the s_z "phase flip" stabilizers, now with elements set to 1 corresponding to the presence of σ_z factors. Here $\frown$ denotes the Fourier dual $H : \mathbb{Z}_2 \to \widehat{\mathbb{Z}}_2$, where

$$H = \frac{1}{\sqrt{2}} \begin{pmatrix} 1 & 1 \\ 1 & -1 \end{pmatrix}$$

is the Hadamard, or 2-dimensional discrete Fourier, transform.

Each of the bit flip stabilizers commutes with the others, and similarly for the phase flip stabilizers. Since $\sigma_x \sigma_z = -\sigma_z \sigma_x$, a bit flip stabilizer commutes with a phase flip stabilizer when the set of qubits on which both act nontrivially is even. Thus the condition that all the stabilizers of a CSS code commute (and therefore have a nonempty fixed set) can be written $P_z P_x^T = 0$. That is, the sequence of maps

$$\mathbb{Z}_2^{s_x} \xrightarrow{P_x^T} \widehat{\mathbb{Z}}_2^n \xrightarrow{P_z} \mathbb{Z}_2^{s_z}$$

satisfies: $\mathrm{im} P_x^T \subset \ker P_z$. The elements of $\ker P_z$ label a basis for the subspace of $(\mathbb{C}^2)^{\otimes n}$ fixed by the σ_z stabilizers, while $\mathbb{Z}_2$-addition by the elements of $\mathrm{im} P_x^T$ defines the orbit of the action of the σ_x stabilizers. Averaging over the orbits gives the subspace simultaneously fixed by *all* the stabilizers. Thus the code subspace is identified with the $\mathbb{Z}_2$-homology $\ker P_z^{\mathrm{T}}/\mathrm{im} P_x^{\mathrm{T}}$ and is $2^{n-s_x-s_z}$ dimensional.*

It is physically natural to realize the topology of CSS codes in a geometrical setting in which there is a *locality* condition requiring each stabilizer to involve only a bounded number of σ_x or σ_z. We will derive such local codes from the Riemannian geometry of a closed smooth manifold M^d. The construction is a straightforward generalization of Kitaev's "toric codes" [12].

Let $\mathcal{C}$ be a piecewise smooth cellulation of M and let $\mathcal{C}^*$ be its dual cellulation. Let C_p denote the $\mathbb{Z}_2$-chain group of p-cells of $\mathcal{C}$. Then the Hilbert space of qubits associated to the p-cells is isomorphic to $(\mathbb{C}^2)^{\otimes n}$ where n is the number of p-cells, which is more functorially written as the functions on the chain group: $\mathbb{C}^{C_p}$. For each $(p-1)$-cell a and $(p+1)$-cell b we define stabilizers A_a and B_b acting on $\mathbb{C}^{C_p}$ with nonidentity tensor factors acting on the qubits associated to the p-cells, π, whose boundary contains a and which lie in the boundary of b, respectively:

$$A_a = \bigotimes_{a\in\partial\pi} \sigma_z^\pi \otimes \bigotimes_{a\notin\partial\pi} \mathrm{id}^\pi \quad \text{and} \quad B_b = \bigotimes_{\pi\in\partial b} \sigma_x^\pi \otimes \bigotimes_{\pi\notin\partial b} \mathrm{id}^\pi.$$

These stabilizers define a CSS code which is the $\mathbb{Z}_2$-homology of the manifold, $H_p = H_p(M; \mathbb{Z}_2)$. The common fixed set of the A operators, $\mathrm{fix}A := \bigcap_{(p-1)-\mathrm{cells}\ a} \mathrm{fix}\ A_a$, is naturally identified with $\mathbb{C}^{Z_p}$ where Z_p denotes the $\mathbb{Z}_2$-cycles in C_p. The operators B_b act on Z_p by addition of ∂b and therefore on $\mathrm{fix}A$ by the induced action on functions. The common fixed space $\mathrm{fix}A \cap \mathrm{fix}B$ (where $\mathrm{fix}B := \bigcap_{(p+1)-\mathrm{cells}\ b} \mathrm{fix}B_b$) under this identification is the space of boundary invariant functions from Z_p to $\mathbb{C}$, i.e., $\mathbb{C}^{H_p}$.

$\mathbb{C}^{H_p}$ is the protected space of a quantum code that protects $\mathrm{rank}(H_p)$ qubits inside n qubits against $\lfloor \frac{t}{2} \rfloor$ errors where t is the minimum of (1) the fewest number of p-cells in an essential $\mathbb{Z}_2$-p-cycle of $\mathcal{C}$, and (2) the fewest number of dual $(d-p)$-cells in an essential $\mathbb{Z}_2$-$(d-p)$-cycle of $\mathcal{C}^*$.

To understand this estimate, a straightforward generalization of the estimate for toric codes [12], we must discuss the error-recovery procedure. We have noted that $\mathrm{fix}A$ is the space of superpositions of cycles, $\mathbb{C}^{Z_p} \subset \mathbb{C}^{C_p} \equiv \mathbb{C}^2 \otimes \cdots \otimes \mathbb{C}^2$, one copy labeled by each p-cell. The basis $(|0\rangle, |1\rangle)$ for $\mathbb{C}^2$ is used to fix the equivalence, $\equiv$. Since $H\sigma_z H^{-1} = \sigma_x$, conjugating by the Hadamard transformation H on each factor transforms each B_b into $A_{\widehat{b}} = \bigotimes_{\widehat{b}\in\partial\theta} \sigma_x^\theta \otimes \bigotimes_{\widehat{b}\notin\partial\theta} \mathrm{id}^\theta$ where $\widehat{b}$ is the dual q-cell,

---

*This perspective on CSS codes is implicit in [12] and has been advocated by Greg Kupperberg.

$q := d - p$, to b in the dual cellulation $\mathcal{C}^*$ and θ runs over $(q+1)$-dual cells. Thus fix$B$ can be interpreted in a new way: using the Fourier basis $\{\frac{1}{\sqrt{2}}(|0\rangle + |1\rangle), \frac{1}{\sqrt{2}}(|0\rangle - |1\rangle)\}$ to define equivalence, $\equiv$, above, a state is in fix$B$ iff it is a superposition of dual cycles. All errors may be projected into σ_x and σ_z errors operating on various qubits [13]. The location of such errors may be deduced from the commuting system of measurements $\{A_a, B_b\}$. If the number of σ_x (σ_z) errors $e_x(e_z)$ satisfies: $e_x < \lfloor \frac{t}{2} \rfloor$ ($e_x < \lfloor \frac{t}{2} \rfloor$) then it will be possible to reconstruct all cycles (dual cycles) of the superposition viewed in the $\{|0\rangle, |1\rangle\}$-basis $(\{\frac{1}{\sqrt{2}}(|0\rangle + |1\rangle), \frac{1}{\sqrt{2}}(|0\rangle - |1\rangle)\}$-basis) by applying σ_z (σ_z) at no more than $\lfloor \frac{t}{2} \rfloor$ p- (q-) cells (dual cells). Doing this cannot change the class (dual class) in H_p^e ($H_q^{e^*}$) since any cycle (dual cycle) will be changed in fewer than $\lfloor \frac{t}{2} - \frac{1}{2} \rfloor + \lfloor \frac{t}{2} \rfloor < t$ cells (dual cells). So after repair, the state is changed by two operators, a null cycle C of σ_xs and by a null dual cycle C^* of σ_zs. Note that nullity implies an even number of intersections $C \cap C^*$; so the two operators commute and their order of application is irrelevant. Nullity further implies that the first operator is a composition of several B_b while the second operator is a composition of several A_a, indexed by the respective coboundary chains. Thus the two operators have no effect on the code space fix$A \cap$ fix$B$ and the repair has been successful.

If a closed Riemannian surface M is given a fine triangulation of bounded geometry—(bounded edge lengths and angles of the triangles)—a condition that ensures locality of the corresponding code based on 1-cell labeled qubits, then up to multiplicative constants, $t \approx 1-\text{systole}(M)$ and $n \approx \text{area}(M)$. Loewner's theorem and its generalizations tell us that no bounded geometry surface code with n qubits can do better than to protect against $\lfloor \frac{t}{2} \rfloor$ worst case errors, where $t \leq C\, n^{1/2}$ and $C = C(V)$ depends only on the valence V of the triangulation. It has been asked whether this square root relation between t and n is intrinsic—coming from the dichotomy of "bit" versus "phase" error—or is merely a feature of "surface codes." We will see that in the manifold context the p- and q-volumes of dual $\mathbb{Z}_2$-cycles can escape (narrowly) from systolic inequalities, and this "freedom" is then mapped back to the world of local quantum codes. We show that higher dimensional codes offer—through the phenomenon of systolic freedom—some slight but larger than constant improvement:

$$t \geq \text{constant} \cdot n^{1/2} \log^{1/2} n \qquad (12.1)$$

in the distance of the code. While an improvement so slight is of only conceptual interest, it is quite open whether better families of metrics can be found that would raise the exponent of n in formula (12.1).

THEOREM 12.3

There is a family of local stabilizer quantum codes of one qubit into n which protect against $\lfloor \frac{t}{2} \rfloor$ worst case errors for $t \geq$ constant$\cdot n^{1/2} \log^{1/2} n$.

Before leaving surface codes, we note that in the theory of Fuchsian groups the trace of a group element α, $\alpha = \left(\begin{smallmatrix} a & b \\ c & d \end{smallmatrix} \right) \in SL(2, \mathbb{Z})$, satisfies

$$\mathrm{tr}(\alpha) = 2\cosh\left(\frac{L(\alpha)}{2}\right)$$

where $L(\alpha)$ is the translation length of α. This formula allows the 1-systoles (and the number of curves representing the 1-systole) of many families of Fuchsian quotients to be computed. For example, in [20] the $N \to \infty$ asymptotics of the 1-systole for the quotients of $\Gamma_{-1,p}(N) \subset SL(\mathbb{Z}, R)$ for prime $p \equiv 3 \pmod 4$ are computed for

$$\Gamma_{-1,p}(N) = \left\{ \left(\begin{matrix} 1 + N(a + b\sqrt{p}) & N(-C + d\sqrt{p}) \\ N(c + d\sqrt{p}) & 1 + N(a - b\sqrt{p}) \end{matrix} \right) : \begin{matrix} \det = 1 \text{ and} \\ a, b, c, d \in \mathbb{Z} \end{matrix} \right\}.$$

Expressing the result in terms of the genus $g = g(N)$ of the quotient $\mathbb{H}^2/\Gamma_{-1,p}(N)$ one obtains:

$$\text{1-systole} > \text{constant} \cdot \frac{4}{3} \log g.$$

Taking fine triangulations with vertex valence bounded by 7, one may translate this geometric result into codes:

THEOREM 12.4

There is a constant $C_0 > 0$ so that for any positive integer n, there exists a 7-local CSS code that can protect $2g$ qubits $(\mathbb{C}^2)^{\otimes 2g}$ by imbedding them into $\lfloor C_0 n g \rfloor$ qubits $(\mathbb{C}^2)^{\otimes \lfloor C_0 n g \rfloor}$ so as to protect against $\log g \, n^{1/2}$ worst case errors.

The "$\log g$" reflects the "economy of scale" common in coding theory: "encryption is cheaper by the dozen."

12.1   Mapping torus constructions

To understand the difficulty of producing systolically free families of metrics with $\mathbb{Z}_2$-coefficients (or any torsion coefficients), consider the following low dimensional picture which readily generalizes to all dimensions. Let $A = D^2 \times [0, \epsilon]/(\rho, \theta, 0) = (\rho, \theta + \pi, \epsilon)$ be the unit 2-disk cross a short interval—a penny—with opposite faces identified by a π-twist. Any factor disk $D^2 \times t$, $t \in [0, \epsilon]$ is least area among essential relative 2-cycles with integer coefficients. The proof can be based on integral geometry or the divergence theorem applied to $\frac{\partial}{\partial t}$, often called a *calibrating field*. However, if we consider $\mathbb{Z}_2$-coefficients, much smaller essential 2-cycles with the topology of a punctured Klein bottle K^- are present, as shown in Figure 12.1.

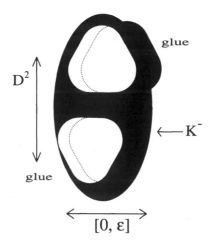

FIGURE 12.1
The topology of a punctured Klein bottle K^-.

Consider the problem of computing a useful lower bound on the (closed) 3-systole of a metric on $M \times S^1$, where M is a closed 3-manifold. There is a resource for this, the "co-area" inequality. In most examples, such as Gromov's original family of $\mathbb{Z}$-systolically free metrics on $S^3 \times S^1$ [14], the arithmetic of the co-area argument *fails* to force systolically free scaling of the $\mathbb{Z}_2$-systoles. We begin by presenting an extended example of a $\mathbb{Z}$-systolically free family of metrics on $S^3 \times S^1$ for which the co-area argument *fails* to prove $\mathbb{Z}_2$-systolic freedom and a higher dimensional tubing construction analogous to the tube drawn in Figure 12.1 shows that this family of metrics is, in fact, $\mathbb{Z}_2$-systolically rigid. The purpose of the example is to introduce the co-area argument first in families of manifolds $\{S_R^3\}$ and to see how the isoperimetric exponent of the S_R^3 "fibers" enters the arithmetic. This will motivate our switch to an example based on arithmetic surfaces whose underlying hyperbolic geometry possesses a linear isoperimetric inequality. This, together with the logarithmic growth (with genus) of the systoles of arithmetic surfaces (see Section 12.0), puts us over the boundary from rigidity to systolic freedom.

Example 12.1

Set $Q_R = S_R^3 \times [0, R^{1/3}]/\text{twist}$, where twist means an $R^{2/3}$ isometric rotation along the Hopf fibers of the 3-sphere of radius R, S_R^3. Again, applying the divergence theorem to the calibrating field $\frac{\partial}{\partial t}$ shows that the factor S_R^3 realizes the $\mathbb{Z} - 3$-systole. The $\mathbb{Z} - 1$-systole is approximately realized both by the generator of $H_1(S^3 \times S^1; \mathbb{Z})$ and by its $R^{1/3}$ power, each of which has length scaling like $R^{2/3}$. Thus, the $\mathbb{Z}$-systoles scale like:

$$\frac{\mathbb{Z} - 4\text{-systole} \sim R^{10/3}}{\mathbb{Z} - 3\text{-systole} \sim R^3 \quad \mathbb{Z} - 1\text{-systole} \sim R^{2/3}}.$$

Since $3 + \frac{2}{3} > 3\frac{1}{3}$, this family is $\mathbb{Z}$-systolically free.

Because the dimension of the ambient manifold is no more than 7 and we are considering a codimension 1 $\mathbb{Z}_2$-class, we may apply a theorem of Federer [12] to represent the $\mathbb{Z}_2 - 3$-systole by an embedded (but not necessarily orientable) 3-manifold $X_R \subset Q_R$. Clearly, the $\mathbb{Z} - 3$-systole, which infimizes over a smaller set, dominates the $\mathbb{Z}_2 - 3$-systole; so we have:

$$\text{vol}(S_R^3) \geq \text{vol}(X_R)$$

or if $x = \text{scaling}\left(\text{vol}(X_R)\right) = \limsup \log_R \text{vol}(X_R)$ is the growth exponent for the $\mathbb{Z}_2 - 3$-systole, we have $3 \geq x$.

The co-area inequality [6] states

$$\int_{t=0}^{R^{1/3}} \text{area}(X_R \cap S_R^3 \times t) dt \leq \text{vol}(X_R),$$

so

$$A_R = \min \text{area}(X_R \cap S_R^3 \times t) \leq \frac{\text{vol}(X_R)}{R^{1/3}} \quad t \in [0, R^{1/3}].$$

To get an inequality on growth rates, set $w = \limsup \log_R A_R$, so that $w \leq x - \frac{1}{3}$.

If the scaling of the $\mathbb{Z} - 3$-systole is by $x < 3$, we can attempt to reach a contradiction by cutting and gluing X_R to an orientable representative Z for the generator of $H_3(Q_R; \mathbb{Z})$. This uses a topological lemma:

LEMMA 12.1

If $Z \subset S^3 \times S^1$ is an embedded 3-manifold representing the nonzero element of $H_3(S^3 \times S^1; \mathbb{Z}_2)$ and $Z \cap (S^3 \times t) = \emptyset$ for some t, then Z is orientable and represents a nontrivial element, in fact a generator if Z is connected, of $H_3(S^3 \times S^1; \mathbb{Z})$.

PROOF of Lemma 12.1 $Z \subset S^3 \times (0,1) \subset S^3 \times S^1$ and must carry the $\mathbb{Z}_2 - 3$-cycle in $S^3 \times (0,1)$. It follows that Z separates the two ends of $S^3 \times (0,1)$, is orientable (use a normal vector field to Z to construct the orientation) and therefore is essential in $H_3(S^3 \times (0,1); \mathbb{Z})$. The lemma follows.   ☐

Let W_R be a t-cross section of X_R with scaling w and let $Y_R \subset S_R^3 \times t$ bound W_R. Using the isoperimetric inequality in S_R^3, which asymptotically is Euclidean 3-space, we see that:

$$y = \text{scaling}\left(\text{vol}(Y_R)\right) \leq \frac{3}{2}w.$$

Now $Z_R = \left(X_R\text{-neighborhood }(W_R)\right) \cup 2\text{-copies }(Y_R)$ is $\mathbb{Z}_2$-homologous to X_R and disjoint from $S_R^3 \times t$; so by Lemma 12.1, Z_R is orientable and must have volume at least as great as the $\mathbb{Z} - 3$-systole. Thus, if $x < 3$ then most of the volume of Z_R must come from the patch Y_R. The scaling of the patch volume is:

$$\frac{3}{2}\left(x - \frac{1}{3}\right) \geq \frac{3}{2}w \geq 3$$

so $x \geq \frac{7}{3}$.

The scaling as estimated (and, in fact, true) for the $\mathbb{Z}_2 - 4$-; 3- and 1-systoles of Q_R as powers of R are:

$$\frac{R^{10/3}}{R^{7/3} \cdot R^{2/3}}.$$

We have only proved $\frac{7}{3}$ is a lower bound, but a construction similar to the "penny" example realizes that exponent. Since $\frac{7}{3} + \frac{2}{3} \leq \frac{10}{3}$, the family Q_R is not $\mathbb{Z}_2$-systolically free. The interval scaling of Q_R, $R^{1/3}$, was chosen to come as close as possible to $\mathbb{Z}_2$-systolic freedom among similar metrics. ☐

Next we turn to the construction of $\mathbb{Z}_2 - (1,2)$-free metrics using mapping cylinders of arithmetic surfaces $\Sigma_g = \Gamma_{-1,p}(N)$ from Section 12.0. It seems reasonable at this point to define a variant *weak c-(p,q)* freedom, where the coefficient $c \cong \mathbb{Z}$ or $\mathbb{Z}_2$. We say a dimension $d = p+q$ has weak c-(p,q) freedom if there is a family of closed Riemannian d-manifolds M_i with nonvanishing $H_j(M_i; c)$, $j = p, q$ so that the systolic ratio is

$$\frac{d - c\text{-systole } (M_i)}{p - c\text{-systole } (M_i) \cdot q - c\text{-systole } (M_j)} \longrightarrow 0.$$

Thus, in the definition of weak freedom, we allow variable topology. The negation of weak freedom is *strong rigidity*. In Section 12.0, we commented that dimension 2 is strongly (1,1) rigid. The reason for introducing the notion of weak freedom is that when we consider (Section 12.3) the quantitative *freedom function* and *weak freedom function*, our examples exhibit substantially less freedom than weak freedom and it is weak freedom that has the more direct relevance to the construction of local quantum codes.

We will use the genus $g = g(N)$ of the surface $\Sigma_g = \mathbb{H}^2/\Gamma_{-1,p}(N)$ as our parameter. We record the important geometric properties of Σ_g:

$$1\text{-systole}(\Sigma_g) > C_1 \log g.$$

Also by a variant of Selberg's theorem (see [6]), the first eigenvalue of the Laplacian (on functions) exceeds some constant $C_2 > 0$, with

$$\lambda_1(\Sigma_g) > C_2. \tag{12.2}$$

By an analysis of the Dirichlet integral [7], (12.2) implies a uniform linear isoperimetric inequality for a null bounding 1-manifold $\gamma \subset \Sigma_g, \gamma = \partial A = \partial B, A \cup_\partial B = \Sigma_g$, area $A \leq$ area B:

$$\text{area}(A) \leq C_3 \text{ length}(\gamma), \tag{12.3}$$

C_3 independent of g.

Now (12.3) leads to a differential inequality on the Morse theory of "distance from a base point $*_g \subset \Sigma_g$":

$$\frac{\mathrm{d}}{\mathrm{d}t} \text{area}\big(\text{Ball}_*(t)\big) = \text{length}\,(\partial_t) \geq \frac{1}{C_3} \text{area}\big(\text{Ball}_*(t)\big),$$

as long as $\text{area}\big(\text{Ball}_*(t)\big) \leq \frac{1}{2}\text{area}\,(\Sigma_g)$. It follows that

$$\text{diameter}\,(\Sigma_g) < C_4 \log g, \tag{12.4}$$

C_4 independent of g. From (12.4) and the Morse theory of the distance function from a base point, we find an embedded wedge of circles W^*, which is the union of descending 1-manifolds spanning $H_1(\Sigma_g; \mathbb{Z}_2)$ with a length estimate (12.5). By the embedded property, we may find $W \subset W^* \subset \Sigma_g$ and using Alexander duality, the complement $\Sigma_g \setminus W$ is a disk, so $H_1(W; \mathbb{Z}_2) \longrightarrow H_1(\Sigma_g; \mathbb{Z}_2)$ is an isomorphism, and

$$\text{radius}_*(W) \leq C_4 \log g. \tag{12.5}$$

Three properties of Σ_g (for an infinite set of gs) will serve as input to our construction:

1. Σ_g has a uniform linear isoperimetric inequality (12.3).

2. There exists an isometry $\tau : \Sigma_g \to \Sigma_g$, with $\text{order}(\tau) \geq C_5(\log g)^{1/2}$.

3. The map $\Sigma_g \to \Sigma_g/\langle \tau(\sigma) \equiv \sigma \rangle =: {}_gS$ is a covering projection to the surface ${}_gS$ (of genus $< g$) and ${}_gS$ has injectivity radius$({}_gS) \geq C_6(\log g)^{1/2}$

where C_4, C_5 and C_6 are positive constants independent of g.

In Lemma 2 of [20] it is proved that:

$$\text{inj. rad.}\big(\mathbb{H}^2/\Gamma_{-1,p}(N)\big) = \mathcal{O}(\log N)$$

and in the proof of Theorem 6 of [20] that $\text{genus}\big(\mathbb{H}^2/\Gamma_{-1,p}(N)\big) =: g(\Sigma_g) =: g(N)$ satisfies:

$$\mathcal{O}(N^2) \leq g(N) \leq \mathcal{O}(N^3)$$

so

$$\text{inj. rad.}(\Sigma_g) = \mathcal{O}(\log g). \tag{12.6}$$

Now choose a sequence of h and g to satisfy $\log g = \mathcal{O}(\log h)^2$ and so that $N(h)$ divides $N(g)$. Thus, we have a covering projection $\Sigma_g \longrightarrow \Sigma_h$. Let α be the shortest essential loop in Σ_h. By (12.6) $\text{length}(\alpha) = \mathcal{O}(\log h)$. Choosing a base point on α, $[\alpha] \in \Gamma_{-1,p}\big(N(h)\big)/\Gamma_{-1,p}\big(N(g)\big)$ satisfies:

$$\text{order}[\alpha] \geq \mathcal{O}(\log h) = \mathcal{O}(\log g)^{1/2},$$

since the translation length of $\alpha = \mathcal{O}(\log g)^{1/2}$ must be multiplied by $\mathcal{O}(\log g)^{1/2}$ before it reaches length $\mathcal{O}(\log g)$, a necessary condition to be an element in the subgroup $\Gamma_{-1,p}(N(g))$.

Let τ be the translation determined by $[\alpha]$. We have just checked condition (2) above: $\operatorname{order}(\tau) > \mathcal{O}(\log g)^{1/2}$. Factor the previous covering as:

$$\Sigma_g \longrightarrow \Sigma_g/\langle\tau\rangle \longrightarrow \Sigma_h$$

and set $\Sigma_g/\langle\tau\rangle =: {}_gS$. Since ${}_gS$ covers Σ_h, we conclude condition (3):

$$\operatorname{inj.\ rad.}({}_gS) \geq \operatorname{inj.\ rad.}(\Sigma_h) \geq \mathcal{O}(\log h) = \mathcal{O}(\log g)^{1/2}.$$

Let $M_g = (\Sigma_g \times \mathbb{R})/\langle(x,t) \equiv (\tau x, t+1)\rangle$ be the Riemannian "mapping torus" of τ. We can also think of $M_g = \Sigma_g \times [0,1]/\langle(x,0) \equiv (\tau x, 1)\rangle$. Our first objective is to describe, with quantitative estimates, a sequence of Dehn surgeries that transforms M_g into a topological $S^2 \times S^1$. Since the number and length of these surgeries is so extravagant, we will describe as an alternative $\mathcal{O}(g)$ surgeries of length $\mathcal{O}(\log g)$ that transform M_g into a $\mathbb{Z}_2$-homology $S^2 \times S^1$. The second sequence is sufficient to exemplify weak freedom. By two theorems of Lickorish [17], we may first write out τ^{-1} in the mapping class group of Σ_g as a product of d_g Dehn twists σ_i along simple loops $\gamma_i \subset \Sigma_g$:

$$\tau^{-1} = \sigma_{d_g} \circ \cdots \circ \sigma_2 \circ \sigma_1$$

and second, performing Dehn surgeries along pushed-in copies of $\{\gamma_i\}$,

$$\left\{ \gamma_1 \times \left(\frac{1}{2} + \frac{1}{3d_g}\right), \gamma_2 \times \left(\frac{1}{2} + \frac{2}{3d_g}\right), \ldots, \gamma_i \times \left(\frac{1}{2} + \frac{i}{3d_g}\right), \ldots, \gamma_{d_g} \times \left(\frac{1}{2} + \frac{1}{3}\right) \right\}$$

obtain a diffeomorphic copy of $\Sigma_g \times [0,1]$ whose product structure, when compared to the original, gives $[\tau^{-1}] : \Sigma_g \times 1 \longrightarrow \Sigma_g \times 1$.

Thus, d_g Dehn surgeries on M_g produce the mapping torus for $\tau^{-1} \circ \tau$, i.e., $\Sigma_g \times S^1$. To allow us to estimate the freedom function in this family Luo has computed (see Appendix) an upper bound $C(g)$ to both the number and length of closed geodesics $\{\gamma_i\}$. Unfortunately, $F(g)$ has the form $F(g) = g^{g^{g^{\cdots g}}}$ ($3g - 3$ many exponents).

To convert $\Sigma_g \times S^1$ to $S^2 \times S^1$ an additional $2g$ Dehn surgeries are needed. Do half (a "sub-kernel") of these surgeries at level $\frac{1}{2} + \frac{1}{6d_g}$ and the dual half at level $\frac{1}{2}$. The result of all $d_g + 2g$ Dehn surgeries is topologically $S^2 \times S^1$, and once these surgeries are metrically specified, we obtain a sequence of Riemannian 3-manifolds $(S^2 \times S^1)_g$. To merely establish $\mathbb{Z}_2$-freedom, we do not need Luo's estimates; the estimate is used in Section 12.3 to quantify the amount of $\mathbb{Z}_2$-freedom.

We now take up the topologically easier problem of simply converting M_g to a $\mathbb{Z}_2$-homology $S^2 \times S^1$. Applying the Gysin sequence [22] we have:

$$H_1(\Sigma_g; \mathbb{Z}_2) \xrightarrow{1-\tau_*} H_1(\Sigma_g; \mathbb{Z}_2) \to H_1(M_g; \mathbb{Z}_2) \to \mathbb{Z}_2 \to 0$$

or

$$0 \to \operatorname{coker}(1 - \tau_*) \to H_1(M_g; \mathbb{Z}_2) \to \mathbb{Z}_2 \to 0.$$

Let $r = \operatorname{rank image}(1 - \tau_*)$. Take from the petals of W (12.5) a basis of loops $W_1, \ldots, W_{2g}$ for $H_1(\Sigma_g; \mathbb{Z}_2)$; these have length $\mathcal{O}(\log g)$. At most r of these loops lie in $\operatorname{image}(1 - \tau_*)$ so there is a subset reordering the index $\{W_1, \ldots, W_{2g-r}\}$ spanning cokernel $(1 - \tau_*)$. It is elementary (see, e.g., [6]) that surgeries (with any framings) applied to $\left\{ W_1 \times \frac{1}{2} + \frac{1}{2(2g-r)}, W_2 \times \frac{1}{2} + \frac{2}{2(2g-r)}, \ldots, W_{2g-r} \times 1 \right\}$ kill coker $(1 - \tau_*)$ and produce a $\mathbb{Z}_2$-homology, $S^2 \times S^1$, which we denote P_g.

In Section 12.2 where $\mathbb{Z}_2$-freedom is established, four metrical properties of these surgeries will be required. They are:

1. *The core curves for the Dehn surgeries are taken to be geodesics in $\Sigma_g \times [0, 1]$ so that the boundaries $\partial T_{i,\epsilon}$ of their ϵ neighborhoods are Euclidean flat. (Flatness follows from translational symmetry.)*

2. *The replacement solid tori $T'_{i,2\epsilon}$ have $\partial T'_{i,\epsilon}$ isometric to $\partial T_{i,\epsilon}$ and are defined as twisted products $D^2 \times [0, 2\pi\epsilon]/\beta$ (the meridians in $T_{i,\epsilon}$ have length $2\pi\epsilon$) where β is an isometric rotation of the disk D^2 adjusted to equal the holonomy obtained by traveling orthogonal to the surgery slopes in $\partial T_{i,\epsilon}$ from $\partial D^2 \times pt$ back to itself.*

3. *The geometry on the disk D^2 above is rotationally symmetric and has a product collar on its boundary as long as the boundary itself, yet $\operatorname{area}(D^2) \le \mathcal{O}\big((\operatorname{length} \partial D^2)^2\big)$.*

4. *Finally, $\epsilon > 0$ is so small that the total volume of all the replacement solid tori, $\cup_i T'_{i,\epsilon}$, is $o(g)$.*

With specifications (1)...(4), Dehn surgery yields a piece-wise smooth Riemannian manifold for which all the relevant notions of p-area are defined. We could work in this category but there is no need to do so since perturbing to a smooth metric will not effect the status of $\mathbb{Z}_2$-systolic freedom (rigidity) in either strong or weak forms.

12.2 Verification of freedom and curvature estimates

We regard the Riemannian manifolds $(S^2 \times S^1)_g$ and P_g as essentially specified in Section 12.1. Technically, there is the parameter ϵ which controls the "thickness" of the Dehn surgeries. On two occasions (Propositions 12.1 and 12.2), we demand this to be sufficiently small at the cost of an increase in the maximum absolute value of the Riemann curvature tensor as a function of g.

To obtain the family P_g, we performed $\mathcal{O}(g)$ surgeries on loops of length $\mathcal{O}(\log g)$. By the "collar theorem" (see Section 3 of [6]), if each loop is represented by a simple geodesic in a Σ_g level it has a collar of length $e^{-\mathcal{O}(\log g)}$ in that level. In the interval direction the surgeries are separated by a distance $\mathcal{O}(g^{-1})$; so together we may find disjoint tubular neighborhoods $T_{i,\epsilon}$ of radius:

$$\epsilon = \left(\frac{1}{g}\right)^{\alpha}, \tag{12.7}$$

for some $\alpha > 1$.

To ensure that Dehn surgery does not substantially increase $\mathrm{vol}(M_g) = \mathrm{vol}(\Sigma_g \times [0,1]) = \mathrm{area}(\Sigma_g) = 2\pi\chi(\Sigma_g) = \mathcal{O}(g)$, we need:

$$\mathcal{O}(g) \cdot \mathrm{area}\ D_i^2 \cdot 2\pi\epsilon \leq \mathcal{O}(g),$$

or from the long collar property (C) of D_i^2:

$$\mathcal{O}(g)(\log g)^2 \cdot \epsilon \leq \mathcal{O}(g)$$

so

$$\epsilon \leq \mathcal{O}(\log g)^{-2}. \tag{12.8}$$

Comparing with line (12.7), we see that this condition (12.8) is less stringent so we may pick $\epsilon = \left(\frac{1}{g}\right)^{\alpha}$, some $\alpha > 1$.

To obtain $(S^2 \times S^1)_g$ with volume $\mathcal{O}(g)$, ϵ must be chosen fantastically small. Again, the collar theorem gives the controlling restriction on ϵ: $\epsilon < e^{-F(g)}$. The volume condition:

$$F(g) \cdot F(g)^2 \cdot 2\pi\epsilon \leq \mathcal{O}(g)$$

is less stringent. We have proved:

PROPOSITION 12.1

Provided $\epsilon < \left(\frac{1}{g}\right)^{\alpha}$, for some $\alpha > 1$ the family P_g has $\mathrm{vol}(P_g) = \mathbb{Z}_2 - 3$-systole$(P_g) = \mathcal{O}(g)$. Provided $\epsilon < e^{-F(g)}$, the family $(S^2 \times S^1)_g$ has $\mathrm{vol}(S^2 \times S^1)_g = \mathbb{Z}_2 - 3$-systole$(S^2 \times S^1)_g = \mathcal{O}(g)$.

The next proposition is more subtle.

PROPOSITION 12.2

Provided $\epsilon < \mathcal{O}\left(\frac{1}{\log g}\right)$, the family P_g has $\mathbb{Z}_2 - 2$-systole $(P_g) = \mathcal{O}(g)$. Provided $\epsilon < g \cdot F(g)^{-2}$, the family $(S^2 \times S^1)_g$ has $\mathbb{Z}_2 - 2$-systole $(S^2 \times S^1)_g = \mathcal{O}(g)$.

Note: The corresponding bounds on ϵ are less strict in Proposition 12.2 than in Proposition 12.1; so, in practice, we must choose ϵ to satisfy the stricter bounds in Proposition 12.1.

PROOF of Proposition 12.2   For this proof only let N_g denote either P_g or $(S^2 \times S^1)_g$. According to [12] a nonoriented minimizer among all nonzero codimension 1 cycles always exists and is smooth provided the ambient dimension is no more than 7. Let $X_g \subset N_g$ denote these minimizers. The argument in the two cases is parallel so no confusion should result from the double use of the symbol X_g. For a contradiction, assume area$(X_g) < \mathcal{O}(g)$.

The Dehn surgeries in Section 12.1 were confined to $\Sigma_g \times \left[\frac{1}{2}, 1\right]$; so the surfaces $\Sigma_g \times t$, $t \in \left(0, \frac{1}{2}\right)$ persist as submanifolds of N_g. By Sard's theorem, for almost all $t_0 \in \left(0, \frac{1}{2}\right)$, $\Sigma_g \times t_0$ intersects X_g transversely. Let W_t, $t \in \left(0, \frac{1}{2}\right)$ denote the intersection. By the co-area formula,

$$\mathcal{O}(g) > \text{area } (X_g) \geq \int_{t=0}^{1/2} \text{length}(W_t)dt.$$

Consequently, for some transverse $t_0 \in \left(0, \frac{1}{2}\right)$,

$$\text{length } (W_{t_0}) < \mathcal{O}(g). \tag{12.9}$$

Since both $\Sigma_g \times t_0$ and X_g represent the nonzero element of $H_2(N_g; \mathbb{Z}_2)$, the complement $N_g \setminus (\Sigma_g \times t_0 \cup X_g)$ can be two-colored into black and white regions (change colors when crossing either surface) and the closure B of the black points, say, is a piecewise smooth $\mathbb{Z}_2$-homology between $\Sigma_g \times t_0$ and X_g.

For homological reasons, the reverse Dehn surgeries $N_g \rightsquigarrow M_g$ to the τ-mapping torus have cores with zero (mod 2) intersection with X_g. This means that the tori $\partial T_{i,\epsilon} = \partial T'_{i,\epsilon}$ each meet X_g in a null homologous, possibly disconnected, 1-manifold $X_g \cap \partial T_{i,\epsilon} \subset \partial T_{i,\epsilon}$. Again, if ϵ is a sufficiently small function of g, we may "cut off" X_g along these tori to form $X'_g = (X_g \setminus \cup_i T_{i,\epsilon}) \cup \delta_i$, where δ_i denotes a bounding surface for $X_g \cap \partial T_{i,\epsilon}$ in $\partial T_{i,\epsilon}$, with negligible increase in area. Note that $X'_g \subset M_g$. In particular, we still have:

$$\text{area}(X'_g) < \mathcal{O}(g), \tag{12.10}$$

provided

$$(\# \text{ surgeries}) \cdot (\text{max length surgery}) \cdot \epsilon < \mathcal{O}(g). \tag{12.11}$$

In the cases $N_g = P_g$ and $(S^2 \times S^1)_g$, (12.11) reduces, respectively, to

$$\mathcal{O}(g) \cdot \mathcal{O}(\log g) \cdot \epsilon < \mathcal{O}(g),$$

and

$$F(g) \cdot F(g) \cdot \epsilon < \mathcal{O}(g).$$

Specifically, we choose δ_i to be the "black" piece of $\partial T_{i,\epsilon}$, i.e., $\delta_i \subset B$. If we set

$$B' = \text{closure}(B \setminus \bigcup_i T_{i,\epsilon})$$

and recall

$$\bigcup_i T_{i,\epsilon} \cap \Sigma'_g \times t_0 = \emptyset,$$

we see that B' is a $\mathbb{Z}_2$-homology from X'_g to $\Sigma_g \times t_0$.

It is time to use property (i), i.e., (12.3): W_{t_0} separates $\Sigma_g \times t_0$ into two subsurfaces meeting along their boundaries: One subsurface sees black on the positive side, the other on its negative side. Thus, the smaller of these two subsurfaces, call it $Y \subset \Sigma_g \times t_0$, must satisfy:

$$\text{area}(Y) \leq C_3 \text{length}(W_{t_0}),$$

where C_3 is independent of g. Combining with (12.9), we have:

$$\text{area}(Y) < \mathcal{O}(g). \tag{12.12}$$

Now modify X'_g to Z_g by cutting along W_{t_0} and inserting two parallel copies of Y. This may be done so that the result is disjoint from $\Sigma_g \times t_0$ but bordant to it by a slight modification B'' of B', with B'' still disjoint from $\Sigma_g \times t_0$. See Figure 12.2 below.

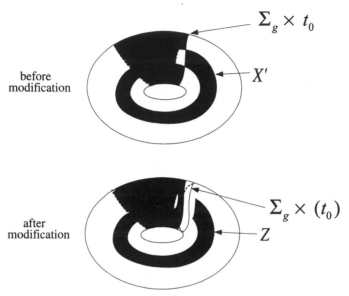

FIGURE 12.2
Modification of X'_g to Z_g.

Combining (12.10) and (12.12),

$$\operatorname{area}(Z_g) < 3 \cdot \mathcal{O}(g) = \mathcal{O}(g).$$

Now reverse the Dehn surgeries and consider:

$$B'' \subset M_g \setminus \Sigma_g \times (t_0) \subset M_g. \qquad (12.13)$$

The middle term of (12.13) is diffeomorphic to $\Sigma_g \times \mathbb{R}$, which is a codimension 0 submanifold of $\mathbb{R}^3$. This proves that B'', and in particular Z, is orientable. But this looks absurd; we have constructed an oriented surface Z oriented-homologous to the fiber $\Sigma_g \times t_0$ of M_g with smaller area.

As in Example 12.1, let $\frac{\partial}{\partial t}$ be the divergenceless flow in the interval direction. Lift Z to $\widetilde{Z}$ in the infinite cyclic cover $\Sigma_g \times \mathbb{R}$ and consider the flow through the lift $\widetilde{B}''$, the lift of B''. The divergence theorem states that the flux through $\widetilde{Z}$ is equal to the flux through $\Sigma_g \times t_0$. Since $\frac{\partial}{\partial t}$ is orthogonal to $\Sigma_g \times t_0$,

$$\text{area}\,(\Sigma_g \times t_0) \leq \text{area}\,(\widetilde{Z}) = \text{area}\,(Z) \qquad (12.14)$$

completing the contradiction. ∎

PROPOSITION 12.3
  $\mathbb{Z}_2 - 1\text{-systole}(P_g) \geq \mathcal{O}(\log g)^{1/2}$ and $\mathbb{Z}_2 - 1\text{-systole}(S^2 \times S_g^1)$ $\geq \mathcal{O}(\log g)^{1/2}$.

PROOF of Proposition 12.3   We actually show that any homotopically essential loop obeys this estimate. The long collar condition, (C) in Section 12.1, implies that any arc in $T'_{i,\epsilon}$ with end points on $\partial T'_{i,\epsilon}$ can be replaced with a shorter arc with the same end points lying entirely within $\partial T'_{i,\epsilon}$. It follows that any essential loop in P_g or $(S^2 \times S^1)_g$ can be homotoped to a shorter loop lying in the complement of the Dehn surgeries.

   Thus, it is sufficient to show that any homotopically essential loop γ in M_g has length $\gamma \geq \mathcal{O}(\log g)^{1/2}$. For a contradiction, suppose the opposite. Since the bundle projection $\pi : M_g \to [0,1]/\langle 0 = 1 \rangle$ is length nonincreasing, $\text{degree}(\pi \circ \gamma) < \mathcal{O}(\log g)^{1/2}$. Lift $\gamma \setminus pt.$ to an arc $\widetilde{\gamma}$ in $\Sigma_g \times \mathbb{R}$. The lift $\widetilde{\gamma}$ joins some point (p,t) to $(\tau^d p, t + d)$ where $d = \text{degree}(\pi \circ \gamma)$. Since $d < \mathcal{O}(\log g)^{1/2}$ and since condition (2) from Section 12.1 requires $\text{order}(\tau) \geq \mathcal{O}(\log g)^{1/2}$, we see that p and $\tau^d p$ differ by a nontrivial covering translation of the cover $\Sigma_g \to {}_g S$. But any non-trivial covering translation moves each point of the total space at least twice the injectivity radius of the base, a quantity guaranteed by (3) of Section 12.1 to be at least $\mathcal{O}(\log g)^{1/2}$. Now using that the projection $\Sigma_g \times \mathbb{R} \to \Sigma_g$ is also length nonincreasing, we see that $\text{length}(\widetilde{\gamma}) \geq \mathcal{O}(\log g)^{1/2}$. Since $\text{length}(\widetilde{\gamma}) = \text{length}(\gamma)$, the same estimate applies to γ. ∎

THEOREM 12.5
  Given the restrictions on $\epsilon > 0$ imposed in Proposition 12.1, the family P_g exhibits weak $\mathbb{Z}_2 - (2,1)$-systolic freedom. The family $(S^2 \times S^1)_g$ exhibits $\mathbb{Z}_2 - (2,1)$-systolic freedom.

PROOF of Theorem 12.5   From Propositions 12.1, 12.2 and 12.3, we have:

$$\frac{\mathbb{Z}_2 - 3\text{-systole}(P_g)}{\mathbb{Z}_2 - 2\text{-systole}(P_g) \cdot \mathbb{Z}_2 - 1\text{-systole}(P_g)} \leq \frac{\mathcal{O}(g)}{\mathcal{O}(g)\,\mathcal{O}(\log g)^{1/2}} \longrightarrow 0,$$

and the same statement holds replacing P_g by $(S^2 \times S^1)_g$. ∎

Let us now estimate the maximum absolute value of curvatures and their first covariant derivatives, maximum $\{|\nabla_h R^\ell_{ijk}|, |R^\ell_{ijk}|\} =: R(\gamma)$ for the two families P_g and $(S^2 \times S^1)_g$. Of course piecewise smooth constructions may have "infinite" curvature at the gluing locus, but after rounding the corners, the residual curvature is dominated by $\mathcal{O}\left(\frac{1}{\ell}\right)^2$ where ℓ is the smallest length scale of the construction, and the first derivatives are dominated by $\mathcal{O}\left(\frac{1}{\ell}\right)^3$. In our case $\ell = \epsilon$. Thus, for the family P_g,

$$R(g) \leq \mathcal{O}(g^{3\alpha}), \tag{12.15}$$

for some $\alpha > 1$, and for $(S^2 \times S^1)_g$,

$$R(g) \leq e^{3F(g)}. \tag{12.16}$$

As in [11], let us consider the $\mathbb{Z}_2$-freedom function of both families. This function quantifies the amount of freedom present in a family. To do this, we homothetically rescale each member of the family to be as small as possible and yet have $R \leq 1$, i.e., all its curvatures and first derivatives of curvature lying in the interval $[-1, 1]$.

This is accomplished by rescaling P_g by $\mathcal{O}(g^{\frac{3}{2}\alpha})$ as suggested by (12.15) and (12.16). We find:

$$\frac{\mathbb{Z}_2 - 3\text{-systole}}{\mathbb{Z}_2 - 2\text{-systole} \cdot \mathbb{Z}_2 - 1\text{-systole}} \leq \frac{\mathcal{O}(g^{3\alpha})\,\mathcal{O}(g)}{\mathcal{O}(g^{2\alpha})\,\mathcal{O}(g) \cdot \mathcal{O}(g^\alpha)\,\mathcal{O}(\log^{1/2} g)}. \tag{12.17}$$

Now we ask for a function which will serve as a lower bound to $\frac{d(n)}{n}$ where n and d are the numerator and denominator (as a function of the numerator) in the right hand side of (12.17). This, by definition, is a lower bound to the $\mathbb{Z}_2$-freedom of the family. An easy computation yields:

$$\frac{d(n)}{n} = \mathcal{O}(\log^{1/2} n),$$

so $\{P_g\}$ has (at least) $\log^{1/2}$ $\mathbb{Z}_2$-systolic freedom.

Rescaling $(S^2 \times S^1)_g$ (according to (12.16)) by $e^{F(g)}$ we find:

$$\frac{\mathbb{Z}_2 - 3\text{-systole}}{\mathbb{Z}_2 - 2\text{-systole} \cdot \mathbb{Z}_2 - 1\text{-systole}} \leq \frac{e^{3F(g)} \; \mathcal{O}(g)}{e^{2F(g)} \; \mathcal{O}(g) \cdot e^{F(g)} \; \mathcal{O}(\log^{1/2} g)}$$

or

$$\frac{d(n)}{n} = \mathcal{O}\left(\log G\left(\frac{1}{3}(\log n)\right) \right)^{1/2}, \qquad (12.18)$$

where $G \colon \mathbb{R}^+ \longrightarrow \mathbb{R}^+$ is some rapidly decaying monotone function that is an inverse to Luo's function F for integral g, $G(F(g)) = g$. Thus the family $(S^2 \times S^1)_g$ has at least this (12.18) absurdly small $\mathbb{Z}_2$-freedom function.

Many examples of $\mathbb{Z}_2$-freedom in all higher dimensions can be obtained from the families P_g and $(S^2 \times S^1)_g$ by taking products with spheres of appropriate radii. Additional surgeries may be done after taking products to adjust the topology of the families if desired. For an application to quantum codes, the simplest construction is to form $\overline{P}_g = P_g \times S^1_{r(g)}$ where S^1_r is the circle of radius $r(g) = g(\log g)^{-1/2}$.

PROPOSITION 12.4
$\mathbb{Z}_2 - 2\text{-systole}(\overline{P}_g) = \mathcal{O}(g)$.

PROOF of Proposition 12.4   Let $[\alpha] \in H_2(\overline{P}_g; \mathbb{Z}_2) \cong \big(H_1(P_g; \mathbb{Z}_2) \otimes H_1(S^1_r; \mathbb{Z}_2)\big) \oplus H_2(P_g; \mathbb{Z}_2)$. Since the projection $\overline{P}_g \longrightarrow P_g$ is area non-increasing, a 2-cycle α must have area $\mathcal{O}(g)$ if its projection in P_g is essential. Suppose for a contradiction that $\text{area}(\alpha) < \mathcal{O}(g)$. Then by the above, $[\alpha] \in H_1(P_g; \mathbb{Z}_2) \otimes H_1(S^1_r; \mathbb{Z}_2)$ the first factor. For generic $\theta \in S^1_r$, $\alpha \cap \text{proj}^{-1}(\theta)$ defines the essential element of $H_1(P_g; \mathbb{Z}_2)$. By Fubini's theorem, $\text{area}(\alpha) \geq \mathbb{Z}_2 - 1\text{-systole} \, (P_g) \cdot \mathbb{Z}_2 - 1\text{-systole} \, (S^1_r) = \mathcal{O}(\log^{1/2} g) \cdot (g \log^{-1/2} g) = \mathcal{O}(g)$, a contradiction. $\blacksquare$

THEOREM 12.6
The family $\overline{P}_g$ exhibits weak $\mathbb{Z}_2$-systolic $(2, 2)$-freedom with freedom function $\mathcal{O}(\log^{1/2})$.

PROOF of Theorem 12.6   This follows immediately from Proposition 12.4 and the fact that $\text{vol} \, (\overline{P}_g) = \mathcal{O}(g^2)(\log g)^{-1/2}$. $\blacksquare$

Remark: Although the family $\overline{P}_g$ suffices to construct local quantum codes (see Section 12.3) that exceed square-root efficiency, for aesthetic reasons one may wish to achieve $\mathbb{Z}_2$-$(2,2)$-freedom on a family whose underlying manifolds are all diffeomorphic to $S^2 \times S^2$. This would remove the word "weak" in Theorem 12.6, but the price is that the quantitative level of $\mathbb{Z}_2$-freedom may slip to an absurdly small amount as in (12.18). This is what happens if we form $(S^2 \times S^2)_g$ from $(S^2 \times S^1)_g \times S_r^1$ by performing two 1-surgeries on the $\pi_1\big((S^2 \times S^1)_g \times S_r^1\big) \cong \mathbb{Z} \oplus \mathbb{Z}$ generators.

THEOREM 12.7

There is a family of Riemannian metrics on $S^2 \times S^2$, $(S^2 \times S^2)_g$, which exhibits $\mathbb{Z}_2$-$(2,2)$-freedom, with freedom function bounded below by $\big(\log G(\tfrac{1}{3} \log d)\big)^{1/2}$. There exists a weak family of Riemannian metrics on closed 4-dimensional manifolds $\big\{ (S^2 \times S^2)_g^h \big\}$ with the $\mathbb{Z}_2$-homology of $S^2 \times S^2$ which exhibit $\mathbb{Z}_2$-$(2,2)$-freedom with freedom function bounded below by $\mathcal{O}(\log d)^{1/2}$.

PROOF of Theorem 12.7   The first statement follows by performing two framed 1-surgeries on each $(S^2 \times S^1)_g \times S_r^1$. The second family is similarly derived from $\overline{P}_g$ by two 1-surgeries on each member. ∎

12.3   Quantum codes from Riemannian manifolds

Let N be a closed Riemannian manifold of dimension d. We assume that the metric on N has been homothetically scaled so that the injectivity radius is at least 1 and the maximum absolute value of any entry in the curvature tensor satisfies $|R_{ijk}^{\ell}| \leq 1$. Also all first covariant derivatives of curvature are assumed bounded, $|\nabla_h R_{ijk}^{\ell}| \leq 1$.

THEOREM 12.8

There are constants $0 < d_1, d_2, d_3 < \infty$ depending only on the dimension d so that N admits a piecewise smooth pair (cellulation, dual triangulation) $= (\mathcal{C}, \mathcal{C}^*)$, so that:

(1) The number of cells of all dimensions in both $\mathcal{C}$ and $\mathcal{C}^*$ satisfies: $|\mathcal{C}| + |\mathcal{C}^*| < d_1 \, vol(M)$.

(2) For every p-cell π of $C \cup C^*$:

$$p - vol\,(\pi) < d_2.$$

(3) For each cell of $C \cup C^*$, both the number of cells in its boundary and the number of cells in its coboundary are less than d_3.

PROOF of Theorem 12.8   The curvative bound implies that there is a fixed ϵ, depending only on dimension d, so that every 4ϵ-ball in M is 1-1 quasi-isometric to the 4ϵ-ball in Euclidean d-space.

Let $V \subset M$ be a maximal set of points so that the ϵ-balls centered at V are disjoint. The 2ϵ-balls centered at V cover M. Roughly, we would like to take C to be the V-Voronoi cellulation centered at V but there are bothersome (open?) technicalities about estimating p-volumes of p-faces. A simple alternative is to let C' be the V-Voronoi cellulation, which after perturbing V we may assume to be generic and piecewise smooth, and to let $C^*_{\text{comb.}}$ be the combinatorial triangulation formally dual to C'. We can construct an embedded, in M, piecewise smooth homeomorph C^* of $C^*_{\text{comb.}}$ by inductively constructing the simplices of C^* by "geodesic sweep out." The vertices of C^* are the set V. The 1-simplices are the shortest center-connecting arcs for pairs of C'-cells that meet along a $d - 1$ face. The 2-simplices (corresponding to triple coincidences of C'-cells of dimension $d - 1$) are defined (unnaturally) by picking out one of the three vertices and sweeping a geodesic segment emanating from that vertex and ending on the opposite edge. Proceeding inductively, the totality of the simplices so constructed is C^*. Define $C_{\text{comb.}}$ to be the combinatorial dual cellulation: barycentrically subdivide $C^*_{\text{comb.}}$ to obtain $\overline{C^*}_{\text{comb.}}$. Then the p-cells of $C_{\text{comb.}}$ are the closed stars of the vertices of $\overline{C^*}_{\text{comb.}}$.

Again using geodesic sweep-out, $C_{\text{comb.}}$ may be piecewise smoothly embedded in M; call the result C. Because of the explicit inductive construction of all the cells of C^* and C, d_2 can be found so that the estimate (2) of Theorem 12.8 holds. Notice that the bound on covariant derivatives of curvature already comes into showing that the 2-simplices, constructed as a family of geodesic arcs, have bounded area.

Now the local combinatorics of C and C^* is dictated by that of C', which is bounded by volume considerations. For v, $v' \in V$, each d-cell D_v of C' has volume at least $(\frac{1}{1.1})^d$ vol(Euclidean ϵ-ball) and can only share a $(d - 1)$-face with another $D_{v'}$ if dist$(v, v') < 4\epsilon$. The number of such potential $(d-1)$-faces per d-cell is bounded by:

$$\frac{(1.1)^d \, \text{vol(Euclidean } 4\epsilon\text{-ball)}}{\left(\frac{1}{1.1}\right)^d \text{vol(Euclidean } \epsilon\text{-ball)}} = (1.1)^{2d} \cdot 4^d.$$

Similarly the number of $(d-k)$-cells of C is bounded by the binomial coefficient $\begin{pmatrix} \lceil (1.1)^{2d} \rceil \cdot 4^d \\ k \end{pmatrix}$. For appropriate constants, d_1 and d_3 conditions (1) and (3) now follow. $\blacksquare$

DEFINITION 12.2   *Given a cellulated, triangulated manifold $(M; C, C^*)$, we may define combinatorial $\mathbb{Z}_2 -$ systoles and dual systoles as follows: the combinatorial $\mathbb{Z}_2 - k$-systole is the fewest number of k-cells of C that form an essential cycle in $H_k(M; \mathbb{Z}_2)$; and the combinatorial $\mathbb{Z}_2 - \ell$-systole is the fewest number of ℓ-cells of C^* that form an essential cycle in $H_\ell(M; \mathbb{Z}_2)$ (constructed from the dual chains).*

Let us translate the result of Section 12.2 into combinatorics using Theorem 12.8. Recall from Theorem 12.6 the family of Riemannian 4-manifolds $(S^2 \times S^2)_g^h$ with $\mathbb{Z}_2 - (4, 2)$-systoles scaling like

$$\left(\mathcal{O}(g^2) \log^{-1/2} g, \mathcal{O}(g) \right)$$

and bounded curvatures. These may be (cellulated, triangulated) in accordance with Theorem 12.8:

$$\left((S^2 \times S^2)_g; C_g; \ C_g^* \right).$$

The result are (cellulations, triangulations) with bounded geometry. No cell or simplex has more than d_3 terms in its boundary or coboundary. We obtain the following scalings on C_g and C_g^*: combinatorial $\mathbb{Z}_2 -$ 2-systole $(C_g) = \mathcal{O}(g)$ and combinatorial $\mathbb{Z}_2 - 2$-systole $(C_g^*) = \mathcal{O}(g)$, and we may bound the number of 2$-$cells:

$$\mathcal{O}\big(2\text{-cells}\,(C_g)\big) \leq \mathcal{O}\big(4\text{-cells}\,(C_g)\big) \leq \mathcal{O}\big(\text{vol}(S^2 \times S^2)_g\big) \leq \mathcal{O}(g^2 \log^{-1/2} g).$$

Now, as described in Section 12.0, we may build a d_3-local-parity-check-CSS code Code$_g$ by assigning qubits to the 2-cells of C_g. Since $H_2(\overline{P}_g; \mathbb{Z}_2) \cong \mathbb{Z}_2 \oplus \mathbb{Z}_2$ these are $\left[\mathcal{O}(g^2 \log^{-1/2} g), 2, \mathcal{O}(g) \right]$ codes, i.e., quantum codes that encode 2 qubits into $\mathcal{O}(g^2 \log^{-1/2} g)$ qubits capable of recovering from $\mathcal{O}(g)$ worst case errors. Introducing the parameter $n = g^2 \log^{-1/2} g$ and $\lfloor \frac{t}{2} \rfloor$ for the number of tolerated worst case errors, we compute the asymptotics of the codes Codes$_g$ to be

$$t = \mathcal{O}\big(n^{1/2} \log^{1/2} n\big).$$

This completes the proof of Theorem 12.3. $\square$

Appendix

The purpose of this appendix is to prove the following result:

THEOREM 12.9

Suppose Σ_g is a closed surface with a hyperbolic metric of injectivity radius r. There exists a computable constant $C(g,r)$ so that each isometry of Σ_g is isotopic to a composition of positive and negative Dehn-twists $D_{c_1}^{\pm 1} \ldots D_{c_k}^{\pm 1}$ where $k \leq C(g,r)$ and the length $l(c_i)$ of c_i is at most $C(g,r)$ for each i.

As a by-product of the proof, we also obtain the following result, which may be of some independent interest in view of the recent work on symplectic 4-manifolds.

Call a self-homeomorphism of the surface *positive* if it is isotopic to a composition of positive Dehn-twists.

THEOREM 12.10

Suppose $\Sigma_{g,n}$ is a compact orientable surface of genus g with n boundary components. Let $\{a_1, \ldots, a_{3g-3+2n}\}$ be a 3-holed sphere decomposition of the surface where $\partial \Sigma_{g,n} = a_{3g-2+n} \cup \cdots \cup a_{3g-3+2n}$. Then each orientation preserving homeomorphism of the surface that is the identity map on $\partial \Sigma_{g,n}$ is isotopic to a composition qp where p is positive and q is a composition of negative Dehn-twists on a_i's.

The basic idea of the proof of Theorem 12.9 suggested by M. Freedman is as follows. Let f be an isometry of the surface. Choose a surface filling system of simple geodesics $\{s_1, \ldots, s_k\}$ whose lengths are bounded (in terms of r and g). Since the lengths of s_i and $f(s_j)$ are bounded, the intersection numbers between any two members of $\{s_1, \ldots, s_k, f(s_1), \ldots, f(s_k)\}$ are bounded. Now the proof of Lickorish's theorem in [17] is constructive and depends only on the intersection numbers between simple loops. Thus, one produces a bounded number of simple loops of bounded lengths so that the composition of positive or negative Dehn-twists on them sends s_i to $f(s_i)$. This shows that f is isotopic to the composition.

The proof below follows Freedman's sketch. We shall choose the surface filling system to be of the form $\{a_1, \ldots, a_{3g-3}, b_1, \ldots, b_{3g-3}\}$ where $\{a_i\}$ forms a 3-holed sphere decomposition of the surface so that $l(a_i) \leq 26(g-1)$ (Ber's theorem) and the b_is have bounded lengths so that $b_i \cap a_j = \emptyset$ for $j \neq i$. Then we establish a controlled version of Lickorish's lemma (Lemma 2 in [17]) by estimating the lengths of loops involved in the Dehn-twists.

We shall use the following notations and conventions. Surfaces are oriented. If a is a simple loop on a surface, D_a denotes the positive Dehn-twist along a and $l(a)$ denotes the length of the geodesic isotopic to a. Two isotopic simple loops a and b will be denoted by $a \cong b$. Given two simple loops a, b, *their geometric intersection number*, denoted by $I(a, b)$, is $\min\{|a' \cap b'| \mid a' \cong a, b' \cong b\}$. It is well known that if a, b are two distinct simple geodesics, then $|a \cap b| = I(a, b)$. We use $|a \cap b| = 2_0$ to denote two simple loops a, b so that $I(a, b) = |a \cap b| = 2$ and their algebraic intersection number is zero.

To prove Theorem 12.9, we begin with the following.

PROPOSITION 12.5

Suppose a and b are homotopically nontrivial simple loops in a hyperbolic surface of injectivity radius r. Then,

(a) *(Thurston).* $I(a, b) \leq \frac{4}{\pi r^2} l(a) l(b)$.

(b) $l(D_a(b)) \leq I(a, b) l(a) + l(b)$.

(c) *For each integer n,* $\frac{\pi r^2 |n| I(a,b)}{4 l(b)} \leq l(D_a^n(b)) \leq |n| I(a, b) l(a) + l(b)$.

(d) *If $|a \cap b| \geq 3$ or $|a \cap b| = 2$ so that the two points of intersection have the same intersection signs, then there exists a simple loop c so that* $l(c) \leq l(a) + l(b)$, $|D_c(b) \cap a| < |b \cap a|$ *and* $l(D_c(b)) \leq 2l(a) + l(b)$.

(e) *There exists a sequence of simple loops $c_1, \ldots, c_k$ so that $k \leq |a \cap b|$,* $l(c_i) \leq (2i - 1)l(a) + l(b)$ *for each i and $D_{c_k} \ldots D_{c_1}(b)$ is either* *disjoint from a, or intersects a at one point, or intersects a at two points of different signs.*

PROOF of Proposition 12.5   Part (a) is essentially in [10], p. 54, Lemma 2. We produce a slightly different proof so that the coefficient is $\frac{4}{\pi r^2}$. Without loss of generality, we may assume that both a and b are simple geodesics. Construct a flat torus as the metric product of two geodesics a and b. The area of the torus is $l(a)l(b)$. Each intersection point of a with b gives a point p in the torus. Now the flat distance between any two of these points p_s is at least the injectivity radius r (otherwise there would be Whitney discs for $a \cup b$). Thus the flat disks of radius $r/2$ around these p_s are pairwise disjoint. This shows that the sum of the areas of these disks is at most $l(a)l(b)$ which is the Thurston's inequality.

To see part (b), we note that the Dehn-twisted loop $D_a(b)$ is obtained by taking $I(a, b)$ many parallel copies of a and resolving all the intersection points between b and the parallel copies (from a to b). Thus the inequality follows.

Part (c) follows from parts (a) and (b). Note that we have used the fact that $I(D_a^n(b), b) = |n|I(a, b)$ (see for instance [18] for a proof, or one also can check directly that there are no Whitney disks for $D_a^n(b) \cup b$).

Part (d) is essentially in Lemma 2 of [17]. Our minor observation is that one can always choose a positive Dehn-twist D_c to achieve the result.

We need to consider two cases.

Case 1. There exist two intersection points x, $y \in a \cap b$ adjacent along in a which have the same intersection signs (see Figure 1). Then the curve c as shown in Figure 1 (with the right-hand orientation on the surface) satisfies all conditions in the part (d). If the surface is left-hand oriented, take $D_c(b)$ to be the loop c.

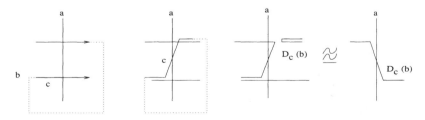

FIGURE 12.3
Right-hand orientation on the plane.

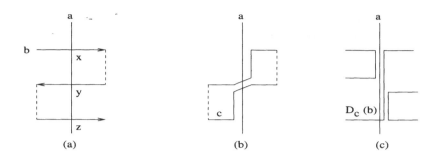

FIGURE 12.4
Right-hand orientation on the plane.

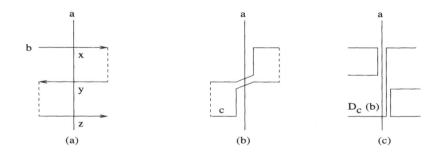

FIGURE 12.5
Left-hand orientation on the plane.

Case 2. Suppose any pair of adjacent intersection points in $a \cap b$ has different intersection signs. Then $|a \cap b| \geq 3$. Take three intersection points $x, y, z \in a \cap b$ so that x, y and y, z are adjacent in a. Their intersection signs alternate. Fix an orientation on b so that the arc from x to y in b does not contain z as shown in Figure 2. If the surface Σ is right-hand oriented as in Figure 2, take c as in Figure 2(b). Then $D_c(b)$ is shown in Figure 2(c). If the surface has the left-hand orientation, then take c as shown in Figure 3(b). The loop $D_c(b)$ is shown in Figure 3(c). One checks easily that the simple loop c satisfies all the conditions.

Part (e) follows from part (d) by induction on $|a \cap b|$. ∎

We shall also need the following well-known lemma in order to deal with disjoint loops and loops intersecting at one point.

LEMMA 12.2

Suppose a and b are two simple loops intersecting transversely at one point. Then,

(a) $D_a D_b(a) \cong a,$

(b) $(D_a D_b D_a)^2$ *sends a to a, b to b and reverses the orientations on both a and b.*

See [2] and [17] for a proof, or one can check it directly. Note that $(D_a D_b D_a)^2$ is the hyper-elliptic involution on the 1-holed torus containing both a and b.

We first give a proof of Theorem 12.10. The proof of Theorem 12.9 follows by making length estimate at each stage of the proof of Theorem 12.10.

PROOF of Theorem 12.10   Let f be an orientation-preserving homeomorphism of $\Sigma_{g,n}$, which is the identity map on the boundary. We shall show that there exists a composition p of positive Dehn-twists so that for each i, $pf^{-1}|_{a_i} = $ id. It follows that pf^{-1} is a product of Dehn-twists on a_i's.

We prove the theorem by induction on the norm $|\Sigma_{g,n}| = 3g - 3 + n$ of the surface (the norm is the complex dimension of the Teichmuller space of complex structures with punctured ends on the interior of the surface). The basic property of the norm is that if Σ' is an incompressible subsurface that is not homotopic to $\Sigma_{g,n}$, then the norm of Σ' is strictly smaller than the norm of $\Sigma_{g,n}$. For simplicity, we assume that the Euler characteristic of the surface is negative (though the proof below also works for the torus).

If the norm of a surface is zero, then the surface is the 3-holed sphere. The theorem is known to hold in this case (see [9]).

If the norm of the surface is positive, we pick a nonboundary component, say a_1, of the 3-holed sphere decomposition as follows. If the genus of the surface $\Sigma_{g,n}$ is positive, a_1 is a nonseparating loop. By Proposition 12.5(e) applied to $a = a_1$ and $b = f^{-1}(a_1)$, we find a sequence of simple loops $c_1, \ldots, c_k$, $k \leq I(a,b)$ so that $a'_1 = D_{c_k} \ldots D_{c_1} f^{-1}(a_1)$ satisfies: either $a'_1 \cap a_1 = \emptyset$, or $|a'_1 \cap a_1| = I(a'_1, a_1) = 1$, or $|a'_1 \cap a_1| = 2_0$. There are two cases we need to consider: (1) both a_1 and a'_1 are separating loops, and (2) both of them are nonseparating.

In the first case, by the choice of a_1, the genus of the surface is zero. First, $I(a_1, a'_1) = 1$ cannot occur due to homological reason. Second, since the homeomorphism $D_{c_k} \ldots D_{c_1} f^{-1}$ is the identity map on the non-empty boundary $\partial \Sigma_{0,n}$, it follows that $I(a_1, a_1') = 2$ is also impossible and a_1' is actually isotopic to a_1. After composing with an isotopy, we may assume that $D_{c_k} \ldots D_{c_1} f^{-1}|_{a_1}$ is the identity map. Now cut the surface open along a_1 to obtain two subsurfaces of smaller norms. Each of these subsurfaces is stablized under $D_{c_k} \ldots D_{c_1} f^{-1}$. Thus the induction hypothesis applies and we conclude the proof in this case.

In the second case that both a_1 and a_1' are nonseparating, then either $|a_1' \cap a_1| = 1$, or there exists a third curve c so that c transversely intersects each of a_1 and a_1' in one point. By Lemma 12.2(a), one of the product h of positive Dehn-twists $D_{a_1'} D_{a_1}$, or $D_c D_{a_1} D_{a_1'} D_c$, will send a_1' to a_1. If the homeomorphism $h D_{c_k} \ldots D_{c_1} f^{-1}$ sends a_1 to a_1 reversing the orientation, by Lemma 12.2(b), we may use six more positive Dehn-twists (on a_1, a_1', or c, a_1) to correct the orientation. Thus, we have constructed a composition of positive Dehn-twists $D_{c_m} \ldots D_{c_1} f^{-1}$ so that it is the identity map on a_1 and $m \leq I(a,b) + 10$. Now cut the surface open along a_1 and use the induction hypothesis. The result follows. ∎

We note that the proof fails if we do not choose a_1 to be a non-separating simple loop in the case the surface is closed of positive genus.

Now we prove Theorem 12.9 by making length estimate on each step above.

PROOF of Theorem 12.9 Let f be an isometry of a hyperbolic closed surface $\Sigma_g = \Sigma_{g,0}$.

We begin with the following result, which gives bound on the lengths of a_i's and c used in the proof of Theorem 12.10.

PROPOSITION 12.6
Suppose Σ_g is a hyperbolic surface of injectivity radius r.

(a) *(Bers) There exists a 3-holed sphere decomposition $\{a_1, \ldots, a_{3g-3}\}$ of the surface so that $l(a_i) \leq 26(g-1)$.*

(b) *If a and b are two nonseparating simple geodesics in a compact hyperbolic surface Σ which is a totally geodesic subsurface in $\Sigma_{g,n}$ so that either $I(a,b) = 0$ or $|a \cap b| = 2_0$, then there exists a simple geodesic c in Σ so that $I(c,a) = I(c,b) = 1$ and $l(c) \leq \frac{8(g-1)r}{\sinh r} + 8r$.*

PROOF of Proposition 12.6   See Buser [6], p. 123 for a proof of part (a).

To see part (b), we first note that there are simple loops x so that $I(x,a) = I(x,b) = 1$ by the assumption on a and b. Let c be the shortest simple loop in Σ satisfying $I(c,a) = I(c,b) = 1$. We shall estimate the length of c as follows. Let $N = \lfloor \frac{l(c)}{2r} \rfloor$ be the largest integer smaller than $\frac{l(c)}{2r}$. Let $P_1 = a \cap c$, P_2, ..., P_N be N points in c so that their distances along c are $d(P_1, P_{i+1}) = 2ri$ and $d(P_i, P_{i+1}) = 2r$. Let B_i be the disc of radius r centered at P_i and B_k be the ball containing $c \cap b$. Then the shortest length property of c shows that the intersections of the interior $\text{int}(B_i) \cap \text{int}(B_j)$ is empty if $1 \leq i < j < k$ or $k < i < j \leq N$. Thus the sum of the areas of the $N - 2$ balls $B_2, \ldots, B_{k-1}, B_{k+1}, \ldots, B_N$ is at most twice the area of the surface $\Sigma_{g,0}$. This gives the estimate required.

∎

Fix a 3-holed sphere decomposition $\{a_1, \ldots, a_{3g-3}\}$ of the hyperbolic surface so that $l(a_i) \leq 26(g-1)$. We may assume that the loops a_i are so labeled that $a_1, a_2, \ldots, a_g$ are nonseparating loops and the rest are separating.

We now show that there exists a computable constant $C' = C'(g,r)$ so that any orientation-preserving isometry f of the hyperbolic surface Σ_g is isotopic to a product qp where q is a product of positive or negative Dehn-twists on a_is and p is a product of at most $C'(g,r)$ many positive Dehn-twists on curves of lengths at most $C'(g,r)$.

We now rerun the constructive proof of Theorem 12.10 by estimating the lengths of loops involved. To begin with, we take $a = a_1$ and $b = f^{-1}(a_1)$ of lengths at most $26g$. By Thurston's inequality, their intersection number $I(a,b)$ is at most $\frac{52^2 g^2}{\pi r^2}$. By Propositions 12.5(e), 12.6(b) and the proof of Theorem 12.10, we produce a finite set of simple loops $\{c_1, \ldots, c_k\}$ so that $k \leq I(a,b) + 10$, the lengths of c_i are bounded in g, r and $f_1 = D_{c_k} \ldots D_{c_1} f^{-1}$ is the identity map on a_1. Now we take $a = a_2$ and $b = f_1(a_2)$ and run the same constructive proof as above in the totally geodesic subsurface $\Sigma_{g-1,2}$ obtained by cutting Σ_g open along a_1. In order for the proof to work, we need to see that the length of b is bounded. Indeed, Proposition 12.5(b) gives the estimate of $l(b)$ in terms of $l(c_i)$, $l(a_2)$, and g, r (here we estimate the intersection number $I(c_i, x)$ in terms of the lengths by Thuston's inequality). Thus, we construct a finite set of simple loops $d_1, \ldots, d_m$ so that m and $l(d_i)$ are bounded in g, r, $d_i \cap a_1 = \emptyset$, and $D_{d_m} \ldots D_{d_1} f_1^{-1}$ is the identity map on $a_1 \cup a_2$. Inductively, we produce the required positive homeomorphism p.

We remark that if the injectivity radius r is at least $\log 2$, then the number $C'(g, r)$ that we obtained is at least $g^{g^{g^{\cdots^{g}}}}$ (there are $3g - 3$ many exponents) in magnitude.

As a consequence, we obtain the following expression for the homeomorphism $p^{-1}f = D_{a_1}^{n_1} \ldots D_{a_{3g-3}}^{n_{3g-3}}$. It remains to show that the exponents n_is are bounded. To this end, for each index i, we pick a geodesic loop b_i which is disjoint from all a_j's for $j \neq i$ and b_i intersects a_i at one point or two points of different signs. A simple calculation involving right-angled hyperbolic hexagon shows that we can choose these b_i to have lengths at most $182(g - 1) - \log(r/4)$. Thus the lengths of curve $p^{-1}f(b_i)$ is bounded (in terms of g and r). By Proposition 12.5(d), the growth of the lengths of loops $D_{a_i}^n(b_i)$ is linear in $|n|$ if $|n|$ is large. Thus we obtain an estimate on the absolute value of the exponents $|n_i|$. This finishes the proof. $\square$

References

[1] M. Berger, À l'ombre de Loewner, *Ann. Sci. École Norm. Sup. (4)* **5** (1972) 241–260.

[2] J. Birman, Mapping class groups of surfaces, in *Braids*, J. Birman and A. Libgober, eds., Proceedings of a summer research conference, Santa Cruz, CA, 13–16 July 1986, Contemp. Math. **78** (Providence, RI: AMS 1988) 13–44.

[3] I. Babenko and M. Katz, Systolic freedom of orientable manifolds, *Ann. Sci. École Norm. Sup. (4)* **31** (1998) 787–809.

[4] I. K. Babenko, M. G. Katz and A. I. Suciu, Volumes, middle-dimensional systoles, and Whitehead products, *Math. Res. Lett.* **5** (1998) 461–471.

[5] W. Browder, *Surgery on Simply-Connected Manifolds*, *Ergebnisse der Mathematik und ihrer Grenzgebiete* **65** (New York: Springer-Verlag 1972).

[6] P. Buser, *Geometry and Spectra of Compact Riemann Surfaces*, Progress in Mathematics **106** (Boston, MA: Birkhäuser 1992).

[7] P. Buser, A note on the isoperimetric constant, *Ann. Sci. École Norm. Sup. (4)* **15** (1982) 213–230.

[8] A. R. Calderbank and P. W. Shor, Good quantum error-correcting codes exist, *Phys. Rev.* A **54** (1996) 1098–1105.

[9] M. Dehn, *Papers on Group Theory and Topology*, translated and edited by J. Stillwell (New York: Springer-Verlag 1987).

[10] A. Fathi, F. Laudenbach and V. Poénaru, *Travaux de Thurston sur les surfaces*, *Astérisque* **66–67** (1979).

[11] M. H. Freedman, $\mathbb{Z}_2$-systolic-freedom, in *Proceedings of the Kirbyfest*, J. Hass and M. Scharlemann, Eds., Geometry and Topology Monographs **2** (Coventry, UK: Geometry and Topology Publications 1999) 113–123.

[12] H. Federer, *Geometric Measure Theory*, *Die Grundlehren der mathematischen Wissenschaften* **153** (New York: Springer-Verlag 1969).

[13] D. Gottesman, *Stabilizer codes and quantum error correction*, `quant-ph/9705052`; CalTech Ph.D. Thesis (1997).

[14] M. Gromov, Systoles and intersystolic inequalities, in *Actes de la Table Ronde de Géométrie Différentielle*, Luminy, 1992, *Sémin. Congr.* **1** (Paris: Soc. Math. France 1996) 291–362.

[15] A. Yu. Kitaev, Fault-tolerant quantum computation by anyons, `quant-ph/9707021`.

[16] M. Katz and A. I. Suciu, Volume of Riemannian manifolds, geometric inequalities, and homotopy theory, in *Tel Aviv Topology Conference: Rothenberg Festschrift (1998)*, Contemp. Math. **231** (Providence, RI: AMS 1999) 113–136.

[17] R. Lickorish, A representation of oriented combinatorial 3-manifolds, *Ann. Math.* **72** (1962) 531–540.

[18] F. Luo, Multiplication of simple loops on surfaces, Rutgers preprint (1999).

[19] C. Pittet, Systoles on $S^1 \times S^n$, *Diff. Geom. Appl.* **7** (1997) 139–142.

[20] P. Schmutz Schaller, Extremal Riemann surfaces with a large number of systoles, in *Extremal Riemann Surfaces*, J. R. Quine and P. Sarnak, Eds., San Francisco, CA, 1995, Contemp. Math. **201** (Providence, RI: AMS 1997) 9–19.

[21] P. W. Shor, Scheme for reducing decoherence in quantum computer memory, *Phys. Rev.* A **52** (1995) R2493–R2496.

[22] E. Spanier, *Algebraic Topology* (New York: McGraw-Hill 1966).

[23] A. M. Steane, Error correcting codes in quantum theory, *Phys. Rev. Lett.* **77** (1996) 793–797.

Quantum Teleportation

Chapter 13

Quantum teleportation

Kishore T. Kapale and M. Suhail Zubairy

Abstract   The amazing properties of quantum systems, such as non-locality, entanglement, superpositions and interference, have revolutionized our thinking about information processing and computing. Quantum teleportation is probably the most notable and dramatic among various concepts developed through the application of quantum mechanics to the information science. In this chapter we review the basic ideas behind quantum teleportation and discuss some interesting proposals for teleportation of various quantum systems. We also discuss the implementation of quantum teleportation based on the well-developed tools of cavity quantum electrodynamics.

13.1   Introduction

Teleportation has a mysterious connotation, owing to the manner in which it is portrayed in science fiction stories. We see the fictional characters beaming themselves from one location to another almost instantaneously. Certainly, instantaneous teleportation of macroscopic objects is an impossibility due to physical limitations like causality, which prohibit

1-58488-282/4/02/$0.00+$1.50
© 2002 by Chapman & Hall/CRC

instantaneous information transfer [1]. Nevertheless, after a careful analysis of the teleportation concept one can discern the key steps involved. Namely, scanning of the original object to obtain sufficient information to be able to recreate it, transmission of this information to the receiver's location and reconstruction using locally available material. At the completion of the process the original object is destroyed and its exact replica is reproduced at the prescribed location. Thus, teleportation, in principle, is different from fax transmission, which leaves the original intact and only creates an approximate replica at the receiving location. Furthermore, the most important ingredient of the teleportation concept is that the actual object is not to traverse the actual distance between the two locations. However, it should be mentioned that some other equivalent object has to cover the distance.

Until recently a common opinion in the scientific community was that teleportation is not possible specially for quantum systems, as according to Heisenberg's uncertainly principle it is impossible to measure all the attributes of a general quantum state exactly and simultaneously. This makes communicating quantum information a very daunting task. Even if we use direct transfer, there is no way to be sure that the quantum state is transferred accurately due to its continual interaction with the surroundings. In such a case teleportation seems to be a promising avenue for communicating a quantum state of a system to a distant location as actual traversing of the system is not involved. However, due to our incapacity to measure the quantum state effectively, there seems no possible way to communicate sufficient information to a distant receiver to enable him to recreate the original quantum state. As it turns out this view was changed radically with a startling discovery by Bennett et al. [2], giving rise to the new field of quantum teleportation. There onwards followed a lot of interesting theoretical proposals [3] and experimental realizations [4] covering a wide range of quantum systems. Here we review several efforts made in the direction of employing quantum mechanical properties to achieve successful teleportation. Just to note, the quantum mechanical principles utilized have also given rise to very interesting fields like quantum computing [5], quantum cryptography [6] and communication [7].

Quantum teleportation primarily relies on quantum entanglement in a form considered by Einstein, Podolsky, and Rosen [8] and various new insights added by J. S. Bell [9, 10]. Quantum entanglement essentially implies an intriguing property that two quantum correlated systems cannot be considered independent even if they are far apart in the sense that measuring the state of one of them allows one to predict something about the state of the other one. Nevertheless, the measurement statistics of each one is no different from the case in which there exists no correlation between them. Such an entangled pair of quantum systems constitutes an important element in any quantum teleportation scheme, and facilitates the interesting feat of faithfully transferring an unknown quantum state to a distant location. The concept should come clear when we discuss the teleportation proposal for a two-level system in Section 13.2.

The essentials of the quantum teleportation scheme can be summarized in three major steps: (1) The sender and receiver acquire one particle each off a pair of entangled particles through a quantum channel. (2) The sender then performs a measurement of a special observable on the combined system of the shared entangled particle and the to-be-teleported particle and sends the classical result of the measurement to the receiver through a classical channel. (3) Depending on the message received, the receiver then performs a unitary operation on the entangled particle in his possession, and teleportation is achieved in the sense that his particle now acquires the state of the input particle.

It is interesting to see that such a scheme does not violate any laws of physics. As communication of classical information is essential to complete the teleportation process, no instantaneous transmission is suggested or required and hence there is no violation of causality. Moreover, the state of the original particle changes in an irrecoverable manner and reappears at the receiving end in complete accord with the no-cloning theorem of quantum mechanics [11, 12]. To elaborate, this no-cloning theorem illustrates a very fundamental feature of quantum information; namely, it can be swapped from one system to another but no duplication or *cloning* is possible. A notable feature of this teleportation scheme is that, in principle, a very small amount of information needs to be communicated in order to accomplish the process.

Our discussion starts with the simplest formal teleportation scheme for a two-state quantum system, and its implementation based on cavity quantum-electrodynamics (QED) in Section 13.2. A proposal to extend this method to a discrete N-state systems and its implementation is covered in Section 13.3. To illustrate the possibility of using the teleportation concept for quantum computing and quantum communication, teleportation of an entangled quantum state is discussed in Section 13.4. In Section 13.5 an extension of teleportation methods to a system of continuous variables is visited. Therein two of the most important proposals based on non local measurements and Wigner functions are considered.

Finally a few remarks are presented in the last section.

We do not cover all the work that has been done in the field in this review. However, we choose a few representative ideas to illustrate and introduce the subject matter. Some of the work that is not covered includes atomic state teleportation schemes of Davidovich et al. [13], Cirac and Parkins [14], Bose et al. [15], and Zheng and Guo [16], and cavity quantum electrodynamics based teleportation schemes of Moussa [17, 18]. We would like to point out, to the reader who is interested in further study of the field, an entanglement purification method proposed by Bennett et al. [19], which allows faithful teleportation through noisy channels, providing tremendous practical promise to the teleportation idea. We would also like to mention the work on teleportation of superpositions of chiral amplitudes by Maierle *et al.* [20], a scheme using bright squeezed light by Ralph and Lam [21] and squeezed vacuum state by Milburn and Braunstein [22], photon number teleportation scheme of Koniorczyk et al. [23], matter-wave teleportation using molecular dissociation and collisions by Opatrny and Kurizki [24], and the teleportation scheme using three-particle entanglement of Karlsson and Bourennane [25]. More recent and notable work includes continuous variable teleportation schemes of Loock, Braunstein and Kimble [26]–[28], a dense coding scheme for continuous variables [29], and a universal formalism for teleportation by Braunstein et al. [30], extending the regime of the quantum states that can be teleported.

13.2   Teleportation of a two-state system

In this section we present a variant of the very first quantum teleportation proposal by Bennett et al. [2], in the context of photon number states of a microwave cavity. We further consider a scheme for implementing the teleportation protocol for this simplest 2-state quantum system using well-established tools of cavity quantum electrodynamics. The goal in mind is to introduce the basic methodology and an experimentally feasible implementation strategy that can be very naturally extended to more complicated systems like an N dimensional quantum state, a two-qubit entangled state and the like.

13.2.1   The formal scheme

The original Bennett et al. proposal was designed to teleport an unknown spin-half system. We resort here to the discussion of photon number states of a microwave cavity, which can be designated as $|0\rangle$ and $|1\rangle$ and correspond to the presence of 0 and 1 photons in the cavity, respectively. However, the treatment considered here is in no way restricted to the photon number states and is equally applicable to any two-state quantum system. To name a few: polarization states of a photon, $|\leftrightarrow\rangle$ and $|\updownarrow\rangle$, states of a spin-half particle $|\uparrow\rangle$ and $|\downarrow\rangle$, the states of photon in a two-arm interferometer, $|a\rangle$ and $|b\rangle$ and ground state and excited state of an atom or ion, $|g\rangle$ and $|e\rangle$, can be very well described by the treatment presented here. As a matter of terminology we label the system to be teleported as 1, and take the entangled pair to be consisting of system 2 and 3. At the conclusion of the teleportation process the state of system 1 is destroyed and is acquired by system 3. The formal scheme is schematically shown in Figure 13.1.

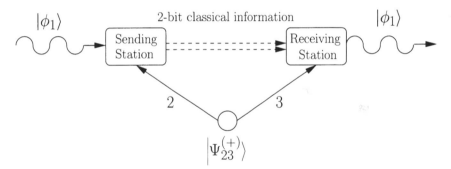

FIGURE 13.1
Schematic view of the quantum teleportation scheme: Sharing an EPR entangled state between the sender and the receiver allows one to teleport an arbitrary, 2-state input quantum state using a classical channel. The quantum channel is denoted by a solid line and the classical one by a dashed line. The same convention is followed in the other figures also.

As mentioned earlier, the first step is the preparation of an entangled state among the cavities 2 and 3

$$\left|\Psi_{23}^{(+)}\right\rangle = \frac{1}{\sqrt{2}}\left(|0\rangle_2 |1\rangle_3 + |1\rangle_2 |0\rangle_3\right), \tag{13.1}$$

which are at the sending and receiving stations, respectively. The notation $|\Psi_{23}^{(+)}\rangle$ for the state signifies one of the four possible maximally entangled states for a system of two entities and will become transparent in further steps. The subscripts to the photon number states correspond to the cavity label. The unknown "input" state of cavity 1 can be taken to be in the most general superposition form

$$|\varphi_1\rangle = \alpha\,|0\rangle_1 + \beta\,|1\rangle_1\,, \tag{13.2}$$

with normalization $|\alpha|^2 + |\beta|^2 = 1$. This is the state that is to be teleported to the receiver. The next step is the joint measurement by the sender on the two cavities, namely, the original unknown cavity 1 and the entangled cavity 2. It is therefore instructive to express the direct product $|\ \rangle_1|\ \rangle_2$ in the complete state of all the three cavities

$$|\Psi_{123}\rangle \equiv |\varphi_1\rangle \otimes \left|\Psi_{23}^{(+)}\right\rangle = \frac{\alpha}{\sqrt{2}}\Big(|0\rangle_1\,|0\rangle_2\,|1\rangle_3 + |0\rangle_1\,|1\rangle_2\,|0\rangle_3\Big)$$
$$+ \frac{\beta}{\sqrt{2}}\Big(|1\rangle_1\,|0\rangle_2\,|1\rangle_3 + |1\rangle_1\,|1\rangle_2\,|0\rangle_3\Big), \tag{13.3}$$

in the famous Bell operator Basis [31]

$$\left|\Psi_{12}^{(\pm)}\right\rangle = \frac{1}{\sqrt{2}}\Big(|0\rangle_1\,|1\rangle_2 \pm |1\rangle_1\,|0\rangle_2\Big),$$
$$\left|\Phi_{12}^{(\pm)}\right\rangle = \frac{1}{\sqrt{2}}\Big(|0\rangle_1\,|0\rangle_2 \pm |1\rangle_1\,|1\rangle_2\Big). \tag{13.4}$$

It is fairly easy to see that the states $\left|\Psi_{12}^{(\pm)}\right\rangle$ and $\left|\Phi_{12}^{(\pm)}\right\rangle$ form a complete orthonormal basis for the combined state of cavities 1 and 2. One can write the complete state in terms of the Bell-basis states as

$$|\Psi_{123}\rangle = \frac{1}{2}\left[\left|\Psi_{12}^{(+)}\right\rangle\Big(\alpha\,|0\rangle_3 + \beta\,|1\rangle_3\Big) + \left|\Psi_{12}^{(-)}\right\rangle\Big(\alpha\,|0\rangle_3 - \beta\,|1\rangle_3\Big)\right.$$
$$\left.+ \left|\Phi_{12}^{(+)}\right\rangle\Big(\beta\,|0\rangle_3 + \alpha\,|1\rangle_3\Big) + \left|\Phi_{12}^{(-)}\right\rangle\Big(-\beta\,|0\rangle_3 + \alpha\,|1\rangle_3\Big)\right]. \tag{13.5}$$

If a measurement is performed at this stage by the sender on the combined state of cavities 1 and 2 in the Bell basis, regardless of the incoming state $|\varphi_1\rangle$ all four outcomes are equally likely. As a result of this measurement, cavity 3, which is the receiving station, gets projected with equal probability into one of the four states as can be easily seen from Equation 13.5. One can write these states in a simple 2-D vector notation as

$$|\varphi_3\rangle \equiv \begin{pmatrix}\alpha\\\beta\end{pmatrix},\ \begin{pmatrix}\alpha\\-\beta\end{pmatrix},\ \begin{pmatrix}\beta\\\alpha\end{pmatrix},\ \text{or}\ \begin{pmatrix}-\beta\\\alpha\end{pmatrix} \tag{13.6}$$

where the first index is the coefficient of the $|0\rangle$ state and the second one is that of $|1\rangle$. As one can see, each of these possible resultant states of cavity 3 is related in a simple way to the original state that was to be teleported. Thus after receiving the classical result of the measurement the receiver can perform a unitary operation to complete the teleportation process. These unitary transformations can be represented in a matrix form as

$$\mathcal{U}_3 \equiv \begin{pmatrix} 1 & 0 \\ 0 & 1 \end{pmatrix}, \begin{pmatrix} 1 & 0 \\ 0 & -1 \end{pmatrix}, \begin{pmatrix} 0 & 1 \\ 1 & 0 \end{pmatrix}, \text{or} \begin{pmatrix} 0 & 1 \\ -1 & 0 \end{pmatrix} \qquad (13.7)$$

corresponding to the states in Equation 13.6, respectively. It is easy to check that these unitary transformations take the appropriate projected state of cavity 3 to the original state of cavity 1, resulting in faithful teleportation.

Once again to emphasize, this process is in complete accord with causality, and the no-cloning theorem of quantum mechanics. Moreover, a two-bit classical channel is essential to communicate the result of measurement between the sending and receiving stations, in the case of a two-state system. A lesser capacity channel would be inadequate for the purpose and it would amount to guessing on the part of the receiver, thus affecting the fidelity of the process immensely. It is imperative to mention that a method proposed by Vaidman et al. [32, 33] based on non local measurements provides an alternative approach, which can be easily extended to the case of continuous degrees of freedom and will be discussed in Section 13.4.

It is interesting to note that considerable progress has been made in experimental realizations of quantum teleportation. The very first proof-of-principle experiment [34] performed at Innsbruck demonstrated teleportation of a polarization state of a photon. This scheme used polarization entangled photon pairs generated through parametric down-conversion, as the EPR pair. The fact to consider is that only 2 out of the 4 Bell states could be detected; thus the maximum efficiency is only 50%. The process, however, validates the quantum teleportation idea experimentally. The first experimental realization was followed by various interesting schemes, a majority of them mainly dealing with the polarization states of photons. More specifically the Rome experiment [35] used the entanglement of the spatial motion of a single photon with its polarization; all the four states associated with the Bell basis could be distinguished, but the input state had to be generated internally and could not come from outside. The experimental scheme of Braunstein and Kimble [36] is of special interest since it established con-

tinuous quantum variable teleportation that we are going to discuss in
Section 13.4, by demonstrating teleportation of a coherent state of an
optical field using squeezed-state entanglement. Another interesting experiment by Pan et al. [37] extended the teleportation concept one step
further by demonstrating swapping of entanglement from a pair of photons to the other one, thus establishing teleportation as a very effective
tool in quantum computing and communication.

Thus, for the simplest two-state system implementation of teleportation, early experiments were unable to accomplish complete Bell basis measurement, which is an essential requirement for efficient teleportation. Recently, Scully and collaborators [38] proposed an ingenious
scheme to perform the Bell state measurement with an ideal success rate
of 100% in the context of the polarization-entangled two-photon states.
It utilizes resonant two-photon absorption of atoms which are suitably
prepared in coherent superposition of hyperfine states, and can be implemented experimentally utilizing non linear optical processes. More
recently, Vitali et al. [39] proposed a scheme based on Kerr nonlinearity
for the Bell state measurement of the similar polarization-entangled two-
photon state. A recent notable experiment by Shih's group [40] achieves
a simultaneous complete Bell state measurement using elegant experimental techniques based on nonlinear interactions introduced by optical
sum frequency generation, thus turning teleportation into an experimental reality with 100% fidelity under ideal circumstances. It was recently
pointed out to us that the teleportation scheme of Solano et al. [41]
for the internal states of trapped ions also provides 100% success under
ideal circumstances. This protocol uses a Bell state analyzer based on
an operation that transforms the Bell basis into a disentangled one unitarily and hence it can be easily implemented using the experimentally
available tools to manipulate trapped ions.

13.2.2   Cavity QED implementation

We discussed how teleportation of a two-state system can be achieved
in principle. However, when it comes to implementation of the process
utilizing the experimentally available tools, one has to look at the details
of all the steps involved, i.e., preparing an entangled state, the Bell
state measurement, and the unitary transformation needed at the end
to complete the process. Here we discuss the implementation strategy to
achieve the above mentioned teleportation scheme for a photon-number
states of a microwave cavity, employing the well-established tools of

cavity quantum electrodynamics.

One starts with a two-level atom in an excited state $|a\rangle$, the ground state of which is denoted by $|b\rangle$. This ancillary atom is to be used to prepare the cavities 2 and 3 in an entangled state as in Equation 13.1. Initially the cavities do not contain any photons; thus the complete state of the system comprised of the atom and the cavities is

$$|a\rangle\,|0\rangle_2\,|0\rangle_3\,. \tag{13.8}$$

The atom is to be passed through the cavities one after the other with specific interaction times and coupling strengths with the cavity modes. This ancillary two-level atom is taken to be in resonance with the cavity mode, meaning the energy difference between the atomic levels matches with the energy of the cavity mode photon. Such an interaction is usually described by the Jaynes–Cummings Hamiltonian [42]. The velocity of the atom can be manipulated to control the interaction times with the cavities. One has to choose the strength of the interaction and the passage time of the atom in such a way that the atom sees a $\pi/2$ pulse in the first cavity and a π pulse in the second cavity. After the passage through the first cavity the atomic states form a coherent superposition and the combined state of the atom and the cavity field becomes

$$\frac{1}{\sqrt{2}}\Big(|a\rangle\,|0\rangle_2 + |b\rangle\,|1\rangle_2\Big)\,|0\rangle_3\,. \tag{13.9}$$

Furthermore, at the end of the passage through the second cavity, the atomic superposition is transferred to the second cavity. Thus the cavity field acquires the required entangled state

$$\frac{1}{\sqrt{2}}\Big(|0\rangle_2\,|1\rangle_3 + |1\rangle_2\,|0\rangle_3\Big), \tag{13.10}$$

and the ancillary atom comes down to the ground state $|b\rangle$.

The next goal is to achieve the Bell-basis measurement for the combined state of cavity 1 and 2. As it will come clear, three ancillary atoms labeled A, B and C are needed to accomplish this. First atom A is passed through both cavities with the intention to measure the total number of photons residing in both cavities taken together. The motivation for this is clear from the observation that the states $|\Psi^{(\pm)}\rangle$ and $|\Phi^{(\pm)}\rangle$ differ in the number of photons. The $|\Psi^{(\pm)}\rangle$ state contains one photon, whereas $|\Phi^{(\pm)}\rangle$ state contains either zero or two photons (see Equation 13.4). Thus the character to be distinguished is whether both the cavities taken together contain an odd or even (including zero) number of photons. The distinction among the individual group states, i.e., the $+$ and $-$ states, remains to be resolved; thus the photon number counting process has to be non destructive to keep the cavity fields intact. This non demolition measurement can be achieved as follows, using a technique called Ramsey interferometry. To start with we provide a very simplified treatment, keeping aside the entanglement of the atom with the cavity fields, and just following the effects on the state of the atom as it passes through the cavities. Once the concept of how one can manipulate or measure cavity fields through atoms is clear, we consider a rigorous treatment taking into account the whole system with all its constituents and entanglement between them.

Atom A is required to be non resonant with the cavity field and its initial state is taken to be the superposition $(|a\rangle_A + |b\rangle_A)/\sqrt{2}$. This atom goes through the cavities with the same interaction time for both of them. Due to the dispersive nature of atomic interaction with the cavity fields, the level $|b\rangle_A$ picks up a relative phase shift which is proportional to the total number of photons in the two cavities and the interaction time without affecting the number of photons in the cavities, i.e., the resulting state of the atom becomes

$$\frac{1}{\sqrt{2}}\left(|a\rangle_A + e^{ip\theta}|b\rangle_A\right), \qquad (13.11)$$

where p is the total number of photons in the two cavities and θ is a parameter that includes the dependence on atom-field coupling strength, detuning, and interaction time [43]. The parameter θ can be controlled easily by changing the interaction time. This modified atom, which is carrying the information about the number of photons in the cavities, is then passed through a resonant classical field, termed a Ramsey field. At this stage interaction time and the coupling parameters are chosen so as to obtain the mixing of states as

$$|a\rangle \rightarrow \frac{1}{\sqrt{2}}(|a\rangle + |b\rangle) \text{ and } |b\rangle \rightarrow \frac{1}{\sqrt{2}}(|a\rangle - |b\rangle). \qquad (13.12)$$

The state of atom A thus becomes

$$e^{ip\theta/2}\left[\cos(p\theta/2)|a\rangle_A - i\sin(p\theta/2)|b\rangle_A\right]. \qquad (13.13)$$

Now one chooses the interaction parameters so as to make $\theta = \pi$. Thus if the cavity 1-2 system is in $|\Psi^{(\pm)}\rangle$ state, which contains 1 photon, the state of atom A is reduced to $|b\rangle_A$ and for the state $|\Phi^{(\pm)}\rangle$, which contains either 0 or 2 photons, the atom emerges in state $|a\rangle_A$. Thus, just measuring the state of the ancillary atom A as it emerges from the two cavities one can determine which group (Ψ or Φ) the cavity state belongs to.

A word of caution is necessary at this point. Strictly speaking one can not really talk about the atomic state until it is measured, as at all stages before the measurement it is entangled with the cavity field. Whereas the treatment above simplifies the discussion and is, in principle, correct as at the end of the measurement, cavity state is unaffected as the photon number measurement was intended to be of a non demolition kind. To illustrate this point we give a more rigorous treatment below in terms of the complete state of all the cavities and the ancillary atom A.

To start with we consider the combined state of atom A and all cavities as

$$\frac{1}{2\sqrt{2}}\left(|a\rangle_A + |b\rangle_A\right)\left[\left|\Psi_{12}^{(+)}\right\rangle\left|\varphi_3^{(1)}\right\rangle + \left|\Psi_{12}^{(-)}\right\rangle\left|\varphi_3^{(2)}\right\rangle\right.$$
$$\left. + \left|\Phi_{12}^{(+)}\right\rangle\left|\varphi_3^{(3)}\right\rangle + \left|\Phi_{12}^{(-)}\right\rangle\left|\varphi_3^{(4)}\right\rangle\right],$$

$$(13.14)$$

in accordance with Equation 13.5. Here we have introduced an abridged notation for the possible states of the third cavity according to

$$\left|\varphi_3^{(1)}\right\rangle = \alpha|0\rangle_3 + \beta|1\rangle_3, \quad \left|\varphi_3^{(2)}\right\rangle = \alpha|0\rangle_3 - \beta|1\rangle_3,$$
$$\left|\varphi_3^{(3)}\right\rangle = \beta|0\rangle_3 + \alpha|1\rangle_3, \quad \left|\varphi_3^{(4)}\right\rangle = -\beta|0\rangle_3 + \alpha|1\rangle_3. \quad (13.15)$$

After the atom passes through both cavities, atomic level $|b\rangle_A$ acquires an appropriate phase as mentioned above according to the number of photons it sees in the cavities, and this reduces the atom-field state to the form

$$\frac{1}{2\sqrt{2}}\left\{|a\rangle_A\left[\left|\Psi_{12}^{(+)}\right\rangle\left|\varphi_3^{(1)}\right\rangle + \left|\Psi_{12}^{(-)}\right\rangle\left|\varphi_3^{(2)}\right\rangle\right.\right.$$
$$\left. + \left|\Phi_{12}^{(+)}\right\rangle\left|\varphi_3^{(3)}\right\rangle + \left|\Phi_{12}^{(-)}\right\rangle\left|\varphi_3^{(4)}\right\rangle\right]$$
$$- |b\rangle_A\left[\left|\Psi_{12}^{(+)}\right\rangle\left|\varphi_3^{(1)}\right\rangle + \left|\Psi_{12}^{(-)}\right\rangle\left|\varphi_3^{(2)}\right\rangle\right.$$
$$\left.\left. - \left|\Phi_{12}^{(+)}\right\rangle\left|\varphi_3^{(3)}\right\rangle - \left|\Phi_{12}^{(-)}\right\rangle\left|\varphi_3^{(4)}\right\rangle\right]\right\}. \quad (13.16)$$

Here we have assumed that the interaction parameters are already chosen in order to have $\theta = \pi$. Thus the state $|b\rangle_A$ acquires a phase

$e^{i\pi} = -1$. The next step of mixing the atomic levels through the Ramsey field (13.12) gives rise to the atom-cavity state

$$\frac{1}{2}\left\{|a\rangle_A \left[\left|\Phi_{12}^{(+)}\right\rangle \left|\varphi_3^{(3)}\right\rangle + \left|\Phi_{12}^{(-)}\right\rangle \left|\varphi_3^{(4)}\right\rangle\right]\right.$$
$$\left. + |b\rangle_A \left[\left|\Psi_{12}^{(+)}\right\rangle \left|\varphi_3^{(1)}\right\rangle + \left|\Psi_{12}^{(-)}\right\rangle \left|\varphi_3^{(2)}\right\rangle\right]\right\}. \qquad (13.17)$$

From a careful observation of the above equation it is clear that, after the measurement, if the atom is found in the state $|a\rangle_A$ the cavity state is one of the $|\Phi^{(\pm)}\rangle$ states, whereas for the atomic state $|b\rangle_A$ the cavity is in one of the $|\Psi^{(\pm)}\rangle$ states. It is interesting to note that this measurement is equivalent to the measurement of the parity of the photon field proposed by Englert *et al.* [44] in the sense that the odd or even character of the total number of photons in cavities 1 and 2 taken together is being probed. Their method corresponds to a single cavity field but can be easily extended to the combined state of two cavities, as required in the present case.

It still remains to resolve the four states completely as we have not separated the $\pm$ states from both the groups. We need two more ancillary atoms B and C to achieve this. We take atom B to be in the ground state $|b\rangle_B$; thus at this stage the state of the whole system becomes

$$\frac{1}{2}|b\rangle_B \left\{|a\rangle_A \left[\left|\Phi_{12}^{(+)}\right\rangle \left|\varphi_3^{(3)}\right\rangle + \left|\Phi_{12}^{(-)}\right\rangle \left|\varphi_3^{(4)}\right\rangle\right]\right.$$
$$\left. + |b\rangle_A \left[\left|\Psi_{12}^{(+)}\right\rangle \left|\varphi_3^{(1)}\right\rangle + \left|\Psi_{12}^{(-)}\right\rangle \left|\varphi_3^{(2)}\right\rangle\right]\right\}. \qquad (13.18)$$

Here we have assumed that the measurement on atom A is not performed yet. This does not change anything that is achieved till now as atom A is untouched in the rest of the process. Now we let atom B interact with the first cavity, cavity 1, with the interaction time such that it sees a π pulse and absorbs a single photon from the cavity if there is any. Such an interaction transforms the state of the system to

$$\frac{1}{2}|0\rangle_1 \left\{|a\rangle_A \left[\left|\varphi_3^{(3)}\right\rangle (|b\rangle_B |0\rangle_2 + |a\rangle_B |1\rangle_2) \left|\varphi_3^{(4)}\right\rangle (|b\rangle_B |0\rangle_2 - |a\rangle_B |1\rangle_2)\right]\right.$$
$$\left. |b\rangle_A \left[\left|\varphi_3^{(1)}\right\rangle (|b\rangle_B |1\rangle_2 + |a\rangle_B |0\rangle_2) \left|\varphi_3^{(4)}\right\rangle (|b\rangle_B |1\rangle_2 - |a\rangle_B |0\rangle_2)\right]\right\}. $$
$$(13.19)$$

At this point we allow the states of atom B to mix according to Equation 13.12, through a classical Ramsey field giving the state

$$\frac{1}{2\sqrt{2}}|0\rangle_1 \left\{|a\rangle_A \left[\left|\varphi_3^{(3)}\right\rangle \left(|a\rangle_B (|0\rangle_2 + |1\rangle_2) - |b\rangle_B (|0\rangle_2 - |1\rangle_2)\right)\right.\right.$$

$$+ \left|\varphi_3^{(4)}\right\rangle \left(|a\rangle_B \left(|0\rangle_2 - |1\rangle_2\right) - |b\rangle_B \left(|0\rangle_2 + |1\rangle_2\right)\right)\Big]$$
$$+ |b\rangle_A \Big[\left|\varphi_3^{(1)}\right\rangle \left(|a\rangle_B \left(|0\rangle_2 + |1\rangle_2\right) + |b\rangle_B \left(|0\rangle_2 - |1\rangle_2\right)\right)$$
$$+ \left|\varphi_3^{(2)}\right\rangle \left(-|a\rangle_B \left(|0\rangle_2 - |1\rangle_2\right) - |b\rangle_B \left(|0\rangle_2 + |1\rangle_2\right)\right)\Big]\Big\}$$

$$(13.20)$$

for the complete system. The detection process is not finished yet. What we have accomplished so far is to eat up a photon in cavity 1 and transform that information to the atom B. We have to do the same with cavity 2 and we follow exactly the same procedure with the third ancillary atom C. Thus the interaction of atom C, initially in the state $|b\rangle_C$, and cavity 2 give the complete state of the system as

$$\frac{1}{2\sqrt{2}} |0\rangle_1 |0\rangle_2 \Big\{ |a\rangle_A \Big[\left|\varphi_3^{(3)}\right\rangle \left(|a\rangle_B \left(|b\rangle_C + |a\rangle_C\right) - |b\rangle_B \left(|b\rangle_C - |a\rangle_C\right)\right)$$
$$+ \left|\varphi_3^{(4)}\right\rangle \left(|a\rangle_B \left(|b\rangle_C - |a\rangle_C\right) - |b\rangle_B \left(|b\rangle_C + |a\rangle_C\right)\right)\Big]$$
$$+ |b\rangle_A \Big[\left|\varphi_3^{(1)}\right\rangle \left(|a\rangle_B \left(|b\rangle_C + |a\rangle_C\right) + |b\rangle_B \left(|b\rangle_C - |a\rangle_C\right)\right)$$
$$- \left|\varphi_3^{(2)}\right\rangle \left(|a\rangle_B \left(|b\rangle_C - |a\rangle_C\right) |b\rangle_B \left(|b\rangle_C + |a\rangle_C\right)\right)\Big]\Big\}.$$

$$(13.21)$$

The mixing of states of atom C as per Equation 13.12 reduces the state to

$$\frac{1}{2} \Big\{ \left|\varphi_3^{(1)}\right\rangle \left(|b\rangle_A |a\rangle_B |a\rangle_C - |b\rangle_A |b\rangle_B |b\rangle_C\right)$$
$$+ \left|\varphi_3^{(2)}\right\rangle \left(|b\rangle_A |a\rangle_B |b\rangle_C - |b\rangle_A |b\rangle_B |a\rangle_C\right)$$
$$+ \left|\varphi_3^{(3)}\right\rangle \left(|a\rangle_A |a\rangle_B |a\rangle_C + |a\rangle_A |b\rangle_B |b\rangle_C\right)$$
$$- \left|\varphi_3^{(4)}\right\rangle \left(|a\rangle_A |a\rangle_B |b\rangle_C + |a\rangle_A |b\rangle_B |a\rangle_C\right)\Big\}.$$

$$(13.22)$$

Thus one can see that a particular sequence of the result of state-measurement for the atoms A, B and C corresponds to one of the Bell states for cavities 1 and 2, and it reduces cavity 3 to the appropriate state, as summarized in the table below.

Now, the last step is to see how one can implement the unitary transformations of Equation 13.7 to transfer the states in Equation 13.6 to

Measurement Result	Cavity 1-2 State	Cavity 3 State
$\lvert b\rangle_A \lvert a\rangle_B \lvert a\rangle_C$ or $\lvert b\rangle_A \lvert b\rangle_B \lvert b\rangle_C$	$\lvert \Psi_{12}^{(+)}\rangle$	$\lvert \varphi_3^{(1)}\rangle$
$\lvert b\rangle_A \lvert a\rangle_B \lvert b\rangle_C$ or $\lvert b\rangle_A \lvert b\rangle_B \lvert a\rangle_C$	$\lvert \Psi_{12}^{(-)}\rangle$	$\lvert \varphi_3^{(2)}\rangle$
$\lvert a\rangle_A \lvert a\rangle_B \lvert a\rangle_C$ or $\lvert a\rangle_A \lvert b\rangle_B \lvert b\rangle_C$	$\lvert \Phi_{12}^{(+)}\rangle$	$\lvert \varphi_3^{(3)}\rangle$
$\lvert a\rangle_A \lvert a\rangle_B \lvert b\rangle_C$ or $\lvert a\rangle_A \lvert b\rangle_B \lvert a\rangle_C$	$\lvert \Phi_{12}^{(-)}\rangle$	$\lvert \varphi_3^{(4)}\rangle$

Table 13.1 Bell-basis measurement for an entangled state of two cavities.

the original state in Equation 13.2 in order to complete the teleportation process. We describe the strategy one by one for all the four possible outcomes for cavity 3 after the Bell-basis measurement: (1) The first possible state of cavity 3, $\alpha \lvert 0\rangle + \beta \lvert 1\rangle$, is precisely the original state; thus no operation is needed. (2) For the second state of $\alpha \lvert 0\rangle - \beta \lvert 1\rangle$ one needs to change the phase of state $\lvert 1\rangle$, which is possible through the non demolition measurement discussed above. We pass an atom in a superposition state $(\lvert a\rangle + \lvert b\rangle)/\sqrt{2}$, and arrange the interaction time such that the ground state acquires the phase factor -1 in the case of a single photon present in the cavity. Detection of an atom in state $\lvert b\rangle$ reduces the cavity to the state $\alpha \lvert 0\rangle + \beta \lvert 1\rangle$, which requires no further operation, whereas if one detects the atom in state $\lvert a\rangle$, one has to send in another atom and repeat the process until the emerging atom is detected in state $\lvert b\rangle$. (3) Now if the cavity is in the state $\beta \lvert 0\rangle + \alpha \lvert 1\rangle$ we just need to swap the probabilities. To achieve this, we pass an atom initially in the ground state so that the cavity field coherence is transferred to the atom leading to the state $\beta \lvert b\rangle + \alpha \lvert a\rangle$. We now operate the atom with a classical field such that the states are switched thus, turning it into the state $\beta \lvert a\rangle + \alpha \lvert b\rangle$, which is then sent through an empty cavity. At the end of the passage the cavity field takes the form $\alpha \lvert 0\rangle + \beta \lvert 1\rangle$ as required. Even though we used an extra cavity here, in principle, it is not really necessary and one can achieve the whole process in a single cavity itself through time-dependent manipulation of the fields. (4) For the last possible option of $-\beta \lvert a\rangle + \alpha \lvert b\rangle$ for the cavity state, both the swapping of probability and the phase shift as described in the last two cases are necessary. This completes all the processes necessary to achieve teleportation with 100% fidelity.

13.3   Discrete N-state quantum systems

We have discussed above how one can achieve teleportation of a quantum state of a two-state system. Even though the treatment was fairly general and applicable to a variety of 2-state systems of both atoms and photons, it is instructive to see the extension of the methodology to a more general quantum state defined in N-dimensional Hilbert space. Herein we discuss such a case and follow the treatment by Zubairy [45], for its underlying simplicity. A mathematically more general approach has been considered by Stenholm and Bardroff [46] and the original Bennett et al. paper [2] also discusses briefly a scheme for the teleportation of a discrete N dimensional system. The method discussed here is an extension of that suggested by Bennett applied to an arbitrary superposition of Fock states, or photon number states of a microwave cavity. We also discuss the extension of the cavity QED implementation scheme considered above for this case.

The state of the radiation field to be teleported is taken to be a coherent superposition of N Fock states, i.e.,

$$|\psi_1\rangle = \sum_{l=0}^{N-1} w_l |l\rangle_1 . \tag{13.23}$$

It is prepared in cavity 1 to start with and at the end of the process it is to reappear in cavity 3. The cavities are to have high-Q values to assure the success of the process. However, it is not a stringent requirement and recently a scheme has been presented that exploits the cavity decay for the teleportation of the atomic state of an atom trapped in a leaky cavity [15]. It should be noted that the method is valid only for $N = 2^n$, with n being an integer. As in the case of 2-state systems one needs three steps to complete the teleportation process.

To start with, additional cavities 2 and 3 are prepared in the quantum-entangled state,

$$|\psi_{23}\rangle = \frac{1}{\sqrt{N}} \sum_{k=0}^{N-1} |N - 1 - k\rangle_2 |k\rangle_3 . \tag{13.24}$$

The combined state of the fields in cavities 1, 2 and 3 can therefore be described by

$$|\Psi_{123}\rangle = \frac{1}{\sqrt{N}} \sum_{j,k=0}^{N-1} w_j \, |j\rangle_1 \, |N-1-k\rangle_2 \, |k\rangle_3 \,. \tag{13.25}$$

The N^2 basis states for the 1-2 system, which are analogous to the Bell-basis states seen earlier, can be written as

$$\left|\psi_{12}^{(n,m)}\right\rangle = \frac{1}{\sqrt{N}} \sum_{j=0}^{N-1} e^{2\pi i n j/N} \, |j\rangle_1 \, |(N-1-j-m) \bmod N\rangle_2 \,, \tag{13.26}$$

where $n, m = 0, 1, \ldots, (N-1)$. The combined state $|\Psi_{123}\rangle$ can then be rewritten as a linear superposition of the basis states $\left|\psi_{12}^{(n,m)}\right\rangle$ of the 1-2 system as follows:

$$|\Psi_{123}\rangle = \sum_{j,k,l=0}^{N-1} e^{-2\pi i l j/N} w_l \left|\psi_{12}^{(j,k)}\right\rangle |(k+l) \bmod N\rangle_3 \,. \tag{13.27}$$

In the second step one makes a measurement on the combined 1-2 system. A detection of the 1-2 system in the state $\left|\psi_{12}^{(j,k)}\right\rangle$ projects the field state in the cavity 3 onto

$$|\psi_3\rangle = \sum_{l=0}^{N-1} e^{-2\pi i l j/N} w_l \, |(k+l) \bmod N\rangle_3 \,. \tag{13.28}$$

The field state in cavity 3 has thus been projected to a state that has all the information about the amplitudes w_l. In the third and final step of the quantum teleportation, a manipulation of the field in cavity 3 needs to be done to bring state (13.28) to form (13.23), which is analogous to the unitary transformations (13.7) of Section 13.2.1.

We outline the essentials of the implementation of these steps of quantum teleportation using the known and experimentally viable methods based on atom-field interaction; the details can be seen in [45]. In some sense the whole treatment here is an extension of the implementation scheme discussed earlier for teleportation of a two-state system. A generalization of the cavity state preparation method by Vogel, Akulin, and Schleich [47] can be utilized to accomplish the first step of entangled state preparation between cavities 2 and 3 [48]. One sends $N-1$ two-level atoms in the excited state through the cavities 2 and 3 which interact with the resonant modes of the electromagnetic field inside the two cavities. The field inside the two cavities is initially in the vacuum state. The interaction times for the atoms with the field can be controlled by controlling the velocity of the atoms. One then has a choice of $2N - 2$ interaction parameters, two for each atom in the cavities 2 and 3. These parameters can be chosen such that if all the atoms are found in the ground state after the passage through the cavities, the entangled state (13.24) is generated. If, however, any one of the $N - 1$ atoms is found to be in the excited state after the passage through the cavities, it is imperative to *empty* the cavities and start over again.

To illustrate the details of the process one can consider a superposition of only four states, i.e.,

$$|\psi_1\rangle = w_0 |0\rangle_1 + w_1 |1\rangle_1 + w_2 |2\rangle_1 + w_3 |3\rangle_1 . \tag{13.29}$$

In this case one needs the entangled state

$$|\psi_{23}\rangle = \frac{1}{2} \left[|3\rangle_2 |0\rangle_3 + |2\rangle_2 |1\rangle_3 + |1\rangle_2 |2\rangle_3 + |0\rangle_2 |3\rangle_3 \right] \tag{13.30}$$

for the cavities 2 and 3. One needs to pass three atoms labeled a, b and c through the cavities. The interaction times needed to generate this state can be calculated such that $g\tau_{a_2} = 5.0915, g\tau_{a_3} = 1.5708, g\tau_{b_2} = 7.2596, g\tau_{b_3} = 1.1107, g\tau_{c_2} = 7.1704$ and $g\tau_{c_3} = 6.7436$, where g signifies the strength of the interaction between the atoms and the cavity and τ_{α_2} and τ_{α_3} are the passage times for the αth atom inside the cavities 2 and 3, respectively. The parameters are chosen such that the probability of detecting the first two atoms in the ground state is unity. The probability of detecting the third atom in the ground state is, however, about 4%. Thus an average of 25 tries are required before the desired entangled state is produced in the cavities 2 and 3. The field inside cavity 1 remains unaffected in the process, as one would require.

In the next step one needs to perform a measurement to determine the state $|\psi_{12}^{(n,m)}\rangle$ of the combined 1-2 system. A careful look at the above expression (13.26) reveals that the subscript m can be obtained from the total number of photons inside the two cavities, whereas n can be inferred from the relative phase of the states under the summation, which happens to be independent of m. Thus the state of the 1-2 system can be determined in two sets of measurements, the first determining m via the total number of photons inside the two cavities, and the second determining n via the relative phase.

Apart from the state with $m = 0$, which contains $N - 1$ photons in all its constituent states, an arbitrary state $|\psi_{12}^{(n,m)}\rangle$ has two possible number of photons, $N - 1 - m$ and $2N - 1 - m$. For example, when $N = 4$, the states $|\psi_{12}^{(0,m)}\rangle$ are given by

$$\left|\psi_{12}^{(0,0)}\right\rangle = \frac{1}{2}\Big(|0\rangle_1 |3\rangle_2 + |1\rangle_1 |2\rangle_2 + |2\rangle_1 |1\rangle_2 + |3\rangle_1 |0\rangle_2\Big)$$

$$\left|\psi_{12}^{(0,1)}\right\rangle = \frac{1}{2}\Big(|0\rangle_1 |2\rangle_2 + |1\rangle_1 |1\rangle_2 + |2\rangle_1 |0\rangle_2 + |3\rangle_1 |3\rangle_2\Big)$$

$$\left|\psi_{12}^{(0,2)}\right\rangle = \frac{1}{2}\Big(|0\rangle_1 |1\rangle_2 + |1\rangle_1 |0\rangle_2 + |2\rangle_1 |3\rangle_2 + |3\rangle_1 |2\rangle_2\Big)$$

$$\left|\psi_{12}^{(0,3)}\right\rangle = \frac{1}{2}\Big(|0\rangle_1 |0\rangle_2 + |1\rangle_1 |3\rangle_2 + |2\rangle_1 |2\rangle_2 + |3\rangle_1 |1\rangle_2\Big). \quad (13.31)$$

Thus the states $|\psi_{12}^{(n,0)}\rangle$, $|\psi_{12}^{(n,1)}\rangle$, $|\psi_{12}^{(n,2)}\rangle$ and $|\psi_{12}^{(n,3)}\rangle$ have three, (two or six), (one or five) and (zero or four) photon numbers, respectively. The number of photons can be determined via Ramsey interferometry, as discussed in Section 13.2.2. We have seen the details of how the Bell-basis measurement can be achieved for a 2-state cavity field. Therein the distinction between odd and even number of photons in the two cavities was sufficient in the first step, whereas the N state Bell-basis measurement is very much involved and we need to separate all the m values. This process, in principle, needs many more ancillary atoms and they have to be sent in with different phase parameters like $\theta = \pi, \pi/2, \pi/4, \pi/8$, etc. Once the distinction between the m values is accomplished, one starts emptying the cavities one photon at a time, as in the simpler two-state case. However, a method based on adiabatic passage, which uses three level atoms, is needed in the present case to achieve this. Implementation of the third step involves transforming the state of cavity 3 as in Equation 13.28 into a basis appropriate to the input state of cavity 1 as in Equation 13.23. As one can see, the difference in these states is only in terms of phase factors and the displacement of photon numbers. As it turns out Ramsey interferometry can be used in this case also. We refer an interested reader to the original article by Zubairy [45] and references therein for more details of the complete implementation scheme.

13.4   Entangled state teleportation

The extension of teleportation ideas to the systems of interest to quantum computing warrants attention. In this context one needs to manipulate many qubit states which are almost always entangled. It is interesting to see how one can teleport entangled atomic or field states. In this section we discuss teleportation of an entangled state of the radiation field first for the two-qubit case and then a more general N-qubit state as per Ikram et al. [49]. Just to note, the notation we use here is a little bit different from the other sections to simplify the discussion.

13.4.1   Two-qubit entangled state

The cavity entangled state is prepared in two separated high Q cavities A_1 and A_2 and it is to be teleported to another pair of high Q cavities C_1 and C_2. The state of the radiation field can be taken to be

$$|\psi(A_1 A_2)\rangle = \sum_{p_1,p_2=0}^{1} \alpha_{p_1 p_2} |p_1, p_2\rangle. \qquad (13.32)$$

It may be pointed out that this scheme corresponds to the teleportation of entangled two-level atomic states also, as the atomic entanglement can be transferred to the two cavities by passing them through the cavities with π pulse. As usual the teleportation of state (13.32) can be carried out in three steps. The schematic view of the process is shown in the Figure 13.2.

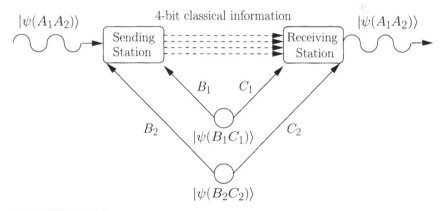

FIGURE 13.2
Quantum teleportation of an entangled state. Two entangled pairs of states $|\psi(B_1 C_1)\rangle$ and $|\psi(B_2 C_2)\rangle$ are shared between the sending and receiving stations to achieve teleportation of a two-bit entangled state $|\psi(A_1 A_2)\rangle$ through a 4-bit classical channel.

In the first step, one needs another set of cavities B_1, B_2 such that the pair of cavities B_1, C_1 and the pair of cavities B_2, C_2 are prepared in entangled states, i.e.,

$$|\psi(B_1C_1)\rangle = \frac{1}{\sqrt{2}}[|0_{B_1}, 1_{C_1}\rangle + |1_{B_1}, 0_{C_1}\rangle], \qquad (13.33)$$

$$|\psi(B_2C_2)\rangle = \frac{1}{\sqrt{2}}[|0_{B_2}, 1_{C_2}\rangle + |1_{B_2}, 0_{C_2}\rangle]. \qquad (13.34)$$

The combined state can therefore be written as

$$|\psi(B_1B_2C_1C_2)\rangle = \frac{1}{2}[|0_{B_1}, 0_{B_2}, 1_{C_1}, 1_{C_2}\rangle + |0_{B_1}, 1_{B_2}, 1_{C_1}, 0_{C_2}\rangle$$
$$+ |1_{B_1}, 0_{B_2}, 0_{C_1}, 1_{C_2}\rangle + |1_{B_1}, 1_{B_2}, 0_{C_1}, 0_{C_2}\rangle].$$
$$(13.35)$$

It is important to note that for the teleportation of a two-qubit entangled quantum state one does not need to prepare an entangled state of four qubits, and two entangled states of two qubits each are sufficient for the purpose. The combined state of the fields in the cavities A_1, A_2, B_1, B_2, C_1 and C_2 is thus given by

$$|\psi(A_1A_2B_1B_2C_1C_2)\rangle = \frac{1}{2}\sum_{p_1,p_2=0}^{1}\alpha_{p_1p_2}|p_1\rangle_{A_1}|p_2\rangle_{A_2}$$
$$\Big[|0_{B_1}, 0_{B_2}, 1_{C_1}, 1_{C_2}\rangle + |0_{B_1}, 1_{B_2}, 1_{C_1}, 0_{C_2}\rangle$$
$$+ |1_{B_1}, 0_{B_2}, 0_{C_1}, 1_{C_2}\rangle + |1_{B_1}, 1_{B_2}, 0_{C_1}, 0_{C_2}\rangle\Big].$$
$$(13.36)$$

Next one needs to define the basis states for the $A_1A_2B_1B_2$ system:

$$|\psi_{j_1,j_2,0,0}(A_1A_2B_1B_2)\rangle = \frac{1}{2}\Big[|0_{A_1}, 0_{A_2}, 1_{B_1}, 1_{B_2}\rangle$$
$$+ e^{i\pi j_2}|0_{A_1}, 1_{A_2}, 1_{B_1}, 0_{B_2}\rangle$$
$$+ e^{i\pi j_1}|1_{A_1}, 0_{A_2}, 0_{B_1}, 1_{B_2}\rangle$$
$$+ e^{i\pi(j_1+j_2)}|1_{A_1}, 1_{A_2}, 0_{B_1}, 0_{B_2}\rangle\Big], (13.37)$$

$$|\psi_{j_1,j_2,0,1}(A_1A_2B_1B_2)\rangle = \frac{1}{2}\Big[|0_{A_1}, 0_{A_2}, 1_{B_1}, 0_{B_2}\rangle$$

$$+e^{i\pi j_2} |0_{A_1}, 1_{A_2}, 1_{B_1}, 1_{B_2}\rangle$$
$$+e^{i\pi j_1} |1_{A_1}, 0_{A_2}, 0_{B_1}, 0_{B_2}\rangle$$
$$+e^{i\pi(j_1+j_2)} |1_{A_1}, 1_{A_2}, 0_{B_1}, 1_{B_2}\rangle \Big], \quad (13.38)$$

$$|\psi_{j_1,j_2,1,0}(A_1 A_2 B_1 B_2)\rangle = \frac{1}{2}\Big[|0_{A_1}, 0_{A_2}, 0_{B_1}, 1_{B_2}\rangle$$
$$+e^{i\pi j_2} |0_{A_1}, 1_{A_2}, 0_{B_1}, 0_{B_2}\rangle$$
$$+e^{i\pi j_1} |1_{A_1}, 0_{A_2}, 1_{B_1}, 1_{B_2}\rangle$$
$$+e^{i\pi(j_1+j_2)} |1_{A_1}, 1_{A_2}, 1_{B_1}, 0_{B_2}\rangle \Big], \quad (13.39)$$

$$|\psi_{j_1,j_2,1,1}(A_1 A_2 B_1 B_2)\rangle = \frac{1}{2}\Big[|0_{A_1}, 0_{A_2}, 0_{B_1}, 0_{B_2}\rangle$$
$$+e^{i\pi j_2} |0_{A_1}, 1_{A_2}, 0_{B_1}, 1_{B_2}\rangle$$
$$+e^{i\pi j_1} |1_{A_1}, 0_{A_2}, 1_{B_1}, 0_{B_2}\rangle$$
$$+e^{i\pi(j_1+j_2)} |1_{A_1}, 1_{A_2}, 1_{B_1}, 1_{B_2}\rangle \Big], \quad (13.40)$$

where $j_1, j_2 = 0, 1$. Thus there are 16 basis states. It should be noted that these basis vectors are analogous to the Bell-basis considered in the case of the two-state system.

The combined state $|\psi(A_1 A_2 B_1 B_2 C_1 C_2)\rangle$ can be rewritten as a linear superposition of the basis states $|\psi_{j_1,j_2,k_1,k_2}(A_1 A_2 B_1 B_2)\rangle$ of the $A_1 A_2 B_1 B_2$ system as follows:

$$|\psi(A_1 A_2 B_1 B_2 C_1 C_2)\rangle = \sum_{j_1,j_2=0}^{1} |\psi_{j_1,j_2,0,0}(A_1 A_2 B_1 B_2)\rangle$$
$$[\alpha_{00} |0_{C_1}, 0_{C_2}\rangle + \alpha_{01} e^{i\pi j_2} |0_{C_1}, 1_{C_2}\rangle$$
$$+\alpha_{10} e^{i\pi j_1} |1_{C_1}, 0_{C_2}\rangle + \alpha_{11} e^{i\pi(j_1+j_2)} |1_{C_1}, 1_{C_2}\rangle]$$
$$+ |\psi_{j_1,j_2,0,1}(A_1 A_2 B_1 B_2)\rangle$$
$$[\alpha_{00} |0_{C_1}, 1_{C_2}\rangle + \alpha_{01} e^{i\pi j_2} |0_{C_1}, 0_{C_2}\rangle$$
$$+\alpha_{10} e^{i\pi j_1} |1_{C_1}, 1_{C_2}\rangle + \alpha_{11} e^{i\pi(j_1+j_2)} |1_{C_1}, 0_{C_2}\rangle]$$
$$+ |\psi_{j_1,j_2,1,0}(A_1 A_2 B_1 B_2)\rangle$$
$$[\alpha_{00} |1_{C_1}, 0_{C_2}\rangle + \alpha_{01} e^{i\pi j_2} |1_{C_1}, 1_{C_2}\rangle$$
$$+\alpha_{10} e^{i\pi j_1} |0_{C_1}, 0_{C_2}\rangle + \alpha_{11} e^{i\pi(j_1+j_2)} |0_{C_1}, 1_{C_2}\rangle]$$

$$+ |\psi_{j_1,j_2,1,1}(A_1 A_2 B_1 B_2)\rangle$$
$$[\alpha_{00} |1_{C_1}, 1_{C_2}\rangle + \alpha_{01} e^{i\pi j_2} |1_{C_1}, 0_{C_2}\rangle$$
$$+ \alpha_{10} e^{i\pi j_1} |0_{C_1}, 1_{C_2}\rangle + \alpha_{11} e^{i\pi(j_1+j_2)} |0_{C_1}, 0_{C_2}\rangle].$$

$$(13.41)$$

In the second step a measurement is performed on the $A_1 A_2 B_1 B_2$ system. A detection of the $A_1 A_2 B_1 B_2$ system in one of the sixteen basis states $|\psi_{j_1,j_2,k_1,k_2}(A_1 A_2 B_1 B_2)\rangle$ projects the field state in the cavities $C_1 C_2$ into

$$|\psi(C_1 C_2)\rangle = \sum_{p_1,p_2=0}^{1} e^{i\pi(j_1 p_1 + j_2 p_2)} \alpha_{p_1 p_2}$$
$$|(k_1 + p_1) \bmod 2\rangle_{C_1} |(k_2 + p_2) \bmod 2\rangle_{C_2}. \quad (13.42)$$

The field state in the cavities $C_1 C_2$ has thus been projected to a state containing all the information about the amplitudes $\alpha_{p_1 p_2}$. In the third and final step of the quantum teleportation, a manipulation of the cavities $C_1 C_2$ needs to be done to bring state (13.42) to form (13.32).

The implementation of this scheme goes exactly parallel to the teleportation of a cavity field as in Section 13.2.2, with an added complication that one has to deal with a double-sized system this time. Namely, one can use two-level atoms to prepare the cavities in the entangled states. Then the measurement and transformation steps can be carried out utilizing Ramsey spectroscopy. The details are involved and can be found in [49].

13.4.2 N-qubit entangled state

It is interesting to consider generalization of the above scheme to a general N-qubit entangled field state. Such an entangled state in N high-Q cavities labeled $A_i(i = 1, 2, \ldots, N)$ can be taken to be

$$|\psi(A_1 \ldots A_N)\rangle = \sum_{n_1,\ldots,n_N=0}^{1} \alpha_{n_1,\ldots,n_N} |n_1, n_2, \ldots, n_N\rangle. \quad (13.43)$$

The goal is to teleport this state to another set of high-Q cavities $C_i(i = 1, 2, \ldots, N)$.

In the first step of the process one needs N pairs of entangled cavities

$$|\psi(B_i C_i)\rangle = \frac{1}{\sqrt{2}} \left[|0\rangle_{B_i} |1\rangle_{C_i} + |1\rangle_{B_i} |0\rangle_{C_i} \right] \quad (13.44)$$

where $i = 1, 2, \ldots, N$. These N entangled states of 2-qubits each can be prepared as earlier by passing two-level atoms initially in an excited state through the two resonant cavities and by setting $\pi/2$ pulse and π pulse, respectively, in the two cavities. The first cavity of each entangled pair belongs to the sender and the second one to the receiver.

At this stage, 2^{2N} basis states for the combined state of cavities $(A_1 \ldots A_N B_1 \ldots B_N)$ can be defined as

$$|\psi_{j_1,\ldots,j_N,k_1,\ldots,k_N}(A_1 \ldots A_N B_1 \ldots B_N)\rangle$$

$$= \sum_{p_1,\ldots,p_N}^{1} \exp\left[i\pi\left(j_1 p_1 + j_2 p_2 + \ldots + j_N p_N\right)\right]$$

$$\times |p_1\rangle_{A_i} |p_2\rangle_{A_2} \cdots |p_N\rangle_{A_N} |(1 - p_1 - k_1) \bmod 2\rangle_{B_1}$$

$$\times |(1 - p_2 - k_2) \bmod 2\rangle_{B_2} \cdots |(1 - p_N - k_N) \bmod 2\rangle_{B_N}$$

$$= \sum_{p_1,\ldots\ldots,p_N}^{1} \prod_{m=1}^{N} \left[e^{i\pi j_m p_m} |p_m\rangle_{A_m} |(1 - p_m - k_m) \bmod 2\rangle_{B_m}\right].$$

$$(13.45)$$

This forms the basis in which the sender will be required to perform the measurement to generate the classical message to be transferred to the receiver. The combined state of the complete system of cavities $(A_1 \ldots A_N, B_1 \ldots B_N, C_1 \ldots C_N)$ in terms of these basis states can be written as

$$|\psi(A_i B_i C_i)\rangle = \sum_{j_1,\ldots,j_N,k_1\ldots,k_N,p_1,\ldots,p_N}^{1} \alpha_{p_1,\ldots,p_N}$$

$$|\psi_{j_1,\ldots j_N,k_1,\ldots k_N}(A_1\ldots A_N B_1\ldots B_N)\rangle$$

$$\times \prod_{m=1}^{N} e^{i\pi j_m p_m} |(p_m + k_m) \bmod 2\rangle_{C_m}. \quad (13.46)$$

A determination of Bell state $|\psi_{j_1,\ldots,j_N,k_1,\ldots,k_N}(A_1\ldots A_N B_1\ldots B_N)\rangle$ projects the state of the field in cavities $C_1, C_2, \ldots, C_N$ in to the entangled state $|\psi(C_1\ldots C_N)\rangle$ as

$$|\psi(C_1\ldots C_N)\rangle = \sum_{p_1,\ldots,p_N}^{1} C_{p_1,\ldots,p_N} \prod_{m=1}^{N} e^{i\pi j_m p_m} |(p_m + k_m) \bmod 2\rangle_{C_m}.$$

$$(13.47)$$

In the final step of the quantum teleportation, a manipulation of the cavities $C_1, C_2, \ldots, C_N$ needs to be done to bring state $|\psi(C_1 C_2 \ldots C_N)\rangle$ to form $|\psi(A_1 A_2 \ldots A_N)\rangle$.

Once again the details of the implementation are a straightforward extension of that of the simple 1-qubit state or that of a 2-qubit entangled state, and can be found in Reference [49].

13.5   Continuous quantum variable states

The schemes considered so far deal with discrete variables to achieve teleportation of a quantum system. Vaidman [32] proposed teleportation of the wave function of a single particle beyond the context of discrete degrees of freedom based on the nonlocal measurement concept of Aharonov et al. [50] and their further work [51, 52]. This proposal can be naturally extended to accomplish teleportation of a quantum system with continuous degrees of freedom, for example, a position or momentum wave function of a particle or a state of an electromagnetic field inside a cavity. The scheme uses an entangled pair to be shared by the sender and receiver as the EPR state with perfect correlations in position and momentum. One can, in principle, use quantum correlations between any two canonically conjugate variables for the purpose. This concept was further extended by Braunstein and Kimble [53] for a general system characterized by dynamical variables with continuous spectra. Their original proposal dealt with finite nonsingular correlations between the entangled systems; we here, however, present a simplified treatment in accord with perfect entanglement, which can be looked upon as a special case of their original consideration. First we discuss the Vaidman proposal for its relevance to the states of continuous degrees of freedom and then the Braunstein–Kimble proposal.

13.5.1   Nonlocal measurements

One starts with two similar systems situated far away from each other and characterized by continuous variables q_1 and q_2 with corresponding conjugate variables p_1 and p_2. A perfect entanglement among these states is an essential requirement. In this context, the "crossed" nonlocal measurements

$$q_1(t_1) - q_2(t_2) = a,$$
$$p_1(t_2) - p_2(t_1) = b, \qquad (13.48)$$

with appropriate results hold the key to the teleportation process leaving the states of the systems after the measurement as

$$\psi_f(q_1) = e^{-ibq_1} \psi_2(q_1 - a),$$
$$\psi_f(q_2) = e^{ibq_2} \psi_1(q_2 + a). \qquad (13.49)$$

As one can see, the state of particle 2 after t_2 is the initial state of particle 1 shifted by $-a$ in q and by $-b$ in p. Similarly the state of particle 1 is the initial state of particle 2 shifted by a in q and b in p. With appropriate back shifts, one has achieved teleportation of the two systems into one another. It is customary to describe how one can achieve the nonlocal measurements of continuous variables as required by Equation 13.48. Similar to the case of the proposal by Bennett et al. [2] one needs a quantum channel to perform these measurements. To accomplish teleportation of a system characterized by a continuous variable, an EPR state with initial correlations like

$$Q_1 + Q_2 = 0,$$
$$P_1 - P_2 = 0, \qquad (13.50)$$

is required. Here Q_1 and Q_2 are continuous variables of the pair of auxiliary particles with corresponding conjugate momenta P_1 and P_2. Next step is to allow two local von Neumann-type interactions described by the Hamiltonian

$$H = g(t - t_1) P_1 q_1 - g(t - t_2) P_2 q_2, \qquad (13.51)$$

which lead to the final state of auxiliary particles such that

$$Q_{1f} + Q_{2f} = q_1(t_1) - q_2(t_2). \qquad (13.52)$$

Thus two local measurements of Q_{1f} and Q_{2f} yield a; another EPR pair is needed for measuring b as required in Equation 13.48. The scheme described here accomplishes two-way teleportation; one can also obtain one-way teleportation similar to the original Bennett proposal. The quantum channel is again taken as the EPR state. It turns out that the

Bell operator measurement for the spin-half system can be generalized to following two consecutive nonlocal measurements of variables

$$q + Q_1 \text{ and } p - P_1, \qquad (13.53)$$

for the case of systems characterized by continuous variables. We refer the interested reader to Reference [33] for the details. The outcomes of the measurements

$$q + Q_1 = a \text{ and } p - P_1 = b \qquad (13.54)$$

accomplish the teleportation of the quantum state $\psi(q)$, up to the known shifts, to the remote particle of the EPR pair, the state of which becomes

$$\psi_f(Q_2) = e^{ibQ_2}\psi(Q_2 + a). \qquad (13.55)$$

The last step consists of appropriate back shifts of the state in Q_2 and P_2 which result in transporting the quantum state $\psi(q)$ at site 1 to the same quantum state $\psi(Q_2)$ of the system at site 2.

13.5.2   Wigner functions

A natural way to deal with the quantum systems characterized by variables with continuous spectra is to use the Wigner function [54]. The unknown input state can be described by the Wigner function $W_1(\alpha_1)$, whereas $W_{23}(\alpha_2, \alpha_3)$ corresponds to the entangled systems 2 and 3. In such a case, the complete system can then be characterized by

$$W_T(\alpha_1, \alpha_2, \alpha_3) = W_1(\alpha_1)\, W_{23}(\alpha_2, \alpha_3) \qquad (13.56)$$

where $\alpha_j = x_j + ip_j$. Here, the real quantities (x_j, p_j) correspond to canonically conjugate variables describing either the position and momentum of the massive particle or the quadrature amplitudes of an electromagnetic field corresponding to the appropriate system. Figure 13.3 summarizes the essentials of the process schematically.

As we can see, beams 1 and 2, characterized by variables α_1 and α_2, are mixed into each other through a partially reflecting mirror. A natural way to describe the mixed state is to construct new variables along paths a and b of the interferometer. The new variables come out to be superpositions of the original ones according to

$$\beta_a = \frac{1}{\sqrt{2}}(\alpha_1 + \alpha_2) = x_a + ip_a,$$

$$\beta_b = \frac{-1}{\sqrt{2}}(\alpha_1 - \alpha_2) = x_b + ip_b. \qquad (13.57)$$

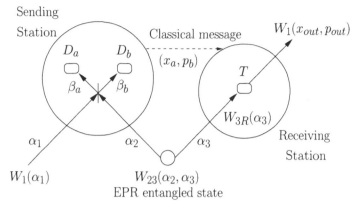

FIGURE 13.3
Continuous variable teleportation. Entangled state of systems represented by continuous variables α_2 and α_3 is shared between the sending and receiving stations. At the sending station the input state governed by α_1 is mixed through a partially reflecting mirror with one of the entangled system changing the characteristic variables of the system to β_a and β_b. Two measurements performed by detectors D_a and D_b generate a classical message (x_a, p_b) which is communicated to the receiver. At the receiving station a transformation is performed on the second entangled component using the classical message, thus accomplishing teleportation of the input state.

Here β_a and β_b are analogous to the original complex variables α_1 and α_2 for the new paths a and b, and the corresponding canonically conjugate variables are (x_a, p_a) and (x_b, p_b). The original pairs of canonically conjugate variables for the systems 1 and 2 can be reexpressed in terms of the coupled variables as

$$x_1 = \frac{1}{\sqrt{2}}\left(x_a - x_b\right),\, p_1 = \frac{1}{\sqrt{2}}\left(p_a - p_b\right),$$
$$x_2 = \frac{1}{\sqrt{2}}\left(x_a + x_b\right),\, p_2 = \frac{1}{\sqrt{2}}\left(p_a + p_b\right). \tag{13.58}$$

Thus, the Wigner function for the complete system in terms of the new coupled variables can be written in the form

$$W_{\mathrm{T}}(\alpha_1, \alpha_2, \alpha_3) = W_1\left(\frac{1}{\sqrt{2}}\left(x_a - x_b\right), \frac{1}{\sqrt{2}}\left(p_a - p_b\right)\right)$$
$$W_{23}\left(\frac{1}{\sqrt{2}}\left(x_a + x_b\right), \frac{1}{\sqrt{2}}\left(p_a + p_b\right); \alpha_3\right). \tag{13.59}$$

Now, on analogous lines with the nonlocal measurement technique one performs measurements of the commuting variables x_a and p_b at the sending station. As after the measurement of variables (x_a, p_b) the system contains no information about the other pair of variables (x_b, p_a), one has to integrate them out, resulting in the reduced Wigner function for system 3,

$$W_{3R}(\alpha_3) = \int dx_b \, dp_a \, W_{\mathrm{T}} \, (x_a, p_a, x_b, p_b, \alpha_2)$$

$$= \int dx_b \, dp_a \, W_1 \, W_{23} \, . \tag{13.60}$$

The classical information obtained through such a measurement is then transmitted to the receiver, depending on which the sender performs an appropriate transformation, along the lines of a nonlocal measurement scheme, to achieve complete teleportation. One needs a specific form for W_{23} to illustrate the transformation that needs to be performed at the receiver's end. The original proposal deals with a general form with variable degree of correlation between the quantum variables. We consider the special case of maximal entanglement between the shared quantum systems, which actually corresponds to an unphysical limit, but it gives the essentials of the scheme in a simplified manner. We start with

$$\begin{aligned} W_{23} \, (\alpha_2, \alpha_3) &= \delta(x_2 + x_3)\delta(p_2 - p_3) \\ &= 2\delta(x_a + x_b + \sqrt{2}x_3)\delta(p_a + p_b - \sqrt{2}p_3), \end{aligned} \tag{13.61}$$

to obtain the reduced Wigner function for system 3

$$W_{3R}(\alpha_3) \equiv W_1(x_3 + \sqrt{2}x_a, p_3 - \sqrt{2}p_b), \tag{13.62}$$

as a result of the measurement on the coupled 1-2 system. Now the following transformation of the variables

$$\frac{1}{\sqrt{2}} (x_a - x_b) \equiv x_{out} = x_3 + \sqrt{2}x_a,$$

$$\frac{1}{\sqrt{2}} (p_a - p_b) \equiv p_{out} = p_3 - \sqrt{2}p_b \tag{13.63}$$

is to be performed by the sender to arrive at

$$W_{3R} \equiv W_1(x_{out}, p_{out}), \tag{13.64}$$

thus accomplishing the teleportation of the input state. It is clear that the transformations performed by the receiver are just appropriate shifts in the variables depending on the classical information obtained from the sender, similar to the proposal based on non local measurement.

It is important to emphasize that the form (13.61) for the entangled state Wigner function corresponds to an ideal case of perfectly squeezed entangled states. In general, one does not have infinite squeezing and hence no maximal entanglement, resulting in less than perfect teleportation. Nevertheless, the scheme holds promise for teleportation of nonclassical states of the electromagnetic field with reasonable fidelity.

13.6 Concluding remarks

To summarize, we have discussed the teleportation proposals for quantum systems of various kinds, namely, a two-state system, a general superposition state in a N-dimensional basis, an entangled quantum state, and quantum states with continuous degrees of freedom. Even though we were not exhaustive in our choice of systems we have covered the most basic ones of them and believe that we provide a good starting point for an interested reader.

Quantum teleportation is an experimental reality and it holds tremendous potential for applications in the fields of quantum communication and computing. It is imperative to mention that the notable idea of dense coding [7], to increase the capacity of communication channels through quantum entanglement, is very similar to quantum teleportation and is quite pertinent to applications in quantum communication. Ideas such as entanglement swapping [37] and two-way teleportation [32], which are based on the quantum teleportation concept, hold a lot of promise in both the fields of quantum computing and communication. However, considerable technical obstacles still remain before quantum teleportation can become a useful means of secure long distance communication. The most significant difficulty arises in maintaining the entanglement for the time required to transfer the classical message. Any large-scale setup would require maintenance of enormous amounts of entangled pairs for practically infinite time in order to teleport a message at an arbitrary time as needed. However, concepts such as entanglement purification and quantum repeaters [55] are steps in the right direction and afford a lot of confidence in the practical applicability of the quantum teleportation concept.

Acknowledgments

We would like to thank Marlan Scully for insightful and valuable interactions. This work is supported by the Office of Naval Research, the Welch Foundation, DARPA, Air Force Research Laboratory and Telecommunication Initiative at TAMU.

References

[1] A. Shimony, in *Proceedings of the International Symposium on Foundations of Quantum Theory* (Physical Society of Japan, Tokyo, 1984).

[2] C. H. Bennett, G. Brassard, C. Crepeau, R. Jozsa, A. Peres, and W. K. Wootters, *Phys. Rev. Lett.* **70**, 1895 (1993).

[3] For various theoretical proposals see Refs. [13]–[30] and [32, 41, 45, 46, 49, 53] below.

[4] See Refs. [34]–[37] and [40] below.

[5] D. Deutsch, *Proc. R. Soc. London Ser.* A **400**, 97 (1985); D. Deutsch and R. Jozsa, *Proc. R. Soc. London Ser.* A **439**, 553 (1992).

[6] S. Wiesner, *Sigact News* **15**, 78 (1983); A. K. Ekert, *Phys. Rev. Lett.* **67**, 661 (1991); C. H. Bennett, *Phys. Rev. Lett.* **68**, 3121 (1992); A. K. Ekert, J. G. Rarity, P. R. Tapster, and G. M. Palma, *Phys. Rev. Lett.* **69**, 1293 (1992).

[7] C. H. Bennett and S. J. Wiesner, *Phys. Rev. Lett.* **69**, 2881 (1992).

[8] A. Einstein, B. Podolsky, and N. Rosen, *Phys. Rev.* **47**, 777 (1935).

[9] J. S. Bell, *Physics (N.Y.)* **1**, 195 (1964).

[10] J. S. Bell, *Speakable and Unspeakable in Quantum Mechanics* (Cambridge Univ. Press, Cambridge, England, 1988), p.196.

[11] D. Dieks, *Phys. Lett.* A **92**, 271 (1982).

[12] W. K. Wootters and W. H. Zurek, *Nature (London)* **299**, 802 (1982).

[13] L. Davidovich, N. Zagury, M. Brune, J. M. Raidmond, and S. Haroche, *Phys. Rev.* A **50**, R895 (1994).

[14] J. I. Cirac and A. S. Parkins, *Phys. Rev.* A **50**, R4441 (1994).

[15] S. Bose, P. L. Knight, M. B. Plenio, and V. Vedral, *Phys. Rev. Lett.* **83**, 5158 (1999).

[16] S.-B. Zheng and G.-C. Guo, *Phys. Lett.* A **232**, 171 (1997).

[17] M. H. Y. Moussa, *Phys. Rev.* A **54**, 4661 (1996).

[18] M. H. Y. Moussa, *Phys. Rev.* A **55** R3287 (1997).

[19] C. H. Bennett, G. Brassard, S. Popescu, B. Schumacher, J. A. Smolin, and W. K. Wootters, *Phys. Rev. Lett.* **76**, 722 (1996).

[20] C. S. Maierle, D. A. Lidar, and R. A. Harris, *Phys. Rev. Lett.* **81**, 5928 (1998).

[21] T. C. Ralph and P. K. Lam, *Phys. Rev. Lett.* **81**, 5668 (1998).

[22] G. J. Milburn and S. L. Braunstein, *Phys. Rev.* A **60**, 937 (1999).

[23] M. Koniorczyk, J. Janszky, and Z. Kis, *Phys. Lett.* A **256**, 334 (1999).

[24] T. Opatrny and G. Kurizki, *Phys. Rev. Lett.* **86**, 3180 (2001).

[25] A. Karlsson and M. Bourennane, *Phys. Rev.* A **58**, 4394 (1998).

[26] P. van Loock and S. L. Braunstein, *Phys. Rev.* A **61**, 010302 (1999).

[27] P. van Loock and S. L. Braunstein, *Phys. Rev. Lett.* **84**, 3482 (2000).

[28] P. van Loock, S. L. Braunstein, and H. J. Kimble, *Phys. Rev.* A **62**, 022309 (2000).

[29] S. L. Braunstein and H. J. Kimble, *Phys. Rev.* A **61**, 042302 (2000).

[30] S. L. Braunstein, G. M. D'Ariano, G. J. Milburn, and M. F. Sacchi, *Phys. Rev. Lett.* **84**, 3486 (2000).

[31] S. L. Braunstein, A. Mann, and M. Revzen, *Phys. Rev. Lett.* **68**, 3259 (1992).

[32] L. Vaidman, *Phys. Rev.* A **49**, 1473 (1994).

[33] L. Vaidman, N. Yoran, *Phys. Rev.* A **59**, 116 (1999).

[34] D. Bouwmeester, J.-W. Pan, K. Mattle, M. Eibl, H. Weinfurter, and A. Zeilinger, *Nature (London)* **390**, 575 (1997).

[35] D. Boschi, S. Branca, F. De Martini, L. Hardy, and S. Popescu, *Phys. Rev. Lett.* **80**, 1121 (1998).

[36] A. Furusawa, J. L. Sorensen, S. L. Braunstein, C. A. Fuchs, H. J. Kimble, and E. S. Polzik, *Science* **282**, 706 (1998).

[37] J.-W. Pan, D. Bouwmeester, H. Weinfurter, and A. Zeilinger, *Phys. Rev. Lett.* **80**, 3891 (1998).

[38] M. O. Scully, B.-G. Englert, and C. J. Bednar, *Phys. Rev. Lett.* **83**, 4433 (1999).

[39] D. Vitali, M. Fortunato, and P. Tombesi, *Phys. Rev. Lett.* **85**, 445 (2000).

[40] Y.-H. Kim, S. P. Kulik, and Y. Shih, *Phys. Rev. Lett.* **86**, 1370 (2001); Y. Shih, *Ann. Phys.* **10**, 19 (2001).

[41] E. Solano, C. L. Cesar, R. L. de Matos Filho, and N. Zagury, *Eur. Phys. J.* D **13**, 121 (2001).

[42] E. T. Jaynes and F. W. Cummings, *Proc. IEEE* **51**, 89 (1963).

[43] See, for example, M. O. Scully and M. S. Zubairy, *Quantum Optics* (Cambridge University Press, Cambridge, 1997), Sec. 19.3.

[44] B.-G. Englert, N. Sterpi, and H. Walther, *Opt. Comm.* **100**, 526 (1993).

[45] M. S. Zubairy, *Phys. Rev.* A **58**, 4368 (1998).

[46] S. Stenholm and P. J. Bardroff, *Phys. Rev.* A **58**, 4373 (1998).

[47] K. Vogel, V. M. Akulin, and W. P. Schleich, *Phys. Rev. Lett.* **71**, 1816 (1993).

[48] M. Ikram, S.-Y. Zhu, and M. S. Zubairy, *Opt. Comm.* **184**, 417 (2000)

[49] M. Ikram, S.-Y. Zhu, and M. S. Zubairy, *Phys. Rev.* A **62**, 022307 (2000).

[50] Y. Aharonov and D. Albert, *Phys. Rev.* D **21**, 3316 (1980); **24**, 359 (1981)

[51] Y. Aharonov, D. Albert, and C. K. Au, *Phys. Rev. Lett.* **47**, 1029 (1981).

[52] Y. Aharonov, D. Albert, and L. Vaidman, *Phys. Rev.* D **34**, 1805 (1986).

[53] S. L. Braunstein and H. J. Kimble, *Phys. Rev. Lett.* **80**, 869 (1998).

[54] E. P. Wigner, *Phys. Rev.* **40**, 749 (1932).

[55] See, for example, D.Bouwmeester, A. Ekert and A. Zeilinger, Eds., *The Physics of Quantum Information* (Springer-Verlag, Berlin, 2000), Chapter 8.

Quantum Secure Communication and Quantum Cryptography

Chapter 14

Communicating with qubit pairs

Almut Beige, Berthold-Georg Englert, Christian Kurtsiefer, and Harald Weinfurter

Abstract   All schemes for quantum cryptography that have been implemented in real-life experiments have in common that single qubits (carried by photons) are used for the exchange of quantum information between the parties. As a consequence, these schemes are *probabilistic*: they involve random processes that decide whether the next qubit will contribute to the shared key sequence or not. This is also an essential element in (almost) all generalizations of the simplest schemes, be it by introducing more choices in the same state space or by enlarging the state space. The situation is completely different in a scheme proposed recently. It uses qubit pairs for the transmission of quantum information (each pair carried by a single photon) and, as a benefit of this doubling, the scheme is *deterministic*: each and every qubit pair sent supplies one key bit. It is, therefore, even possible to communicate confidential messages securely and directly, that is: without the need for establishing a shared key first. We present a detailed analysis of the security of the qubit-pair schemes for cryptography and communication under eavesdropping attacks in which the quantum information in transit is intercepted and a substitute message is forwarded to conceal the interception.

1-58488-282/4/02/$0.00+$1.50
© 2002 by Chapman & Hall/CRC

14.1   Introduction

The seminal paper by Bennett and Brassard [1] has triggered wide-spread interest in quantum cryptography—that is, secure distribution of a key for encrypting and deciphering, protected against unnoticeable eavesdropping by the laws of quantum physics. Roughly speaking, the eavesdropper cannot overcome the fundamentally indeterministic nature of quantum phenomena and cannot avoid the disturbances that measurements on quantum objects bring about.

By now there are fair numbers of generalizations of the Bennett–Brassard scheme. Arguably the most intriguing one is Ekert's scheme of 1991 [2], which exploits the strong quantum correlations in entangled states of the Einstein–Podolsky–Rosen kind [3], inasmuch as an eavesdropper's interference results in a nonviolation of Bell's celebrated inequality [4].

Other generalizations extend the original scheme by either making use of more states in the same state space or by enlarging the state space. Examples are given by [5] and [6, 7], respectively. The monographs by Lo, Popescu, and Spiller [8], and by Bouwmeester, Ekert, and Zeilinger [9] deal with these matters and contain many useful references. This can also be said about the review article by Gisin, Ribordy, Tittel, and Zbinden [10].

We shall here describe and analyze a recently proposed scheme for quantum cryptography [11, 12] that is not just another generalization of these kinds but is markedly different from the Bennett–Brassard scheme and its variants. These standard schemes are *probabilistic* in the sense that one never knows if a quantum-coded bit of information in transmission will be turned into useful classical information upon detection. The novel scheme, by contrast, is *deterministic*: all quantum information sent will surely be useful classical information in the end. That this can be done despite the fundamental lack of determinism of quantum phenomena mentioned above, and without compromising security, is a pleasant surprise. Further, this deterministic feature of the scheme offers the option of secure communication without first establishing a shared key.

The new scheme is a late spin-off of a puzzling observation by Vaidman, Aharonov, and Albert in 1987 [13]. We discuss various features of it in Section 14.2 and thereby set the stage for presentation of the schemes for deterministic quantum cryptography in Section 14.4 and direct secure communication in Section 14.6. As a preparation of these treatments, we first give an account of the 1984 Bennett–Brassard scheme in Section 14.3. Idealized optical setups that realize the cryptography schemes in a single-photon fashion are the subject matter of Section 14.5.

14.2 The mean king's problem

14.2.1 The Vaidman–Aharonov–Albert puzzle

In 1987, Vaidman, Aharonov, and Albert published a paper [13] with the somewhat provocative title "How to ascertain the values of σ_x, σ_y, and σ_z of a spin-$\frac{1}{2}$ particle." It reports the solution of what has later become known as *The Mean King's Problem*. With permission from Verlag der Zeitschrift für Naturforschung, we quote from Reference [14]:

> A shipwrecked physicist gets stranded on a far-away island that is ruled by a mean king who loves cats and hates physicists since the day when he first heard what happened to Schrödinger's cat. A similar fate is awaiting the stranded physicist. Yet, mean as he is, the king enjoys defeating physicists on their own turf, and therefore he maliciously offers an apparently virtual chance of rescue.
>
> He takes the physicist to the royal laboratory, a splendid place where experiments of any kind can be performed perfectly. There the king invites the physicist to prepare a certain silver atom in any state she likes. The king's men will then measure one of the three cartesian spin components of this atom — they'll either measure σ_x, σ_y, or σ_z without, however, telling the physicist which one of these measurements is actually done. Then it is again the physicist's turn, and she can perform any experiment of her choosing. Only after she's finished with it, the king will tell her which spin component had been measured by his men. To save her neck, the physicist must then state correctly the measurement result that the king's men had obtained.
>
> Much to the king's frustration, the physicist rises to the challenge — and not just by sheer luck: She gets the right answer any time the whole procedure is repeated. How does she do it?

Readers who figure it out themselves rather than reading on will enjoy the lesson they learn about the wonderful things entanglement can do for you.

The mean king understands, of course, that the physicist cannot get the answer right always if she just prepares the silver atom in a chosen spin state before the king's men measure σ_x, σ_y, or σ_z and performs a control measurement of the atom's spin later. For example, if she prepares the atom in a spin state with a definite value of σ_x and measures σ_z later, then the physicist can tell the outcome of the intermediate measurement by the royal experimenters if they measured σ_x or σ_z. But if they measured σ_y, she can only guess and is bound to guess wrong in half of these σ_y cases when the procedure is repeated a couple of times.

In this example, the physicist's odds for giving the right answer are $\frac{5}{6} = 83.3\%$ in a single trial, and only 2.6% for 20 repetitions. She can do better than that by preparing the atom in a spin state with a definite value of $\sigma_x + \sigma_y$ and a control measurement of σ_z. Then, she still knows the correct answer with certainty if the king's men measured σ_z, but she must rely on guesswork for both σ_x and σ_y. Nevertheless, her total odds for guessing right increase substantially, namely to $\frac{2}{3} + \frac{1}{6}\sqrt{2} = 90.2\%$ (single trial) and 12.8% (20 repetitions), respectively. But the physicist cannot achieve larger odds than these 90.2%, unless she exploits the power of quantum entanglement.

Indeed, with the aid of quantum entanglement — a major theme, in fact, of Schrödinger's *Generalbeichte* (general confession) of 1935 [15], which also contains the (in)famous cat example that the king is so angry about — she achieves odds of 100% — that is, the stranded physicist can always infer correctly the measurement result obtained by the king's men. To this end, she prepares the privileged silver atom such that its spin degree of freedom, the qubit in question, is entangled with another qubit in a suitable way; and later she performs a judiciously chosen control measurement on this 2-qubit system. Mindful of what follows below, we present this solution to the mean king's problem along the lines of References [14, 16], where recent generalizations of the problem are reported, rather than following the original exposition of Reference [13].

14.2.2   The stranded physicist's solution

We describe the auxiliary qubit by Pauli operators τ_x, τ_y, τ_z — which are, of course, just analogs of σ_x, σ_y, σ_z, respectively — and we write, for example, $|\uparrow_x\downarrow_z\rangle$ for the ket vector associated with the (pure) state that has $\sigma_x = 1$ and $\tau_z = -1$. The initial state prepared by the stranded physicist is then*

$$
\begin{aligned}
|\Psi_0\rangle &= 2^{-\frac{1}{2}}\left(|\uparrow_x\downarrow_x\rangle - |\downarrow_x\uparrow_x\rangle\right) \\
&= 2^{-\frac{1}{2}}\left(|\uparrow_y\downarrow_y\rangle - |\downarrow_y\uparrow_y\rangle\right) \\
&= 2^{-\frac{1}{2}}\left(|\uparrow_z\downarrow_z\rangle - |\downarrow_z\uparrow_z\rangle\right),
\end{aligned}
\tag{14.1}
$$

the familiar antisymmetric Bell state.** If the king's men then measure σ_x, the resulting state reduction amounts to

$$
|\Psi_0\rangle \xrightarrow{\ \sigma_x\,=?\ } \left\{
\begin{array}{l}
|\uparrow_x\downarrow_x\rangle \ \text{if} \ \sigma_x = +1 \ \text{is found} \\
|\downarrow_x\uparrow_x\rangle \ \text{if} \ \sigma_x = -1 \ \text{is found}
\end{array}
\right\} = 2^{-\frac{1}{2}}\left(|\Psi_x\rangle \pm |\Psi_0\rangle\right),
\tag{14.2}
$$

and likewise we have

$$
|\Psi_0\rangle \xrightarrow{\ \sigma_y\,=\,\pm 1\ } 2^{-\frac{1}{2}}\left(|\Psi_y\rangle \pm |\Psi_0\rangle\right)
\tag{14.3}
$$

and

$$
|\Psi_0\rangle \xrightarrow{\ \sigma_z\,=\,\pm 1\ } 2^{-\frac{1}{2}}\left(|\Psi_z\rangle \pm |\Psi_0\rangle\right)
\tag{14.4}
$$

for measurements of σ_y and σ_z. Here we meet the three symmetric Bell states,

$$
\begin{aligned}
|\Psi_x\rangle = \sigma_x |\Psi_0\rangle = -\tau_x |\Psi_0\rangle &= 2^{-\frac{1}{2}}\left(|\uparrow_x\downarrow_x\rangle + |\downarrow_x\uparrow_x\rangle\right) \\
&= 2^{-\frac{1}{2}}\mathrm{i}\left(|\uparrow_y\uparrow_y\rangle + |\downarrow_y\downarrow_y\rangle\right) \\
&= 2^{-\frac{1}{2}}\left(-|\uparrow_z\uparrow_z\rangle + |\downarrow_z\downarrow_z\rangle\right),
\end{aligned}
$$

$$
|\Psi_y\rangle = \sigma_y |\Psi_0\rangle = -\tau_y |\Psi_0\rangle = 2^{-\frac{1}{2}}\left(|\uparrow_y\downarrow_y\rangle + |\downarrow_y\uparrow_y\rangle\right)
$$

---

*The cyclic symmetry displayed here is not available if one employs the standard phase conventions, in which $|\uparrow_x\rangle = 2^{-\frac{1}{2}}(|\uparrow_z\rangle + |\downarrow_z\rangle)$, $|\downarrow_x\rangle = 2^{-\frac{1}{2}}(|\uparrow_z\rangle - |\downarrow_z\rangle)$ and $|\uparrow_y\rangle = 2^{-\frac{1}{2}}(|\uparrow_z\rangle + \mathrm{i}\,|\downarrow_z\rangle)$, $|\downarrow_y\rangle = 2^{-\frac{1}{2}}(|\uparrow_z\rangle - \mathrm{i}\,|\downarrow_z\rangle)$, because they fail to respect the invariance of the Pauli algebra under cyclic permutations, $\sigma_x \to \sigma_y \to \sigma_z \to \sigma_x$. The most natural phase conventions that respect this symmetry are stated by $|\uparrow_x\rangle = -\frac{1}{2}(1 - \mathrm{i})(|\uparrow_z\rangle + |\downarrow_z\rangle)$, $|\downarrow_x\rangle = \frac{1}{2}(1 + \mathrm{i})(|\uparrow_z\rangle - |\downarrow_z\rangle)$ and the relations obtained by cyclic permutations. These were used to establish the y and z versions of $|\Psi_0\rangle$ in (14.1) and also those of $|\Psi_x\rangle$ etc. in (14.5).

**It may help to think of the auxiliary τ qubit as being under the physicist's protection whereas the σ qubit (the spin of the silver atom) is accessible to the king's men. But that is not necessary; the auxiliary qubit could be another binary degree of freedom of the same silver atom. By the rules of the game, the king's men are not allowed to do anything to it.

$$= 2^{-\frac{1}{2}}i\left(|\uparrow_z\uparrow_z\rangle + |\downarrow_z\downarrow_z\rangle\right)$$

$$= 2^{-\frac{1}{2}}\left(-|\uparrow_x\uparrow_x\rangle + |\downarrow_x\downarrow_x\rangle\right),$$

$$|\Psi_z\rangle = \sigma_z|\Psi_0\rangle = -\tau_z|\Psi_0\rangle = 2^{-\frac{1}{2}}\left(|\uparrow_z\downarrow_z\rangle + |\downarrow_z\uparrow_z\rangle\right)$$

$$= 2^{-\frac{1}{2}}i\left(|\uparrow_x\uparrow_x\rangle + |\downarrow_x\downarrow_x\rangle\right)$$

$$= 2^{-\frac{1}{2}}\left(-|\uparrow_y\uparrow_y\rangle + |\downarrow_y\downarrow_y\rangle\right). \quad (14.5)$$

Together with the antisymmetric state $|\Psi_0\rangle$, they constitute an orthonormal basis in the 4-dimensional Hilbert space of the two qubits.

We note in passing that all Bell states have many Schmidt decompositions to choose from (two-parametric sets, in fact); three each are given in (14.1) and (14.5). For all non-Bell (pure) 2-qubit states, the Schmidt decomposition is unique; see the remarks after (2.43) in Chapter 2.

Now, to rise to the king's challenge the physicist must perform a follow-up measurement that distinguishes four mutually orthogonal 2-qubit states (the maximal number), which are characterized by the crucial property that they are orthogonal to one of the outcomes of each of the three possible measurements. Then, upon being told which spin component was measured and knowing the outcome of her control measurement, she can surely state the king's men's measurement result correctly, because the other outcome is excluded by the characteristic orthogonality property.

All together there are eight states of this kind — two sets of four: $|B_1\rangle$, $|B_2\rangle$, $|B_3\rangle$, $|B_4\rangle$ and $|C_1\rangle$, $|C_2\rangle$, $|C_3\rangle$, $|C_4\rangle$. In a compact, self-explaining matrix notation they are explicitly given by

$$\left.\begin{array}{c}\left(|B_1\rangle, |B_2\rangle, |B_3\rangle, |B_4\rangle\right)\\ \left(|C_1\rangle, |C_2\rangle, |C_3\rangle, |C_4\rangle\right)\end{array}\right\} = \left(\pm|\Psi_0\rangle, |\Psi_x\rangle, |\Psi_y\rangle, |\Psi_z\rangle\right)$$

$$\times \frac{1}{2}\begin{pmatrix} 1 & 1 & 1 & 1 \\ 1 & 1 & -1 & -1 \\ 1 & -1 & 1 & -1 \\ 1 & -1 & -1 & 1 \end{pmatrix}. \quad (14.6)$$

This 4×4 matrix is both unitary and Hermitean and therefore it is its own inverse. The same is true for the matrix that relates the two measurement bases to each other,

$$\begin{pmatrix} \langle C_1| \\ \langle C_2| \\ \langle C_3| \\ \langle C_4| \end{pmatrix} = -\frac{1}{2}\begin{pmatrix} -1 & 1 & 1 & 1 \\ 1 & -1 & 1 & 1 \\ 1 & 1 & -1 & 1 \\ 1 & 1 & 1 & -1 \end{pmatrix}\begin{pmatrix} \langle B_1| \\ \langle B_2| \\ \langle B_3| \\ \langle B_4| \end{pmatrix}. \quad (14.7)$$

Since

$$\left(\pm|\Psi_0\rangle,|\Psi_x\rangle,|\Psi_y\rangle,|\Psi_z\rangle\right)$$
$$\rightarrow \tfrac{1}{2}\left(1+\sigma_x\tau_x+\sigma_y\tau_y+\sigma_z\tau_z\right)\left(\pm|\Psi_0\rangle,|\Psi_x\rangle,|\Psi_y\rangle,|\Psi_z\rangle\right)$$
$$=\left(\mp|\Psi_0\rangle,|\Psi_x\rangle,|\Psi_y\rangle,|\Psi_z\rangle\right) \tag{14.8}$$

interchanges the B and C bases, they are related to each other by this swapping transformation, which amounts to $\sigma_k \leftrightarrow \tau_k$ for $k = x, y, z$.

The four B states are mutually orthogonal, and so are the four C states,

$$\langle B_j|B_k\rangle = \delta_{jk}, \quad \langle C_j|C_k\rangle = \delta_{jk}, \tag{14.9}$$

and one set is complementary to the other,

$$|\langle B_j|C_k\rangle| = \tfrac{1}{2}. \tag{14.10}$$

The matrices of the resulting transition probability amplitudes are

$$\begin{pmatrix} \langle\uparrow_x\downarrow_x| \\ \langle\downarrow_x\uparrow_x| \\ \langle\uparrow_y\downarrow_y| \\ \langle\downarrow_y\uparrow_y| \\ \langle\uparrow_z\downarrow_z| \\ \langle\downarrow_z\uparrow_z| \end{pmatrix} \left(|B_1\rangle,|B_2\rangle,|B_3\rangle,|B_4\rangle\right)$$

$$= 2^{-\frac{1}{2}}\begin{pmatrix} 1 & 1 & 0 & 0 \\ -1 & 1 & 0 & 0 \\ 1 & 0 & 1 & 0 \\ -1 & 0 & 1 & 0 \\ 1 & 0 & 0 & 1 \\ -1 & 0 & 0 & 1 \end{pmatrix} \frac{1}{2}\begin{pmatrix} 1 & 1 & 1 & 1 \\ 1 & 1 & -1 & -1 \\ 1 & -1 & 1 & -1 \\ 1 & -1 & -1 & 1 \end{pmatrix} = 2^{-\frac{1}{2}}\begin{pmatrix} 1 & 1 & 0 & 0 \\ 0 & 0 & -1 & -1 \\ 1 & 0 & 1 & 0 \\ 0 & -1 & 0 & -1 \\ 1 & 0 & 0 & 1 \\ 0 & -1 & -1 & 0 \end{pmatrix},$$

$$\tag{14.11}$$

which is relevant if the physicist decides to measure the B basis, and (remember the swapping transformation)

$$\begin{pmatrix} \langle\uparrow_x\downarrow_x| \\ \langle\downarrow_x\uparrow_x| \\ \langle\uparrow_y\downarrow_y| \\ \langle\downarrow_y\uparrow_y| \\ \langle\uparrow_z\downarrow_z| \\ \langle\downarrow_z\uparrow_z| \end{pmatrix} \left(|C_1\rangle,|C_2\rangle,|C_3\rangle,|C_4\rangle\right) = 2^{-\frac{1}{2}}\begin{pmatrix} 0 & 0 & -1 & -1 \\ 1 & 1 & 0 & 0 \\ 0 & -1 & 0 & -1 \\ 1 & 0 & 1 & 0 \\ 0 & -1 & -1 & 0 \\ 1 & 0 & 0 & 1 \end{pmatrix}, \tag{14.12}$$

The mean king's	The stranded physicist detects							
men measure	B_1	B_2	B_3	B_4	C_1	C_2	C_3	C_4
σ_x	+1	+1	−1	−1	−1	−1	+1	+1
σ_y	+1	−1	+1	−1	−1	+1	−1	+1
σ_z	+1	−1	−1	+1	−1	+1	+1	−1

Table 14.1 Measurement results inferred by the physicist. The ±1 entries state the outcome that the mean king's men must have obtained at their intermediate measurement of σ_x or σ_y or σ_z provided that the physicist found the respective B state or C state in her control measurement. She can choose at her discretion which of the two bases she wants to measure; either one will do.

which applies if she chooses the C basis. They contain just the right number of null entries at the right places to make the seemingly impossible happen. For example, if she found B_3 in her control measurement and is told that the king's men had measured σ_y at the intermediate time, then she knows that their measurement result must have been +1 because, as the bold-face **0** entry in (14.11) tells us, $|\downarrow_y\uparrow_y\rangle$ is orthogonal to $|B_3\rangle$ and, therefore, they could not have obtained −1.

These matters are summarized in Table 14.1. Note, in particular, that the columns display all $2^3 = 8$ combinations of $+/-$. And, since for each combination the corresponding B or C state is unique, there are no further states of this kind available. The two bases of (14.6) and (14.7) are the only ones. Note also that, whenever the physicist detects $|B_1\rangle$ or $|C_1\rangle$, she knows the outcome of the king's men's intermediate spin measurement (+1 or −1, respectively) even before being told which component they measured.

The secret of the physicist's success is that she forces the king's men to leave a second quantum record of their measurement result in addition to the final spin state of the selected silver atom. For example, after finding $\sigma_y = 1$, they hand over the 2-qubit state $|\uparrow_y\downarrow_y\rangle$, which is the $\uparrow_y$ record of the measurement result plus its copy (or rather anti-copy) $\downarrow_y$ stored in the state of the auxiliary qubit.

Could the physicist use true copies instead of anti-copies? No, because that would require a B basis that discriminates between the states of pairs such as $|\uparrow_x\uparrow_x\rangle / |\downarrow_x\downarrow_x\rangle$; but there is no B basis of this kind. The state that is orthogonal to, say, $|\uparrow_x\uparrow_x\rangle$, $|\downarrow_y\downarrow_y\rangle$, and $|\downarrow_z\downarrow_z\rangle$ — needed for the $- + +$ sign sequence — is $|\Psi_0\rangle$ of (14.1). This state is, however, orthogonal to *all* ↑↑ and *all* ↓↓ states, so it is never detected but should nevertheless be used for all $+/-$ sequences. Clearly, therefore, copies would not work; in fact, there is no initial state to be used instead of $|\Psi_0\rangle$ that would produce copies.

Certain combinations of copies and anti-copies are all right. They would obtain, for example, if the physicist prepared the 2-qubit system in the Bell state $|\Psi_z\rangle = \sigma_z |\Psi_0\rangle$ rather than $|\Psi_0\rangle$. The procedure is then as described above, modulo the unitary transformation effected by σ_z.

The physicist makes use of entanglement twice in her solution to the problem posed by the mean king. First, she prepares the entangled state $|\Psi_0\rangle$. Second, when measuring the B basis, say, she determines joint eigenstates of the binary 2-qubit observables

$$\frac{1}{2}\left(\sigma_x - \tau_x - \sigma_y\tau_z - \sigma_z\tau_y\right) ,$$
$$\frac{1}{2}\left(\sigma_y - \tau_y - \sigma_z\tau_x - \sigma_x\tau_z\right) ,$$
$$\frac{1}{2}\left(\sigma_z - \tau_z - \sigma_x\tau_y - \sigma_y\tau_x\right) , \tag{14.13}$$

each of them being the product of the other two. (The trio corresponding to the C basis is obtained by the swapping transformation, of course.) For example, the respective eigenvalues $+1$, -1, -1 identify $|B_2\rangle$ (or $|C_2\rangle$, respectively). This characterization of the measurement bases is quite analogous to stating that the four Bell states in (14.1) and (14.5) are joint eigenstates of the trio $\sigma_x\tau_x$, $\sigma_y\tau_y$, $\sigma_z\tau_z$.

This second exploitation of entanglement is the crucial step in any actual demonstration experiment that realizes the mean king's problem. Fortunately, it is possible with optical instruments to measure any 2-qubit basis — or, equivalently, any such trio of binary observables — if both qubits are carried by a single photon in the form of a polarization qubit and an interferometer qubit. How this is done in detail is described in Reference [17], where a photon-polarization version of the mean king's problem is formulated and a corresponding experiment is proposed. A similar, and essentially equivalent, setup is in place and data of satisfactory quality are expected to be available shortly [18].

14.2.3   The mean king's second challenge

As a further preparation for the eventual discussion of communication with qubit pairs, we consider *The Mean King's Second Problem*:

> Being much frustrated by the physicist's apparent good
> luck in guessing right the outcome of the intermediate spin

measurement performed by his men, the mean king decides to change the rules of the game. Rather than measuring σ_x, σ_y, or σ_z, the king's men will now rotate the atom's spin by $180°$ about one of the six axes specified by the 3-vectors $\vec{e}_x \pm \vec{e}_y$, $\vec{e}_y \pm \vec{e}_z$, $\vec{e}_z \pm \vec{e}_x$. Again, the physicist may prepare any initial state of her choosing and make a control measurement of her liking without, however, knowing which rotation is performed until after her control measurement is over.

Only then, the king will tell her if the rotation axis was in the xy plane, or the yz plane, or the zx plane, and the physicist is challenged to identify the actual axis correctly. Again, the physicist frustrates the king by giving the right answer any time the procedure is repeated. How does she do it this time?

Here, too, you will enjoy figuring it out yourself before reading on.

It is immediately clear that the physicist cannot succeed if she manipulates the atom's spin alone. The best guessing odds she could achieve this way are $\frac{5}{6} = 83.3\%$, less than the 90.2% in the mean king's original problem. These 83.3% obtain, for instance, when an eigenstate of $\sigma_x + \sigma_y + \sigma_z$ is prepared and the control measurement determines the final value of this observable.

Accordingly, the physicist must take advantage of entanglement again, and she does so in exactly the same manner as before: prepare the Bell state $|\Psi_0\rangle$ of (14.1) and measure the B basis of (14.6) (or the C basis, or perhaps both in random alternation).

This is how it works. The effect on $|\Psi_0\rangle$ of the six rotations in question is respectively given by

$$|\Psi_0\rangle \rightarrow \left|[xy]_\pm\right\rangle \equiv 2^{-\frac{1}{2}}\left(\sigma_x \pm \sigma_y\right)|\Psi_0\rangle$$
$$= 2^{-\frac{1}{2}}\left(|\Psi_x\rangle \pm |\Psi_y\rangle\right),$$
$$|\Psi_0\rangle \rightarrow \left|[yz]_\pm\right\rangle \equiv 2^{-\frac{1}{2}}\left(\sigma_y \pm \sigma_z\right)|\Psi_0\rangle$$
$$= 2^{-\frac{1}{2}}\left(|\Psi_y\rangle \pm |\Psi_z\rangle\right),$$
$$|\Psi_0\rangle \rightarrow \left|[zx]_\pm\right\rangle \equiv 2^{-\frac{1}{2}}\left(\sigma_z \pm \sigma_x\right)|\Psi_0\rangle$$
$$= 2^{-\frac{1}{2}}\left(|\Psi_z\rangle \pm |\Psi_x\rangle\right). \tag{14.14}$$

These three state pairs constitute the six sums and differences of Bell states that do not involve $|\Psi_0\rangle$; the other six, which do, are the three

Plane containing rotation axis	Stranded physicist detects			
	B_1 or C_1	B_2 or C_2	B_3 or C_3	B_4 or C_4
xy	+	−	−	+
yz	+	+	−	−
zx	+	−	+	−

Table 14.2 Rotation axes inferred by the physicist to meet the mean king's second challenge. If she detected, for example, state $|B_2\rangle$ and is then told that the rotation axis was in the zx plane, she infers that the actual axis is specified by the 3-vector $\vec{e}_z - \vec{e}_x$.

pairs of (14.2)–(14.4). For the three new pairs, the analog of (14.11) reads

$$
\begin{pmatrix}
\langle[xy]_+| \\
\langle[xy]_-| \\
\langle[yz]_+| \\
\langle[yz]_-| \\
\langle[zx]_+| \\
\langle[zx]_-|
\end{pmatrix}
\left(|B_1\rangle, |B_2\rangle, |B_3\rangle, |B_4\rangle \right) = 2^{-\frac{1}{2}}
\begin{pmatrix}
1 & 0 & 0 & 1 \\
0 & 1 & -1 & 0 \\
1 & -1 & 0 & 0 \\
0 & 0 & 1 & -1 \\
1 & 0 & -1 & 0 \\
0 & -1 & 0 & 1
\end{pmatrix},
$$

$$(14.15)$$

and the same 6×4 matrix shows up in the analog of (14.12) because $|\Psi_0\rangle$ is orthogonal to all six $|[\cdot\cdot]_\pm\rangle$ states. The resulting analog of Table 14.1 is Table 14.2.

The king, utterly frustrated by the physicist's continuing success and furiously upset anew by her colleagues' recent abuse of Heisenberg's dog [19], now proceeds to the final challenge: At the intermediate stage his men will either perform one of the three spin measurements or one of the six rotations, and the physicist must infer correctly the measurement result or the rotation axis, respectively. In reply, the physicist just smiles ...

14.2.4 A different perspective

In her control measurement at the final stage of the experiment, the physicist distinguishes the joint eigenstates of the B trio (14.13) from each other, or those of the C trio obtained by swapping the qubits in (14.13). Before this last step the situation is this: a certain 2-qubit state

state detected	B_1	B_2, C_1, C_2, C_3, or C_4	B_3	B_4
educated guess	+	+ or −	−	−
odds	100%	50%	66.7%	66.7%

Table 14.3 Guessing strategy and odds for guessing right if all 6 state pairs of (14.16) are prepared with equal *a priori* probability. The overall odds are $\frac{17}{24} = 70.8\%$ if the physicist measures only the B basis. They are 50% if she detects solely C states, and $\frac{29}{48} = 60.4\%$ if both bases are measured equally frequently.

has been prepared, and later the physicist will be told to which one of the six state pairs

$$2^{-\frac{1}{2}}\left(|\Psi_0\rangle \pm |\Psi_x\rangle\right) ,\ 2^{-\frac{1}{2}}\left(|\Psi_0\rangle \pm |\Psi_y\rangle\right) ,\ 2^{-\frac{1}{2}}\left(|\Psi_0\rangle \pm |\Psi_z\rangle\right) ,$$
$$2^{-\frac{1}{2}}\left(|\Psi_x\rangle \pm |\Psi_y\rangle\right) ,\ 2^{-\frac{1}{2}}\left(|\Psi_y\rangle \pm |\Psi_z\rangle\right) ,\ 2^{-\frac{1}{2}}\left(|\Psi_z\rangle \pm |\Psi_x\rangle\right) \qquad (14.16)$$

it belongs, and then she will be able to identify the state in question: she will know with certainty whether it was the "+" partner of the pair or the "−" partner.

Now imagine that the experiment is first repeated many times and that the physicist is given the "which pair" information only after all measurements are completed. Then, prior to receiving this *classical information*, she has available the *quantum information* represented by the sequence of her measurement results. When she finds B_1, she knows for sure that "+" was the case. For all other results she can just make an educated guess about this +/− binary alternative — this +/− bit — based on the fair assumption that all 12 states are prepared equally likely. Her guessing strategy and the odds for guessing right are summarized in Table 14.3.

We recognize here all the essential ingredients of a quantum cryptography scheme: quantum information with (partly) unclear significance until some classical information becomes available and, very important for the security of the shared key, the possibility of acquiring the quantum information by measuring in either one of two complementary bases.

We shall return to this *cryptography with qubit pairs* in Section 14.4 after recalling, in Section 14.3, the essentials of the standard BB84 scheme, introduced by Bennett and Brassard in 1984 [1] (see also [20]), and later generalized in various ways, in particular by Ekert in 1991 [2].

The qubit-pair scheme suggested by the analysis of the mean king's problem — and thus ultimately originating in the 1987 paper by Vaidman, Aharonov, and Albert [13] — is not such a generalization of BB84, because it is *deterministic*. This is to say that each and every state sent and detected will eventually contribute a key bit whereas, in marked contrast, 50% (or more) of the states sent and detected in BB84 yield quantum information that is of no use. In addition, a random process decides which cases will be the useful ones. This *probabilistic* feature of BB84 (and its generalizations) is not present in the qubit-pair scheme of Section 14.4. Not surprisingly, then, one can also employ qubit pairs for direct communication with ensured security; this is the subject matter of Section 14.6. The qubit-pair communication scheme described there has a very striking property indeed: it is perfectly secure although the encryption key becomes known publicly.

14.3 BB84: cryptography with single qubits

As always, the protagonists are *Alice* and *Bob* — she sends, he receives, and they communicate over a public channel in addition. "Public" just means that there is no precaution taken that would prevent an eavesdropper (canonically: female *Eve* or, less popular, male *Evan*) from accessing the exchanged information — it is as if it were broadcast. Any security analysis must, therefore, proceed from the two main premises that Eve/Evan knows all that is public and that she/he is able to perform any manipulation allowed by the laws of physics.

For simplicity, we shall assume that the transmission between Alice and Bob is noiseless and that their preparation devices (Alice) and measurement apparatus (Bob) are perfect. Then, all transmission errors originate in some interference by Eve/Evan. Obviously, such assumptions are not warranted for any actual device built to implement BB84 or the qubit-pair scheme of Section 14.4, but for a general security analysis they are acceptable. In real experiments one must determine the background noise level and then blame all excess noise on the eavesdropper. If there is too much noise to begin with, its fluctuations will hide the excess noise and then the procedure cannot be relied upon. The quantitative analysis is easiest if there is no background noise at all, and we shall deal with this idealized situation solely.

14.3.1 Description of the scheme

In the simplest version of the BB84 scheme, Alice sends individual qubits to Bob that are prepared in either one of the four states

$$|1_+\rangle = |\uparrow_z\rangle , \qquad |2_+\rangle = |\uparrow_x\rangle ,$$
$$|1_-\rangle = |\downarrow_z\rangle , \qquad |2_-\rangle = |\downarrow_x\rangle \qquad\qquad (14.17)$$

with equal likelihood. States $|1_\pm\rangle$ are "of type 1", states $|2_\pm\rangle$ are "of type 2". Sending and receiving $|1_+\rangle$ or $|2_+\rangle$ will establish a "+" bit, and $|1_-\rangle$ or $|2_-\rangle$ yields a "−" bit. For each qubit arriving, Bob measures either σ_z or σ_x, alternating randomly between the two. A σ_z measurement distinguishes between $|1_+\rangle$ and $|1_-\rangle$, but it cannot tell $|2_+\rangle$ and $|2_-\rangle$ apart. And, likewise, a σ_x measurement is fitting for type-2 qubits but not for those of type 1. Thus, after Alice has sent the necessary number of qubits to Bob, she will have to ask him (publicly) which measurements he made. Those qubits where the measurement does not match the type — σ_z for type 2 or σ_x for type 1 — are then discarded. On average every second qubit sent by Alice is thus of no use in the end. This public *qubit matching* is an essential ingredient in ensuring the security of the process. It is clear that neither Alice nor Bob can know beforehand which qubits will be selected by the matching. Only the probabilistic fact that about 50% of them will be measured in the right basis — the other 50% in the wrong one — is certain at the outset.

After discarding the unmatched qubits, Bob will pick a sufficiently large number of matched qubits and reveal (publicly) what he measured for them: "+" or "−". Alice compares this with what she sent; and if no errors are detected, she concludes that Eve did not listen in. She then tells Bob that the random $+/-$ sequence of the rest of the matched qubits is a secure cryptographic key and that he should expect an encrypted message shortly. If, however, one or more of Bob's measurement results are inconsistent with what Alice sent — such as detecting "−" for $|1_+\rangle$ — then they discard the whole bit sequence and start all over again. If necessary, they must repeat the procedure a couple of times until they get an error-free transmission.

Once a secure key of sufficient length (one key bit for each message bit) is established, the sending of a confidential message follows the usual procedure. Alice encrypts the message by multiplication:

message bit	+	+	−	−
key bit	+	−	+	−
product bit	+	−	−	+

and so produces a random $+/-$ string that she communicates to Bob (publicly). Bob multiplies it again with the key string bit by bit, and so reconstructs the message. As long as Eve has no knowledge of the key bit sequence, this communication is perfectly secure.

Eve's interference is revealed by the errors she causes unavoidably. We shall take a closer look at that below and are content with a cursory account here. In simple terms, these errors occur because Eve does not know the type of the next qubit nor which measurement Bob will perform later. Imagine, for example, that Eve measures σ_x and forwards $|\uparrow_x\rangle$ or $|\downarrow_x\rangle$ to Bob if she finds $\sigma_x = +1$ or $\sigma_x = -1$, respectively. Then, if Bob also measures σ_x and the qubit is one of the matched ones, Eve will have the same information as Bob without giving rise to a detection error. But if Bob measures σ_z (and this matches with Alice's sending a type-1 qubit) then there is a 50% chance that Bob gets $\sigma_z = +1$ for $|\downarrow_z\rangle$ sent or $\sigma_z = -1$ for $|\uparrow_z\rangle$. Tersely: There is a probability of 25% that Eve's interference causes an error. Then, the chance that her eavesdropping remains unnoticed upon comparing 100 test bits, say, is $(\frac{3}{4})^{100} = 3 \times 10^{-13}$, virtually zero.

This line of reasoning exploits implicitly the random nature of the process. It must not be known beforehand if the next qubit is of type 1 or type 2 (otherwise Eve could choose her measurement fittingly), nor must it be known which measurement Bob will perform (otherwise Eve could just do the same measurement), nor must it be known which qubits will be used for the comparison (otherwise Eve would leave those qubits untouched and only look at the other ones).

14.3.2 Eavesdropping: minimizing the error probability

The error probability of 25% refers to the particular situation in which Eve measures σ_x and forwards to Bob a substitute qubit in the σ_x state she detected. We shall now take a more systematic look at eavesdropping strategies of this intercept–resend kind. It is true that there are other things that Eve can do, such as entangling every qubit in transmission with an auxiliary qubit (an *ancilla*) and later extract information by a suitable measurement on all the ancillae,* but we shall be content with a detailed study of intercept-resend attacks. Rather than repro-

---

*If the arguments that Xiang-bin put forward recently [21] survive scrutiny, it may not be necessary to worry about such "coherent attacks" because, if he is right, they are not more efficient than bit-by-bit attacks of the eavesdropper. It is too early to tell, though.

ducing the complete analysis of all conceivable eavesdropping scenarios given elsewhere (see [22], for example), it is our objective to establish a benchmark with which we can compare the efficiency of intercept–resend attack in the qubit-pair scheme of Section 14.4.

This, then, is what Eve does: she performs a measurement that distinguishes the eavesdropping states $|E_1\rangle$ and $|E_2\rangle$ from each other and forwards to Bob a replacement qubit in state $|F_1\rangle$ or $|F_2\rangle$, respectively. The question that Alice and Bob need to have answered is: what is the minimal error probability (for matched qubits) resulting from Eve's interference? It tells them how many test bits they need to verify to exclude eavesdropping (of the intercept–resend kind) with the desired degree of certainty.

We deal with this problem in two steps. Step 1: for given $|E_1\rangle$ and $|E_2\rangle$, determine the forwarded states $|F_1\rangle$ and $|F_2\rangle$ that minimize the error probability. Step 2: now choose those $|E_1\rangle$ and $|E_2\rangle$ that yield the smallest one of all such minimal error probabilities. This "minimum of minima" is the number that Alice and Bob want to know.

Step 1: suppose Eve finds $|E_1\rangle$. This happens with probability

$$\tfrac{1}{4}|\langle E_1|1_+\rangle|^2 + \tfrac{1}{4}|\langle E_1|1_-\rangle|^2$$

$$+\tfrac{1}{4}|\langle E_1|2_+\rangle|^2 + \tfrac{1}{4}|\langle E_1|2_-\rangle|^2 \tag{14.18}$$

since all four states $|1_\pm\rangle$, $|2_\pm\rangle$ are sent equally frequently. Then state $|F_1\rangle$ is forwarded. With matched bases, Bob will detect an error if he gets $|1_-\rangle$ when Alice sent $|1_+\rangle$, for example. The summation over all cases gives the error probability

$$\tfrac{1}{4}|\langle E_1|1_+\rangle|^2|\langle F_1|1_-\rangle|^2 + \tfrac{1}{4}|\langle E_1|1_-\rangle|^2|\langle F_1|1_+\rangle|^2$$

$$+\tfrac{1}{4}|\langle E_1|2_+\rangle|^2|\langle F_1|2_-\rangle|^2 + \tfrac{1}{4}|\langle E_1|2_-\rangle|^2|\langle F_1|2_+\rangle|^2$$

$$= \tfrac{1}{4} - \tfrac{1}{8}\langle F_1|\sigma_z|F_1\rangle\langle E_1|\sigma_z|E_1\rangle - \tfrac{1}{8}\langle F_1|\sigma_x|F_1\rangle\langle E_1|\sigma_x|E_1\rangle$$

$$= \tfrac{1}{4} - \tfrac{1}{8}\langle F_1|\big[\langle E_1|\sigma_z|E_1\rangle\sigma_z + \langle E_1|\sigma_x|E_1\rangle\sigma_x\big]|F_1\rangle \ . \tag{14.19}$$

Now, an operator of the form $z\sigma_z + x\sigma_x$ with real coefficients z, x has eigenvalues $\pm\sqrt{z^2 + x^2} = \pm|z + \mathrm{i}x|$. Therefore, the minimal value of the right-hand side in (14.19), with respect to varying the forwarded state $|F_1\rangle$, is given by

$$\tfrac{1}{4} - \tfrac{1}{8}|\langle E_1|(\sigma_z + \mathrm{i}\sigma_x)|E_1\rangle| \ . \tag{14.20}$$

Together with the analogous expression for $|E_2\rangle$, the minimal error probability associated with distinguishing $|E_1\rangle$ from $|E_2\rangle$ is

$$\text{error}(|E_1\rangle, |E_2\rangle) = \frac{1}{2} - \frac{1}{8} \sum_{k=1,2} |\langle E_k| (\sigma_z + i\sigma_x) |E_k\rangle| \, . \qquad (14.21)$$

This ends step 1. And step 2 just consists of recognizing that*

$$\text{error}(|E_1\rangle, |E_2\rangle) \geq \frac{1}{2} - \frac{1}{8} \text{tr}\{|\sigma_z + i\sigma_x|\} = \frac{1}{2} - \frac{1}{8} \times 2 = 25\% \,, \quad (14.22)$$

where the inequality is an elementary example of what is known as the "Peierls inequality" in statistical physics.** The equal sign holds if $|E_1\rangle$, $|E_2\rangle$ are eigenstates of $\sigma_x \cos\varphi + \sigma_z \sin\varphi$ with any value of the angle parameter φ. For $\varphi = 0$, we have the situation described at the end of Section 14.3.1. As we know now, it is one of the error-minimizing eavesdropping strategies (of the intercept–resend kind).

If Alice and Bob wish a larger minimal error probability than these 25%, to make it easier to detect an eavesdropper, they can decide to use

$$|3_+\rangle = |\uparrow_y\rangle \,, \qquad |3_-\rangle = |\downarrow_y\rangle \qquad (14.23)$$

as a third state pair [5]. Indeed, this raises the minimal error probability to $\frac{1}{3} = 33.3\%$, but a price must be paid: the fraction of matched qubits drops from $\frac{1}{2}$ to $\frac{1}{3}$. Thus, Alice and Bob must balance the increase of the minimal error probability by a factor of $\frac{4}{3}$ against the decrease in efficiency by a factor of $\frac{2}{3}$.

14.3.3 Eavesdropping: maximizing the raw information

Let us now consider this question: if Eve did not care at all whether her listening-in could be detected or not, how well can she guess if a "+" state is transmitted or a "−" state before she gets any of the classical information that Alice and Bob exchange? Admittedly, this is of rather little relevance for BB84. But by studying it first in the technically

---

*A reminder: the modulus $|A|$ of an operator A is equal to $\sqrt{A^\dagger A}$ by definition. It differs from $|A^\dagger| = \sqrt{AA^\dagger}$ unless A is a normal operator. For the example at hand, we have $|\sigma_z \pm i\sigma_x| = 1 \mp \sigma_y$.

**See, in particular, Thirring's textbook [23]. The most important properties of operators with a finite trace of their modulus (called *trace class operators* by some, *nuclear operators* by others) are discussed in many books on Hilbert space theory such as References [23, 24, 25, 26]. In a language that is perhaps more easily accessible to physicists, all properties needed here can be found in Reference [27].

simple context of BB84, we prepare the ground for readdressing this problem for the qubit-pair scheme in Section 14.4, when it will be quite important.

So, now Eve aims at keeping $|1_+\rangle$ and $|2_+\rangle$ apart from $|1_-\rangle$ and $|2_-\rangle$ without caring for the further discrimination between the two "+" states or the two "−" states. Her statistical operator for a qubit in transmission is

$$\rho = \tfrac{1}{4}\big(|1_+\rangle\,\langle 1_+| + |1_-\rangle\,\langle 1_-| + |2_+\rangle\,\langle 2_+| + |2_-\rangle\,\langle 2_-|\big)$$

$$= \tfrac{1}{2}\rho^{(+)} + \tfrac{1}{2}\rho^{(-)} \tag{14.24}$$

with the respective statistical operators for the "+" and "−" sectors given by

$$\left.\begin{aligned}\rho^{(+)} &= \tfrac{1}{2}\big(|1_+\rangle\,\langle 1_+| + |2_+\rangle\,\langle 2_+|\big)\\[4pt]\rho^{(-)} &= \tfrac{1}{2}\big(|1_-\rangle\,\langle 1_-| + |2_-\rangle\,\langle 2_-|\big)\end{aligned}\right\} = \tfrac{1}{2} \pm \tfrac{1}{4}\big(\sigma_z + \sigma_x\big)\,. \tag{14.25}$$

Eve needs to distinguish $\rho^{(+)}$ from $\rho^{(-)}$. This problem is familiar from studies concerning quantitative aspects of wave–particle duality (see [27] and references cited therein), and the known answer is this: Eve's maximal likelihood for guessing the bit right is

$$\mathcal{L} = \tfrac{1}{2} + \tfrac{1}{4}\mathrm{tr}\left\{\big|\rho^{(+)} - \rho^{(-)}\big|\right\}\,, \tag{14.26}$$

and she can acquire this optimum of bit knowledge by measuring a (nondegenerate) observable whose eigenstates are also eigenstates of $\rho^{(+)} - \rho^{(-)}$.

For the statistical operators $\rho^{(\pm)}$ in (14.25), this gives

$$\mathcal{L} = \tfrac{1}{2} + \tfrac{1}{8}\mathrm{tr}\left\{|\sigma_z + \sigma_x|\right\} = \tfrac{1}{2} + \tfrac{1}{4}\sqrt{2} = 85.4\%\,, \tag{14.27}$$

and Eve could just measure $\sigma_z + \sigma_x$ to achieve these guessing odds. By comparison, Bob, who measures either σ_z or σ_x, has guessing odds of only 75%, because his measurement is not optimal for this purpose. As we know from the considerations in Section 14.3.2, a measurement of $\sigma_z + \sigma_x$ also minimizes the error probability; it is optimal for both of Eve's objectives.

Here, too, Alice and Bob can decide to use the third pair $|3_\pm\rangle$ of (14.23) in addition. This would reduce Eve's optimal guessing odds to $\tfrac{1}{2} + \tfrac{1}{6}\sqrt{3} = 78.9\%$ and Bob's to 66.7%.

14.4 Cryptography with qubit pairs

14.4.1 Description of the scheme

Now we turn to the cryptographic scheme suggested by the mean king's problem that we alluded to above in Section 14.2.4. For an implementation, Alice and Bob must first pick an orthonormal set of 2-qubit states, which we denote by $|\Phi_1\rangle, \ldots, |\Phi_4\rangle$. These Φ states will play the role of the Bell states $|\Psi_0\rangle, \ldots, |\Psi_z\rangle$ in Section 14.2. All other states will be defined in terms of the Φ basis. As far as the general considerations are concerned, any choice of states for the Φ basis will do — different choices being related by unitary transformations — but a particular Φ basis might be more practical than another for an actual physical realization of the scheme. We shall address this issue in Section 14.5; for the more immediate purposes all Φ basis are equivalent.

Then there are six pairs of 2-qubit states, the analogs of the state pairs of (14.16), here in lexicographic order:

$$|1_\pm\rangle = 2^{-\frac{1}{2}}\left(|\Phi_1\rangle \pm |\Phi_2\rangle\right), \qquad |4_\pm\rangle = 2^{-\frac{1}{2}}\left(|\Phi_2\rangle \pm |\Phi_3\rangle\right),$$
$$|2_\pm\rangle = 2^{-\frac{1}{2}}\left(|\Phi_1\rangle \pm |\Phi_3\rangle\right), \qquad |5_\pm\rangle = 2^{-\frac{1}{2}}\left(|\Phi_2\rangle \pm |\Phi_4\rangle\right), \qquad (14.28)$$
$$|3_\pm\rangle = 2^{-\frac{1}{2}}\left(|\Phi_1\rangle \pm |\Phi_4\rangle\right), \qquad |6_\pm\rangle = 2^{-\frac{1}{2}}\left(|\Phi_3\rangle \pm |\Phi_4\rangle\right).$$

These are the states that Alice sends to Bob to establish the $+/-$ sequence of the shared key. And Bob detects them in his measurement bases $|B_1\rangle, \ldots, |B_4\rangle$ and $|C_1\rangle, \ldots, |C_4\rangle$, related to the Φ basis in the same manner as the B and C bases of Section 14.2 are related to the Bell basis there, see (14.6),

$$\left.\begin{array}{l}\left(|B_1\rangle, |B_2\rangle, |B_3\rangle, |B_4\rangle\right) \\ \left(|C_1\rangle, |C_2\rangle, |C_3\rangle, |C_4\rangle\right)\end{array}\right\} = \left(\pm|\Phi_1\rangle, |\Phi_2\rangle, |\Phi_3\rangle, |\Phi_4\rangle\right)$$

$$\times \frac{1}{2}\begin{pmatrix} 1 & 1 & 1 & 1 \\ 1 & 1 & -1 & -1 \\ 1 & -1 & 1 & -1 \\ 1 & -1 & -1 & 1 \end{pmatrix}, \qquad (14.29)$$

and (14.7), (14.9), (14.10) continue to hold. It will be convenient to represent kets by complex 4-component columns and bras by 4-component

Type of qubit pair sent by Alice	State detected by Bob							
	B_1	B_2	B_3	B_4	C_1	C_2	C_3	C_4
1	+	+	−	−	−	−	+	+
2	+	−	+	−	−	+	−	+
3	+	−	−	+	−	+	+	−
4	+	−	−	+	+	−	−	+
5	+	−	+	−	+	−	+	−
6	+	+	−	−	+	+	−	−

Table 14.4 Bit values inferred by Bob. For example, when detecting B_3 for a type-4 qubit pair, (14.31) states that the actual state must have been 4_- so that "−" is the value for this bit.

rows, as exemplified by

$$\langle B_3| = \frac{1}{2}\left(1,-1,1,-1\right)\begin{pmatrix}\langle\Phi_1|\\\langle\Phi_2|\\\langle\Phi_3|\\\langle\Phi_4|\end{pmatrix} \cong \frac{1}{2}\left(1,-1,1,-1\right),$$

$$|4_+\rangle = \left(|\Phi_1\rangle,|\Phi_2\rangle,|\Phi_3\rangle,|\Phi_4\rangle\right)\frac{1}{\sqrt{2}}\begin{pmatrix}0\\1\\1\\0\end{pmatrix} \cong \frac{1}{\sqrt{2}}\begin{pmatrix}0\\1\\1\\0\end{pmatrix} \quad (14.30)$$

and put to use in

$$\langle B_3|4_+\rangle = \frac{1}{2}\left(1,-1,1,-1\right)\frac{1}{\sqrt{2}}\begin{pmatrix}0\\1\\1\\0\end{pmatrix} = 0, \quad (14.31)$$

and operators will be represented by the corresponding 4×4 matrices.

Having detected one of the B states or C states, and being told the type $(1, 2, \ldots,$ or $6)$ of the transmitted qubit pair, Bob infers the actual kind ("+" or "−") with the aid of Table 14.4, which combines Tables 14.1 and 14.2 into one, with the necessary changes.

Alice and Bob can make use of two or more state pairs for the purpose of cryptography — that is, for establishing a secure shared key. In particular, there is the "simplest version" of Reference [11] where only the two pairs of type 1 and 2 are used — or, equivalently, most other combinations, such as types 5 and 6, but not 1 and 6, or 2 and 5, or 3 and 4, because they would easily fall prey to eavesdropping. Evan would just need to measure a basis consisting of, for instance, the four mutually orthogonal states $|1_\pm\rangle$, $|6_\pm\rangle$ and forward to Bob the state found.

More relevant than the simple 2-pair version, however, is the scheme in which the four types 1, 2, 5, and 6 are used. In Section 14.5 below we shall argue that a fair comparison with BB84 should be based on such a 4-pair scheme. Other versions with 3, 5, or all 6 state pairs are of much lesser interest, and we shall be content with discussing the 2-pair and 4-pair cases.

In either version, much of the procedure is fully analogous to BB84: Alice sends qubit pairs to Bob in a random sequence of states of the types agreed upon. He detects them either in the B basis or the C basis, switching randomly between his two measurement bases. Then he picks a random sample of sufficiently many events, reports (publicly) which state he detected for each of them, and she checks if his results are consistent with what she sent. If they are, fine; if not, back to square one. In case of success, Alice announces (publicly) the type of each qubit pair, and Bob determines the bit values in accordance with Table 14.4. From then on they share a secure key, and the confidential message to be exchanged can be encrypted and deciphered.

We must not fail to emphasize the crucial difference between this qubit-pair scheme and the single-qubit BB84 scheme and its generalizations. There is no necessity for matching Alice's basis (of preparation) with Bob's basis (for detection). Both bases that he is using in random alternation *always* match her bases. Therefore, she knows in advance that the next qubit pair transmitted will provide useful quantum information when he detects it and will surely contribute a key bit (or a control bit) later, after revealing the classical information of its type.

This *deterministic nature* is not shared by BB84 and its variants. The only other proposal on record with deterministic features is the one of Goldenberg and Vaidman [28], in which the eavesdropper does not have full access to the qubits in transmission. In marked contrast, the qubit-pair scheme that we are discussing here does not rely on measures of such a somewhat artificial kind. Indeed, it is as different from the Goldenberg-Vaidman proposal as it is from BB84.

The deterministic nature of the qubit-pair scheme enables Alice to communicate a given bit sequence to Bob without having to fear that some of the bits will remain unrecognizable to him forever. This makes one wonder if the scheme could not be used for sending a message directly and securely without encrypting it first. Indeed, this is possible; but one needs to modify the scheme because the versions presented so far do not prevent Evan, the eavesdropper, from acquiring pretty good knowledge of the bits in transmission before Alice and Bob become aware of his interference. We shall deal with this question in a quantitative manner in Section 14.4.3, after first determining the minimal error probabilities in Section 14.4.2, and shall present the modified scheme that does not suffer from this problem in Section 14.6.

14.4.2   Eavesdropping: minimizing the error probability

The determination of the minimal error probability that arises from Evan's interception and resending follows the general line of thought of Section 14.3.2. Assume that Evan has found state $|E\rangle$ and wants to forward that state $|F\rangle$ that is least likely to cause a traitorous detector response at Bob's end. The probability for getting $|E\rangle$ in the first place is

$$\frac{1}{2N} \sum_n \left[|\langle E|n_+\rangle|^2 + |\langle E|n_-\rangle|^2 \right] , \qquad (14.32)$$

where N is the number of pairs used and the summation is over the pair types considered — that is, $N = 2$, $n = 1, 2$ for the 2-pair version and $N = 4$, $n = 1, 2, 5, 6$ for the 4-pair version. Clearly, (14.32) is the analog of (14.18).

We now construct the analog of (14.19). Suppose state $|n_+\rangle$ is the state in transmission. Then those B states, for which $\langle n_+|B_k\rangle = 0$, contribute to the error probability in proportion to $|\langle F|B_k\rangle|^2$, their probability of responding to the forwarded state $|F\rangle$. Now, for each B state we have

$$\text{either} \quad \left\{ \begin{matrix} |\langle n_+|B_k\rangle|^2 = \frac{1}{2} \\ \text{and} \\ |\langle n_-|B_k\rangle|^2 = 0 \end{matrix} \right\} \quad \text{or} \quad \left\{ \begin{matrix} |\langle n_+|B_k\rangle|^2 = 0 \\ \text{and} \\ |\langle n_-|B_k\rangle|^2 = \frac{1}{2} \end{matrix} \right\} , \qquad (14.33)$$

the second of which giving rise to an error if $|n_+\rangle$ is in transmission. Both cases of (14.33) are accounted for by the factor

$$2|\langle n_-|B_k\rangle|^2 = 1 - 2|\langle n_+|B_k\rangle|^2 . \qquad (14.34)$$

So, the total contribution from $|n_+\rangle$ and $|B_k\rangle$ is

$$\frac{1}{2N}|\langle E|n_+\rangle|^2 \frac{1}{2}|\langle F|B_k\rangle|^2\left(1-2|\langle n_+|B_k\rangle|^2\right),\qquad(14.35)$$

where the first two factors are the probability of getting $|E\rangle$ copied from (14.32), the $\frac{1}{2}$ is the relative frequency that the B basis is measured (and not the C basis), and the last two factors are as just explained. The total error probability is now obtained by adding the contribution from $|n_-\rangle$, summing over n and k, and adding the corresponding terms referring to the C basis. The outcome

$$\begin{Bmatrix}\text{error probability}\\\text{for given}\\|E\rangle\text{ and }|F\rangle\end{Bmatrix}=\frac{1}{4N}\sum_n\langle E|\,P_n\,|E\rangle$$

$$-\frac{1}{4N}\sum_n\langle E|\,Q_n\,|E\rangle\,\langle F|\,Q_n\,|F\rangle\qquad(14.36)$$

is the analog of (14.19). Here,

$$\begin{matrix}P_n\\Q_n\end{matrix}\Biggr\}=|n_+\rangle\,\langle n_+|\pm|n_-\rangle\,\langle n_-|\,,\qquad(14.37)$$

and on the way to (14.36) one exploits

$$|\langle E|n_\pm\rangle|^2=\tfrac{1}{2}\langle E|\left(P_n\pm Q_n\right)|E\rangle\,,\qquad(14.38)$$

the analogous statements for $|\langle B_k|n_\pm\rangle|^2$ and $|\langle C_k|n_\pm\rangle|^2$ along with $\langle B_k|\,P_n\,|B_k\rangle=\langle C_k|\,P_n\,|C_k\rangle=\frac{1}{2}$, and the identity

$$\sum_{k=1}^4\left(|B_k\rangle\,\langle B_k|\,Q_n\,|B_k\rangle\,\langle B_k|+|C_k\rangle\,\langle C_k|\,Q_n\,|C_k\rangle\,\langle C_k|\right)=Q_n\,,\quad(14.39)$$

which one verifies by inspection.

Given the eavesdropping state $|E\rangle$, the forwarding state $|F\rangle$ that minimizes the right-hand side of (14.36) is the eigenstate of $\sum_n\langle E|\,Q_n\,|E\rangle\,Q_n$ with the largest eigenvalue. With the aid of the matrix representation associated with (14.30), we have ($q_n\equiv\langle E|\,Q_n\,|E\rangle$ for short)

$$\sum_{n=1,2,5,6}q_nQ_n\,\widehat{=}\,\begin{pmatrix}0&q_1&q_2&0\\q_1&0&0&q_5\\q_2&0&0&q_6\\0&q_5&q_6&0\end{pmatrix}\qquad(14.40)$$

with eigenvalues

$$\pm \tfrac{1}{2} |q_1 + \mathrm{i} q_2 + \mathrm{i} q_5 - q_6| \pm \tfrac{1}{2} |q_1 + \mathrm{i} q_2 - \mathrm{i} q_5 + q_6| \,, \qquad (14.41)$$

all four sign combinations occurring. So, the largest eigenvalue is

$$|q_1 + \mathrm{i} q_2| = |\langle E| (Q_1 + \mathrm{i} Q_2) |E\rangle| \qquad (14.42)$$

for the 2-pair scheme, and

$$\tfrac{1}{2} |\langle E| (Q_1 + \mathrm{i} Q_2 + \mathrm{i} Q_5 - Q_6) |E\rangle|$$
$$+ \tfrac{1}{2} |\langle E| (Q_1 + \mathrm{i} Q_2 - \mathrm{i} Q_5 + Q_6) |E\rangle| \qquad (14.43)$$

for the 4-pair scheme.

Upon replacing the $\langle F| \cdots |F\rangle$ expectation value in (14.36) by these eigenvalues and summing over a complete set $|E_1\rangle$, ..., $|E_4\rangle$ of eavesdropping states, we arrive at

$$\begin{Bmatrix} \text{minimal error} \\ \text{probability for} \\ \text{given } E \text{ basis} \end{Bmatrix} = \frac{1}{2} - \frac{1}{8} \sum_{j=1}^{4} |\langle E_j| (Q_1 + \mathrm{i} Q_2) |E_j\rangle| \qquad (14.44)$$

for the 2-pair scheme, and

$$\begin{Bmatrix} \text{minimal error} \\ \text{probability for} \\ \text{given } E \text{ basis} \end{Bmatrix} = \frac{1}{2} - \frac{1}{32} \sum_{j=1}^{4} |\langle E_j| (Q_1 + \mathrm{i} Q_2 + \mathrm{i} Q_5 - Q_6) |E_j\rangle|$$

$$- \frac{1}{32} \sum_{j=1}^{4} |\langle E_j| (Q_1 + \mathrm{i} Q_2 - \mathrm{i} Q_5 + Q_6) |E_j\rangle|$$

$$(14.45)$$

for the 4-pair scheme. These are the respective analogs of (14.21). As in the steps from (14.21) to (14.22), we can invoke the Peierls inequality to find the minimum of (14.44) with respect to varying the E basis,

$$\begin{Bmatrix} \text{minimal error} \\ \text{probability for} \\ \text{2-pair scheme} \end{Bmatrix} = \tfrac{1}{2} - \tfrac{1}{8} \mathrm{tr} \{|Q_1 + \mathrm{i} Q_2|\} = \tfrac{1}{2} - \tfrac{1}{8} \times 2\sqrt{2} = 14.6\% \,.$$

$$(14.46)$$

(Of course, the trace is now over the 4-dimensional 2-qubit space.) Applying the Peierls inequality to both summations in (14.45) establishes

$$\begin{Bmatrix} \text{minimal error} \\ \text{probability for} \\ \text{4-pair scheme} \end{Bmatrix} \geq \tfrac{1}{2} - \tfrac{1}{32} \mathrm{tr} \{|Q_1 + \mathrm{i} Q_2 + \mathrm{i} Q_5 - Q_6|\}$$

$$- \tfrac{1}{32} \mathrm{tr} \{|Q_1 + \mathrm{i} Q_2 - \mathrm{i} Q_5 + Q_6|\}$$

$$= \tfrac{1}{2} - \tfrac{1}{32} \times 4 - \tfrac{1}{32} \times 4 = 25\% \,. \qquad (14.47)$$

To demonstrate that this lower bound is in fact the minimum of the right-hand side in (14.45), we need to show that the same E basis maximizes both sums. Now, recall that the equal sign holds in

$$\sum_{\nu} |\langle \nu | X | \nu \rangle| \leq \operatorname{tr}\{|X|\} \tag{14.48}$$

if the orthonormal $|\nu\rangle$'s constitute the eigenket basis of the unitary operator U that appears in*

$$X = U|X| = |X^{\dagger}|U . \tag{14.49}$$

So it boils down to finding two commuting unitary operators U_+ and U_- such that

$$(Q_1 + iQ_2) \pm (iQ_5 - Q_6) = U_{\pm}|(Q_1 + iQ_2) \pm (iQ_5 - Q_6)| . \tag{14.50}$$

In view of the matrix representations

$$(Q_1 + iQ_2) \pm (iQ_5 - Q_6) \mathrel{\hat{=}} \begin{pmatrix} 0 & 1 & i & 0 \\ 1 & 0 & 0 & \pm i \\ i & 0 & 0 & \mp 1 \\ 0 & \pm i & \mp 1 & 0 \end{pmatrix} \tag{14.51}$$

and

$$|(Q_1 + iQ_2) \pm (iQ_5 - Q_6)| \mathrel{\hat{=}} \begin{pmatrix} 1 & 0 & 0 & \mp i \\ 0 & 1 & i & 0 \\ 0 & -i & 1 & 0 \\ \pm i & 0 & 0 & 1 \end{pmatrix} , \tag{14.52}$$

this challenge is met by

$$U_+ \mathrel{\hat{=}} \begin{pmatrix} 0 & 1 & 0 & 0 \\ 1 & 0 & 0 & 0 \\ 0 & 0 & 0 & -1 \\ 0 & 0 & -1 & 0 \end{pmatrix} \quad \text{and} \quad U_- \mathrel{\hat{=}} \frac{1}{2} \begin{pmatrix} 1 & 1 & i & i \\ 1 & 1 & -i & -i \\ i & -i & -1 & 1 \\ i & -i & 1 & -1 \end{pmatrix} , \tag{14.53}$$

for example. Accordingly, one E basis for which the right-hand side of (14.45) is equal to the lower bound of (14.47) consists of the common eigenstates of these U_+ and U_-, and this observation justifies the replacement $\geq \;\rightarrow\; =$ in (14.47).

---

*In the present finite-dimensional (actually 4-dimensional) situation there is always such a U, although it may not be unique.

In summary, then, in the 2-pair version of the deterministic qubit-pair scheme for quantum cryptography, the minimal error probability resulting from eavesdropping of the intercept–resend kind is 14.6%, and it is 25% for the 4-pair version. The latter probability is equal to the minimal error probability obtained in (14.22) for the standard 2-pair version of the probabilistic BB84 scheme.

14.4.3   Eavesdropping: maximizing the raw information

If it is Evan's objective to determine as efficiently as possible if a bit in transmission is "+" or "−", he is facing the problem that Eve had to solve in Section 14.3.3. Accordingly, his maximum likelihood for guessing the bit right is given by Eve's result of (14.26), with the necessary changes. The trace is now in the 4-dimensional 2-qubit space, and the relevant statistical operators $\rho^{(\pm)}$ are given by

$$\rho^{(+)} = \frac{1}{N} \sum_n |n_+\rangle \langle n_+| , \qquad \rho^{(-)} = \frac{1}{N} \sum_n |n_-\rangle \langle n_-| , \qquad (14.54)$$

so that

$$\mathcal{L} = \frac{1}{2} + \frac{1}{4N} \mathrm{tr} \left\{ \left| \sum_n Q_n \right| \right\} . \qquad (14.55)$$

With $q_1 = q_2 = 1$, $q_5 = q_6 = 0$ or $q_1 = q_2 = q_5 = q_6 = 1$ in (14.40) for the 2-state or the 4-state version, respectively, the eigenvalues of (14.41) are at hand for the evaluation of the trace in (14.55). We find

$$\mathcal{L} = \begin{cases} \frac{1}{2} + \frac{1}{4}\sqrt{2} = 85.4\% & \text{for } N = 2 \text{ and } n = 1, 2, \\ \frac{3}{4} = 75\% & \text{for } N = 4 \text{ and } n = 1, 2, 5, 6. \end{cases} \qquad (14.56)$$

Incidentally, using all six pairs of (14.28) does not reduce $\mathcal{L}$ below the 75% value of the 4-state version.

Although these guessing odds of 75% are smaller than those found in Sec. 14.3.3 for the BB84 scheme, they are forbiddingly large when it comes to using the 4-state qubit-pair scheme for direct communication. A substantial modification would be required.

As we shall see in Section 14.6, it is possible indeed to redesign the scheme such that $\mathcal{L} = 50\%$, which are the guessing odds that one obtains by flipping a coin; genuine bit knowledge is only available if $\mathcal{L} > 50\%$. For example, $\mathcal{L} = 75\%$ could mean that Evan knows $3 : 1$ odds for each bit or that he is sure about half of the bits and must flip a coin for the other half. Which actual situation exists depends on the measurement he performs. It is quite conceivable that he can choose between those two extreme cases and also intermediate ones, each realizing overall odds of 75%. So, to really deny to Evan any information about bits in transit from Alice to Bob, one must employ a scheme of the kind presented in Section 14.6, where $\mathcal{L} = 50\%$ is enforced.

14.5 Idealized single-photon schemes

In this section we describe how the schemes for secure key distribution can be realized experimentally by quantum-optical means. In addition to a source of entangled photon pairs (SEPP) and single-photon detectors, these schemes just employ linear optical elements, such as half-transparent mirrors (HTMs), polarizing beam splitters (PBSs), half-wave plates (HWPs), and phase shifters (PSs).

14.5.1 BB84 scheme with two state pairs

An idealized implementation of BB84, a scheme proposed by Bennett, Brassard, and Mermin in 1992 [29], is sketched in figure 14.1. It uses the polarization qubit of single photons. Having the photon polarized vertically (v) or horizontally (h) is the binary alternative of the first state pair,

$$|1_+\rangle = |\mathsf{v}\rangle \ , \qquad |1_-\rangle = |\mathsf{h}\rangle \ , \qquad (14.57)$$

and their symmetric (s) and antisymmetric (a) superpositions,*

$$|2_+\rangle = |\mathsf{s}\rangle \equiv 2^{-\frac{1}{2}}\big(|\mathsf{v}\rangle + |\mathsf{h}\rangle\big) \ , \quad |2_-\rangle = |\mathsf{a}\rangle \equiv 2^{-\frac{1}{2}}\big(|\mathsf{v}\rangle - |\mathsf{h}\rangle\big) \ , \quad (14.58)$$

make up the second state pair. In physical terms, s and a are linear polarizations halfway between v and h, as indicated by the orientation of the pairs of two-sided arrows:

---

*We are using the usual phase conventions here, not those of the first footnote on page 363.

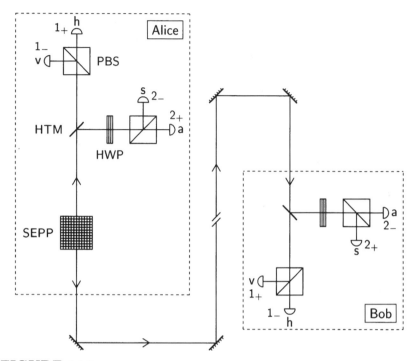

FIGURE 14.1
Single-photon realization of the standard BB84 scheme that uses two
state pairs.

that are meant to symbolize the electric field vectors of photons flying
toward the reader. Thus, contact with Section 14.3 is established by

$$\sigma_z = |v\rangle \langle v| - |h\rangle \langle h| \ ,$$
$$\sigma_x = |s\rangle \langle s| - |a\rangle \langle a| \ , \tag{14.59}$$

and the eigenstates of

$$\sigma_y = \mathrm{i}\sigma_x\sigma_z = \frac{|v\rangle + \mathrm{i}\,|h\rangle}{\sqrt{2}} \frac{\langle v| - \mathrm{i}\,\langle h|}{\sqrt{2}} - \frac{|v\rangle - \mathrm{i}\,|h\rangle}{\sqrt{2}} \frac{\langle v| + \mathrm{i}\,\langle h|}{\sqrt{2}} \tag{14.60}$$

are the two circular polarizations.

Alice has a source of entangled photon pairs (SEPP) at her disposal
that emits 2-photon states of the form

$$2^{-\frac{1}{2}}\left(|v,h\rangle - |h,v\rangle\right) = 2^{-\frac{1}{2}}\left(|a,s\rangle - |s,a\rangle\right) , \tag{14.61}$$

where $|v, h\rangle$, for instance, is the state where the photon moving up in Figure 14.1 is vertically polarized and the one moving down horizontally.* This is, of course, a Bell state of the kind that the stranded physicist prepares in Section 14.2.2; see (14.1).

The up-moving photon encounters a half-transparent mirror (HTM) where it is transmitted or reflected. If transmitted, it passes through a polarizing beam splitter (PBS) that deflects vertically polarized photons to one detector and lets horizontally polarized photons reach another detector. As soon as the detector for h photons responds, Alice knows that the down-moving photon is of the v kind so that a $|1_+\rangle$ state is on its way to Bob. Likewise, a click of the v detector signals that $|1_-\rangle$ has been prepared. Similarly, if the HTM reflects the up-moving photon, Alice determines if it is of the s kind or the a kind. This is achieved by having it pass through a half-wave plate (HWP) that effects the polarization changes $s \rightarrow v$, $a \rightarrow h$ so that the detectors behind the subsequent PBS respond if the HTM reflected a s or a a photon, respectively. After finding that her photon is of a polarization, Alice knows that a $|2_+\rangle = |s\rangle$ state is transmitted to Bob, and it is a $|2_-\rangle = |a\rangle$ state if she gets a click of the s detector.

The equipment at Bob's end is essentially identical with what Alice uses for the polarization analysis. There is one PBS with detectors for measuring in the v/h basis, and a HWP plus PBS for the s/a basis. The random switching between the two measurement bases is done by the HTM that the photon gets to first.

Quite obviously, an experimental setup of this kind would realize the BB84 scheme. We call it "idealized" because we ignore the experimental imperfections that are always present in real life. For example, there could be polarization changing influences on the photons on the road from Alice to Bob, and the HTMs might not be exactly *half*-transparent for all polarizations. Further, most detectors for single photons are not highly efficient, so a substantial number of photons will escape detection. This is not of much concern here because Alice and Bob can just disregard those cases where only one of the photons emitted by the SEPP is detected.

Experiments along these lines have actually been performed. They have shown that secure key distribution is possible indeed, even over a distance of several kilometers [33]. Some experiments use primary

---

*Bright sources of such states are in fact available. Roughly speaking, they use nonlinear crystals to convert blue photons into polarization-entangled pairs of red photons. Consult References [30, 31, 32], for example, for technical details.

sources (the SEPP in Figure 14.1) of polarization-entangled photon pairs
[34, 35]; others use sources of photons entangled in the emission times
[36, 37]. They are superior to the many demonstrations with attenuated
laser pulses (see, for example, the textbook accounts in References [8]
and [9]), since the setup as used by Alice in Figure 14.1 already deliv-
ers single-photon pulses with very good approximation. However, the
technical requirements are much more involved and expensive. With the
development of all-solid-state sources, this situation might change in the
future. When using parametric down-conversion as the source of entangled photon pairs, there are other degrees of freedom that also exhibit
a high degree of correlation and entanglement. For example, momentum conservation enables the generation of momentum-entangled photon
pairs and allows the design of sources (using two or even only one nonlinear crystal) as they would be needed for the new schemes described
in the following.

14.5.2   Qubit-pair scheme with four state pairs

The idealized experimental scheme that implements a particularly
simple form of the qubit-pair scheme with four state pairs is sketched
in Figure 14.2. It has the remarkable and perhaps unexpected property
that all measurements that Alice performs for preparation and Bob for
detection identify product states. In addition, the eight different 2-qubit
states transmitted ($|n_{\pm}\rangle$ for $n = 1, 2, 5, 6$) are product states as well. The
only qubit-pair states that are entangled are the auxiliary states of the Φ
basis, which is not measured but just useful for the theoretical analysis.

As in the BB84 scheme of Figure 14.1, Alice sends single photons
to Bob in the scheme of Figure 14.2 as well. But in addition to the
polarization qubit, the photons now carry a second qubit. Its basic
alternative is the option of propagating along the right transmission line
(R) or the left one (L). The analogs of the s/a superpositions of the v/h
polarization alternative are the symmetric (S) and antisymmetric (A)
superpositions

$$\left.\begin{array}{c} |S\rangle \\ |A\rangle \end{array}\right\} = 2^{-\frac{1}{2}}\left(|R\rangle \pm |L\rangle\right). \qquad (14.62)$$

The unitary transition

$$\begin{array}{ll} |R\rangle \rightarrow |S\rangle\,, & |S\rangle \rightarrow |R\rangle\,, \\ |L\rangle \rightarrow |A\rangle\,, & |A\rangle \rightarrow |L\rangle\,, \end{array} \qquad (14.63)$$

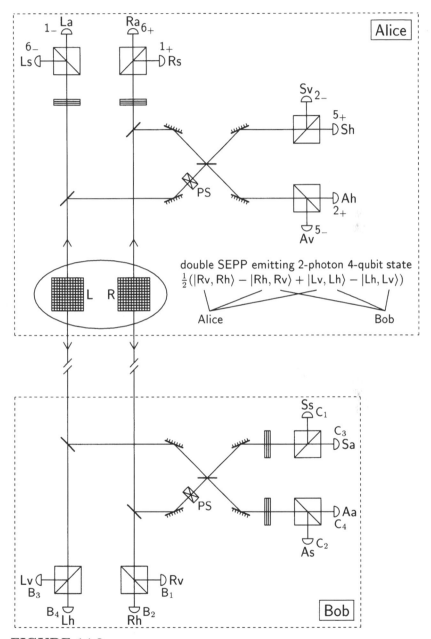

FIGURE 14.2
Single-photon realization of the qubit-pair scheme using four state pairs.

is achieved by a HTM in conjunction with a suitably set phase shifter (PS) that makes sure that the relative phases are the correct ones. More generally, *any* unitary transformation on the qubit pair consisting of the v/h and R/L alternatives of a single photon can be realized by a standard, balanced Mach–Zehnder interferometer supplemented by polarization-changing elements; see Reference [17] for details.

In the scheme of Figure 14.2, the Φ basis of Section 14.4 is given by

$$\begin{pmatrix} \langle\Phi_1| \\ \langle\Phi_2| \\ \langle\Phi_3| \\ \langle\Phi_4| \end{pmatrix} = -\frac{1}{\sqrt{2}} \begin{pmatrix} 0 & 1 & -1 & 0 \\ 0 & 1 & 1 & 0 \\ 1 & 0 & 0 & -1 \\ 1 & 0 & 0 & 1 \end{pmatrix} \begin{pmatrix} \langle Rs| \\ \langle Ra| \\ \langle Ls| \\ \langle La| \end{pmatrix} \tag{14.64}$$

with $|Rs\rangle$, for example, symbolizing a "s polarized photon on the right." Equation (14.29) tells us what Bob's measurement bases are:

$$\left(|B_1\rangle, |B_2\rangle, |B_3\rangle, |B_4\rangle \right) = \left(-|Rv\rangle, |Rh\rangle, |Lv\rangle, |Lh\rangle \right) ,$$

$$\left(|C_1\rangle, |C_2\rangle, |C_3\rangle, |C_4\rangle \right) = \left(-|Ss\rangle, |As\rangle, |Sa\rangle, |Aa\rangle \right) . \tag{14.65}$$

Accordingly, he measures in the B basis by two PBSs plus detectors, one set each for R and L, just doubling the equipment he uses in Figure 14.1 for the discrimination between v and h.

Similarly, for measuring in the C basis, Bob doubles the HWP/PBS combination that tells s from a in Figure 14.1. And to implement the transition of (the second column in) (14.63), he makes the photons pass through a PS/HTM combination on their way to the HWP/PBS combination. As in the BB84 setup of Figure 14.1, the random switching between Bob's measurement bases is here also achieved by the random decision taken at the HTMs that either deflect the down-moving photon to the right, where the C basis is measured, or let it continue downward where B states are detected.

The state pairs $|n_\pm\rangle$, $n = 1, 2, 5, 6$ needed in the four-pair scheme to which (14.47) refers, for instance, are found by applying (14.28) to (14.64). We get

$$\begin{aligned} |1_+\rangle &= -|Ra\rangle , & |2_+\rangle &= -|Av\rangle , \\ |1_-\rangle &= |Ls\rangle , & |2_-\rangle &= |Sh\rangle , \\ |5_+\rangle &= -|Sv\rangle , & |6_+\rangle &= -|Rs\rangle , \\ |5_-\rangle &= |Ah\rangle , & |6_-\rangle &= |La\rangle . \end{aligned} \tag{14.66}$$

Note how this is related to what the stranded physicist does in Section 14.2: if we regard the qubit states paired in $|1_\pm\rangle$ as anti-copies of each other, then we have paired copies in $|6_\pm\rangle$ and similarly for $|2_\pm\rangle$ and $|5_\pm\rangle$.

Alice's double SEPP emits photon pairs in the polarization-entangled state (14.59), whereby emission on the right and on the left occurs equally likely. A well-defined phase between "both photons on the right" and "both photons on the left" ensures that the two-photon state is given by *

$$\tfrac{1}{2}\big(|\mathsf{Rv,Rh}\rangle - |\mathsf{Rh,Rv}\rangle + |\mathsf{Lv,Lh}\rangle - |\mathsf{Lh,Lv}\rangle\big)$$

$$= \tfrac{1}{2}\big(|\mathsf{Ra,Rs}\rangle - |\mathsf{Rs,Ra}\rangle + |\mathsf{La,Ls}\rangle - |\mathsf{Ls,La}\rangle\big)$$

$$= \tfrac{1}{2}\big(|\mathsf{Sv,Sh}\rangle - |\mathsf{Sh,Sv}\rangle + |\mathsf{Av,Ah}\rangle - |\mathsf{Ah,Av}\rangle\big), \qquad (14.67)$$

where $|\mathsf{Ra,Rs}\rangle$, for example, refers to the situation "Alice's up-moving photon is on the right and a polarized, Bob's down-moving photon is on the right and s polarized." This 2-photon state is the tensor product of the antisymmetric polarization Bell state (14.59) and the symmetric R/L Bell state

$$2^{-\frac{1}{2}}\big(|\mathsf{R,R}\rangle + |\mathsf{L,L}\rangle\big) = 2^{-\frac{1}{2}}\big(|\mathsf{S,S}\rangle + |\mathsf{A,A}\rangle\big). \qquad (14.68)$$

Upon detecting Ra, Alice knows that an Rs photon is on the way to Bob so that state $|6_+\rangle$ is in transmission. This is one of the four cases associated with her measurement at the top of Figure 14.2, where the alternatives R/L and s/a are distinguished. The other three prepare $|1_+\rangle$, $|1_-\rangle$, and $|6_-\rangle$. Similarly, states $|2_\pm\rangle$ and $|5_\pm\rangle$ are produced when Alice's photon is detected by one of the four detectors for up-moving photons deflected to the right, where she distinguishes between S and A as well as v and h.

Clearly, the setup of Figure 14.2 would realize the 2-qubit scheme with four state pairs just as well as the setup of Figure 14.1 realizes the 1-qubit BB84 scheme with two state pairs. In Figure 14.1, the photon carries one qubit from Alice to Bob, in Figure 14.2 it carries two qubits, and — fair enough — this doubling requires a doubling of the experimental equipment. This is why a fair comparison between the two schemes should refer to the 2-state-pair version of BB84 and the 4-state-pair version of the qubit-pair scheme.

Alice and Bob get a key bit for every photon sent in Figure 14.2, whereas they need two photons (on average) in Figure 14.1. One might, therefore, be tempted to state that the single-photon 2-qubit scheme is

---

*Roughly speaking, such states can be produced by irradiating coherently two non-linear crystals with blue photons that get converted into pairs of red photons. A pertinent experiment is reported in Reference [38].

twice as efficient — and that is true if there is a limit to the rate at
which the detectors can count single photons. To produce a key of a
certain length would take half as long in Figure 14.2 as in Figure 14.1.
But if one states the efficiency in terms of key bits established per qubit
transmitted, it is one key bit for two qubits in both schemes. The main
difference between the schemes remains, however; in figure 14.2 Alice
is sure that the next qubit pair she sends will contribute a key bit —
this is the *deterministic nature* of the qubit-pair scheme. By contrast,
in Figure 14.1 the next two qubits sent may produce two key bits, one
key bit, or none at all (with relative frequencies of 25%, 50%, and 25%,
respectively *), and there is nothing you can do about this *probabilistic
nature.*

14.6   Direct communication with qubit pairs

In this final section, we present a scheme that enables Alice and Bob to
communicate directly, without the need to establish a shared secure key
first. In some sense, this direct-communication scheme is a modification
of the qubit-pair cryptography scheme with four state pairs that we
discussed in Section 14.4. It also requires the use of four state pairs
$|1_\pm\rangle, \ldots, |4_\pm\rangle$, but these states have properties different from the ones
in Section 14.4 so that Evan cannot obtain any information about bits
in transmission from Alice to Bob.

The scheme described here is the most symmetric variant. A simpler
but less symmetric version, in which the protection against an eavesdropper needs more effort, is laid out in [12], where one can also find an
idealized single-photon scheme, much like the ones in Section 14.5.

14.6.1   Description of the scheme

We pick up the story where we left it at the end of Section 14.4.3.
In order to prevent Evan from gaining any knowledge about bits in
transmission, we must choose the states $|n_\pm\rangle$ such that $\rho^{(+)} = \rho^{(-)}$,

---

*On average this is a 50% efficiency, and only a few variants of BB84 try to go beyond
that [39, 40].

Type of qubit pair	State detected by Bob							
sent by Alice	1_+	2_+	3_+	4_+	1_-	2_-	3_-	4_-
1	+	−	−	−	−	+	+	+
2	−	+	−	−	+	−	+	+
3	−	−	+	−	+	+	−	+
4	−	−	−	+	+	+	+	−

Table 14.5 Bit values inferred by Bob in the scheme defined by (14.71) and (14.72).

that is:

$$\sum_n |n_+\rangle \langle n_+| = \sum_n |n_-\rangle \langle n_-| \tag{14.69}$$

or, equivalently,

$$\sum_n Q_n = 0 \tag{14.70}$$

with Q_n as introduced in (14.37). One way of achieving this, which keeps a lot of the symmetry displayed by the matrices in (14.29) and (14.7), is given by

$$\begin{pmatrix} \langle 1_+| \\ \langle 2_+| \\ \langle 3_+| \\ \langle 4_+| \end{pmatrix} = \frac{1}{\sqrt{3}} \begin{pmatrix} 0 & 1 & 1 & 1 \\ -1 & 0 & 1 & -1 \\ -1 & -1 & 0 & 1 \\ -1 & 1 & -1 & 0 \end{pmatrix} \begin{pmatrix} \langle 1_-| \\ \langle 2_-| \\ \langle 3_-| \\ \langle 4_-| \end{pmatrix} \equiv \mathsf{T} \begin{pmatrix} \langle 1_-| \\ \langle 2_-| \\ \langle 3_-| \\ \langle 4_-| \end{pmatrix} \tag{14.71}$$

in conjunction with

$$|B_n\rangle = |n_+\rangle \quad \text{and} \quad |C_n\rangle = |n_-\rangle \quad \text{for} \quad n = 1, 2, 3, 4. \tag{14.72}$$

Thus, Bob's measurement bases are identical with Alice's preparation bases. Condition (14.69) is met because the "+" states constitute an orthonormal basis in the 2-qubit space, and so do the "−" states.

Upon detecting state $|B_3\rangle = |3_+\rangle$, say, Bob does not know if Alice sent $|3_+\rangle$ or either one of $|1_-\rangle$, $|2_-\rangle$, $|4_-\rangle$. The respective probabilities are $\frac{1}{2}$ and three times $\frac{1}{6}$, so that he has no way of guessing if she sent "+" or "−". This is, of course, the essence of (14.69). But as soon as she announces the type sent ($n = 1, 2, 3,$ or 4), he can infer the bit value with certainty. The matters are summarized in Table 14.5.

With equipment in place that enables Alice to send any of the $|n_\pm\rangle$ states of her liking and Bob to detect the incoming qubit pair in the B

basis of "+" states or the C basis of "−" states, alternating between his two measurement bases in a random fashion, this is how she communicates a confidential $+/-$ message string to him:

Step 1: Pick key sequence, form message sequence

Alice picks the key sequence, a random sequence of 1, 2, 3, 4 identifying the pair types. She matches the $+/-$ message sequence with the key sequence and so forms a sequence of $n_\pm$ state identifiers. Only she knows the key sequence.

Step 2: Add control bits

Alice intersperses this $n_\pm$ sequence with a fair number of additional state identifiers, of random values and at random positions, to be used as control bits. Only she knows the positions and values of the control bits.

Step 3: Send and receive

Alice sends the 2-qubit state sequence thus defined to Bob, who detects the qubit pairs in the "+" basis or the "−" basis, randomly switching between the two. Only he knows which basis is measured.

Step 4: Check security

Alice tells Bob which bits are the control bits (that were added in Step 2) and he announces (publicly) the states in which they were detected. She then checks if his measurement results are consistent with what she sent. If there are inconsistencies, such as detecting $|3_+\rangle$ or $|1_-\rangle$ when $|3_-\rangle$ was sent, they return to Step 1 and try again. Otherwise they continue with Step 5.

Step 5: Announce key sequence

Alice announces (publicly) the key sequence of Step 1. Then Bob identifies the message sequence in accordance with Table 14.5.

An illustrating example is given in Table 14.6.

We emphasize the crucial difference between this *secure direct communication* and the standard *encrypted communication*. In the latter, one establishes a secret shared key, *never* to be made public, and then sends an encrypted message over a public channel. The bits in transmission are not correlated at all with the message bits. In fact, since the encryption key consists of a random bit sequence, the encrypted message is an equally random bit string.

Alice's key	1	3	4	4	1	2	1	3	3 $\cdots$
Message	+	$\boxed{+}$	−	−	−	+	$\boxed{-}$	+	− $\cdots$
States sent	1_+	3_+	4_-	4_-	1_-	2_+	1_-	3_+	$3_- \cdots$
Bob detects	1_+	1_-	4_-	2_+	2_+	4_-	4_+	3_+	$3_- \cdots$

Table 14.6 Direct confidential communication. Alice chooses a random key sequence of $1, 2, 3, 4$ (first row) and matches it with the bit sequence of the message (second row) interspersed with randomly located control bits (boxed) to determined the sequence of states to be sent (third row). Bob obtains a sequence of detected states (fourth row). The control bits are used to test for the presence of an eavesdropper. After Alice reveals the random sequence of the first row, Bob can then reconstruct the message of the second row.

By contrast, the message bits transmitted in the 5-step qubit-pair scheme just described have their actual values — Alice sends $|2_+\rangle$, say, and that really means "+". And the $1, 2, 3, 4$ key sequence of step 1 is *always* made public in Step 5, after Step 4 establishes that Evan did not listen in.

The security of the scheme derives from (1) the impossibility of gaining any knowledge about even a single message bit before Step 5, and (2) the errors that arise unavoidably if the qubit pairs in transmission are intercepted. In Step 4, the values of the control bits are made public when Alice and Bob look for inconsistencies as evidence for Evan's possible interference, but none of the message bits are revealed. Note that one cannot use (some of) the message bits for the security check — one really needs the additional control bits — because that would eventually make public some part of the message, contrary to the objective of keeping each and every message bit confidential.

The scheme possesses property (1) by construction; this is the essence of (14.69) or (14.70). Property (2) is the statement that the minimal error probability is appreciable. Let us address this issue now.

14.6.2 Minimal error probability

The calculation of the minimal error probability resulting from intercept–resend attacks by Evan follows the pattern of Sections 14.3.2 and 14.4.2. For the benefit of technical simplification, it is expedient to first introduce auxiliary states $|\chi_1\rangle$, $\ldots$, $|\chi_4\rangle$ that play a role similar to the one of $|\Phi_1\rangle$, $\ldots$, $|\Phi_4\rangle$ in Section 14.4. The unitary 4×4 matrix T in

(14.71) is anti-Hermitean,

$$-T = T^\dagger = T^{-1} , \tag{14.73}$$

so that

$$T^2 = -1_{4\times4} \quad \text{and} \quad T = \tfrac{1}{2}\left(1_{4\times4} + T\right)^2 = -\tfrac{1}{2}\left(1_{4\times4} - T\right)^2 . \tag{14.74}$$

We exploit this observation in defining the orthonormal χ states in accordance with

$$\begin{pmatrix} \langle\chi_1| \\ \langle\chi_2| \\ \langle\chi_3| \\ \langle\chi_4| \end{pmatrix} = 2^{-\frac{1}{2}}\left(1_{4\times4} - T\right)\begin{pmatrix} \langle1_+| \\ \langle2_+| \\ \langle3_+| \\ \langle4_+| \end{pmatrix} = 2^{-\frac{1}{2}}\left(1_{4\times4} + T\right)\begin{pmatrix} \langle1_-| \\ \langle2_-| \\ \langle3_-| \\ \langle4_-| \end{pmatrix} . \tag{14.75}$$

Then, representing operators by their χ-basis matrices, we have

$$\sum_{n=1}^{4} q_n Q_n \cong \begin{pmatrix} q_1 & 0 & 0 & 0 \\ 0 & q_2 & 0 & 0 \\ 0 & 0 & q_3 & 0 \\ 0 & 0 & 0 & q_4 \end{pmatrix} T - T \begin{pmatrix} q_1 & 0 & 0 & 0 \\ 0 & q_2 & 0 & 0 \\ 0 & 0 & q_3 & 0 \\ 0 & 0 & 0 & q_4 \end{pmatrix} . \tag{14.76}$$

Equation (14.36) carries over, so that

$$\left\{ \begin{array}{c} \text{error probability} \\ \text{for given} \\ |E\rangle \text{ and } |F\rangle \end{array} \right\} = \frac{1}{8} - \frac{1}{16}\sum_n \langle E| Q_n |E\rangle \langle F| Q_n |F\rangle \tag{14.77}$$

after taking $\sum_n P_n = 2$ into account. Upon determining the largest eigenvalue of the operator in (14.76) we get the analog of (14.45),

$$\left\{ \begin{array}{c} \text{minimal error} \\ \text{probability for} \\ \text{given } E \text{ basis} \end{array} \right\} = \frac{1}{2} - \frac{1}{16\sqrt{6}}\sum_{j=1}^{4}\left(\mu_1^{(j)} + \mu_2^{(j)} + \mu_3^{(j)} \right.$$

$$\left. + \left[3\left(\mu_1^{(j)}\mu_2^{(j)} + \mu_2^{(j)}\mu_3^{(j)} + \mu_3^{(j)}\mu_1^{(j)}\right)\right]^{\frac{1}{2}} \right)^{\frac{1}{2}} \tag{14.78}$$

with

$$\mu_1^{(j)} = \langle E_j| \left(Q_1 - Q_2 - Q_3 + Q_4\right) |E_j\rangle^2 ,$$

$$\mu_2^{(j)} = \langle E_j| \left(Q_1 - Q_2 + Q_3 - Q_4\right) |E_j\rangle^2 ,$$

$$\mu_3^{(j)} = \langle E_j| \left(Q_1 + Q_2 - Q_3 - Q_4\right) |E_j\rangle^2 . \tag{14.79}$$

Unfortunately, this does not seem to lend itself to another application of the Peierls inequality. In fact, we do not know for sure the minimal value of (14.78) with respect to varying the E basis, but we surmise that it is $\frac{1}{6} = 16.7\%$. The evidence in support of this *conjecture* is as follows.

The symmetry between the "+" basis, the "−" basis, and the χ basis identifies them as the natural candidates for bases that yield stationary values of the right-hand side of (14.78). Indeed, we find

$$\mu_1^{(j)} = \mu_2^{(j)} = \mu_3^{(j)} = \begin{cases} 0 \text{ for } |E_j\rangle = |\chi_j\rangle \ , \\ \frac{16}{9} \text{ for } |E_j\rangle = |j_\pm\rangle \ , \end{cases} \quad (14.80)$$

for $j = 1, \ldots, 4$, so that the χ basis gives the absolutely largest value of $\frac{1}{2} = 50\%$, whereas the "+" basis and the "−" basis both give $\frac{1}{6} = 16.7\%$. That they are equivalent in this respect is clear from the invariance of (14.79) under the interchange $|n_+\rangle \leftrightarrow |n_-\rangle$ — that is, $Q_n \to -Q_n$, for $n = 1, 2, 3, 4$.

Now let us look at the vicinity of the "+" basis, say, and consider

$$|E_j\rangle = \mathrm{e}^{\mathrm{i}\varepsilon G} |j_+\rangle \quad (14.81)$$

with a self-adjoint generator G and a real parameter ε. Up to order ε^2, this gives

$$\left\{ \begin{array}{c} \text{right-hand side} \\ \text{of (14.78)} \\ \text{near "+" basis} \end{array} \right\} = \frac{1}{6} + \frac{\varepsilon^2}{64} \sum_{n=1}^{4} \langle n_+| G(1 - Q_n)G |n_+\rangle \geq \frac{1}{6} \quad (14.82)$$

where $G(1 - Q_n)G$ is nonnegative since G is self-adjoint and

$$1 - Q_n = \sum_{m(\neq n)} |m_+\rangle \langle m_+| + |n_-\rangle \langle n_-| \geq 0 \ . \quad (14.83)$$

Upon writing the n-th expectation value in (14.82) as

$$\langle n_+| G(1 - Q_n)G |n_+\rangle = \left[\langle n_+| G^2 |n_+\rangle - \langle n_+| G |n_+\rangle^2 \right] \\ + |\langle n_+| G |n_-\rangle|^2 \ , \quad (14.84)$$

where the $[\cdots]$ term is the variance of G in state $|n_+\rangle$, we learn that the equal sign in the inequality of (14.82) can only hold if all four $|n_+\rangle$ states are eigenkets of G. But such a G would just give rise to phase factors in (14.81) so that the E basis would be identical with the "+" basis. Therefore, we have '>' in (14.82) for all bases in the infinitesimal neighborhood of the "+" basis, which is to say that the "+" basis yields a local minimum of the error probability (and the "−" basis yields another equivalent one). Our conjecture thus amounts to asserting that this minimum is global. In addition to the mathematical arguments given above, it is supported by a considerable body of numerical evidence [12].

14.7   Acknowledgments

BGE thanks Yakir Aharonov and Lev Vaidman for illuminating discussions and wishes to express his gratitude for the hospitable environment provided by Gerald Badurek and Helmut Rauch at the Atominstitut in Vienna, where part of this work was done with financial support from the Technical University in Vienna. AB and BGE are grateful for the kind hospitality extended to them at the Erwin–Schrödinger–Institut in Vienna. CK and HW acknowledge support by project QuComm (IST-1999-10033) of the European Union.

References

[1] C. H. Bennett and G. Brassard, in *Proc. IEEE Int. Conf. on Computers, Systems, and Signal Processing, Bangalore* (IEEE, New York 1984) pp. 175–179.

[2] A. Ekert, *Phys. Rev. Lett.* **67** (1991) 661–663.

[3] A. Einstein, B. Podolsky, and N. Rosen, *Phys. Rev.* **47** (1935) 777–780.

[4] J. S. Bell, *Physics* **1** (1964) 195–200.

[5] D. Bruß, *Phys. Rev. Lett.* **81** (1998) 3018–3021.

[6] M. Bourennane, A. Karlsson, and G. Björk, *Phys. Rev. A* **64** (2001) art. 012306 (5 pages).

[7] N. J. Cerf, M. Bourennane, A. Karlsson, and N. Gisin, LANL preprint *quant-ph*/0107130 (2001).

[8] H.-K. Lo, S. Popescu, and T. Spiller, *Introduction to Quantum Computation and Information* (World Scientific, Singapore 1998).

[9] D. Bouwmeester, A. Ekert, and A. Zeilinger, *The Physics of Quantum Information* (Springer-Verlag, Berlin 2000).

[10] N. Gisin, G. Ribory, W. Tittel, and H. Zbinden, LANL preprint *quant-ph*/0101098 (2001).

[11] A. Beige, B.-G. Englert, C. Kurtsiefer, and H. Weinfurter, LANL preprint *quant-ph*/0101066 (2001).

[12] A. Beige, B.-G. Englert, C. Kurtsiefer, and H. Weinfurter, LANL preprint `quant-ph/0111106` (2001); to appear in *Acta Physica Polonica*.

[13] L. Vaidman, Y. Aharonov, and D. Z. Albert, *Phys. Rev. Lett.* **58** (1987) 1385–1387.

[14] Y. Aharonov and B.-G. Englert, *Z. Naturforsch.* **56a** (2001) 16–19. We thank Verlag der Zeitschrift für Naturforschung for the kind permission to reproduce the Mean King's Problem in Section 14.2.1.

[15] E. Schrödinger, *Die Naturwissenschaften* **23** (1935) 807–812, 823–828, 844–849; English translation by J. D. Trimmer, *Proc. Am. Philos. Soc.* (1980) **124**, 323–338; the latter reprinted in [41].

[16] B.-G. Englert and Y. Aharonov, *Phys. Lett. A* **284** (2001) 1–5.

[17] B.-G. Englert, C. Kurtsiefer, and H. Weinfurter, *Phys. Rev. A* **63** (2001) art. 032303 (10 pages).

[18] C. Kurtsiefer and H. Weinfurter, unpublished.

[19] J. A. Bergou and B.-G. Englert, *J. Mod. Opt.* **45** (1998) 701–711.

[20] C. H. Bennett, *Phys. Rev. Lett.* **68** (1992) 3121–3124.

[21] W. Xiang-bin, LANL preprint *quant-ph*/0110089 (2001).

[22] C. A. Fuchs, N. Gisin, R. B. Griffiths, C.-S. Niu, and A. Peres, *Phys. Rev. A* **56** (1997) 1163–1172.

[23] W. Thirring, *A Course in Mathematical Physics, Vol. 4: Quantum Mechanics of Large Systems* (Springer-Verlag, Vienna 1980).

[24] N. Akhieser and I. M. Glazman, *Theory of Linear Operators in Hilbert Space* (Ungar, New York 1961).

[25] M. Reed and B. Simon, *Methods of Modern Mathematical Physics, Vol. 1* (Academic Press, New York 1972).

[26] J. Glimm and A. Jaffe, *Quantum Physics. A Functional Integral Point of View* (Springer-Verlag, Berlin 1987).

[27] B.-G. Englert and J. A. Bergou, *Opt. Commun.* **179** (2000) 337–355.

[28] L. Goldenberg and L. Vaidman, *Phys. Rev. Lett.* **75** (1995) 1239–1343.

[29] C. H. Bennett, G. Brassard, and N. D. Mermin, *Phys. Rev. Lett.* **68** (1992) 557–559.

[30] P. G. Kwiat, K. Mattle, H. Weinfurter, A. Zeilinger, A. V. Sergienko, and Y. H. Shih, *Phys. Rev. Lett.* **75** (1995) 4337–4341.

[31] P. G. Kwiat, E. Waks, A. G. White, I. Appelbaum, and P. H. Eberhard, *Phys. Rev. A* **60** (1999) 773–776.

[32] C. Kurtsiefer, M. Oberparleiter, and H. Weinfurter, *Phys. Rev. A* **64** (2001) art. 023802 (4 pages).

[33] R. J. Hughes, G. G. Luther, G. L. Morgan, C. G. Peterson, and C. Simons, Quantum cryptography over underground optical fibers, in *Advances in Cryptology — Proceedings of Crypto '96* (Springer, Berlin 1996).

[34] T. Jennewein, C. Simon, G. Weihs, H. Weinfurter, and A. Zeilinger, *Phys. Rev. Lett.* **84** (2000) 4729–4732.

[35] D. S. Naik, C. G. Peterson, A. G. White, A. J. Berglund, and P. G. Kwiat, *Phys. Rev. Lett.* **84** (2000) 4733–4736.

[36] W. Tittel, J. Brendel, H. Zbinden, and N. Gisin, *Phys. Rev. Lett.* **84** (2000) 4737–4740.

[37] G. Ribordy, J. Brendel, J.-D. Gautier, N. Gisin, and H. Zbinden, *Phys. Rev. A* **63** (2001) art. 012309 (12 pages).

[38] A. Zeilinger, M. A. Horne, H. Weinfurter, and M. Żukowski, *Phys. Rev. Lett.* **78** (1997) 3031–3034.

[39] H.-K. Lo, H. F. Chau, M. Ardehali, LANL preprint *quant-ph* /0011056 (2000).

[40] A. Cabello, *Phys. Rev. Lett.* **85** (2000) 5635–5638.

[41] J. A. Wheeler and W. H. Zurek, *Quantum Theory and Measurement* (Princeton University Press, Princeton 1983).

Commentary on Quantum Computing

Chapter 15

Transgressing the boundaries of quantum computation: a contribution to the hermeneutics of the NMR paradigm

Stephen A. Fulling

Abstract   This personal essay recounts a misadventure in attempting to find a quantum algorithm with a "general purpose" flavor. Exact arithmetic with large integers can be conducted modulo a list of relatively prime numbers, the answer being reconstructed in conventional terms only at the last moment. Implementing this process on a molecular level would lead not to a quantum computation, but to a massively parallel classical computation. It has been observed that something similar is going on, in practice though not in principle, in the nuclear magnetic resonance implementations of the Grover search algorithm. Certain other NMR computations, which are frankly of an "ensemble" nature, are even closer to my proposal, though more practical. Comparing and contrasting these situations may cast light on the current semantic disputes over what constitutes a "quantum" computation.

1-58488-282/4/02/$0.00+$1.50
© 2002 by Chapman & Hall/CRC

15.1 Review of NMR quantum computing

The first laboratory demonstrations of quantum computational algorithms were achieved by means of (liquid) nuclear magnetic resonance (NMR). (See [10] for a sober review of the state of the art.) The ideal envisioned in this paradigm is the following. A large molecule contains numerous nuclei with spin, which comprise the qubits of the computer. Because they have different masses and molecular environments, the nuclei are individually addressable by properly tuned magnetic pulses.

The reality is that (understandably) so far only very small numbers of qubits have been implemented (e.g., 3 qubits, enabling a Grover search among 8 objects [20]). Perhaps more worrisome, the experiments are performed on a macroscopic sample containing *many* molecules; it can be argued that this misses the main point of quantum parallelism. This criticism has been made more precise [5, 14, 18] as follows. The internal state of a single molecule is *pseudopure*, deviating only slightly from the density matrix M representing maximal mixture:

$$\rho = (1 - \epsilon)M + \epsilon|\Psi\rangle\langle\Psi|, \qquad \epsilon \sim \frac{N}{2^N}$$

for N qubits. Linden and Popescu [14] show that getting the "answer" reliably then requires on the order of ϵ^{-1} repetitions of the computation (or performing it all at once on that many molecules); this does not appear substantially different from a massively parallel classical computation. The earlier, more technical version [5] of the criticism emphasized the issue of *separability*:

> We [prove] that all mixed states of N qubits in a sufficiently small neighborhood of the maximally mixed state are separable (unentangled)[,...] a mixture of product states.... [A]ll states so far used in NMR for quantum computations ... are separable.... [These results] suggest that current NMR experiments are not true quantum computations, since no entanglement appears in the physical states at any stage.... To reach a firm conclusion, much more needs to be understood about what it means for a computation to be a "quantum" computation.

The proponents of NMR computing naturally regard the glass as half full rather than half empty. One of the most informative research papers in this area is that of Jones and Mosca [11], which says that

> NMR quantum computers differ from other implementations in one important way: there is not one single quantum computer, but rather a statistical ensemble of them. ... Some algorithms ... produce a superposition of states (relative to the natural NMR computational basis) as their final result, and in such cases the behavior of an ensemble quantum computer will be quite different [from a "conventional" quantum computer].

In their experiment a Grover-related counting procedure leads to the density matrix

$$\rho = \frac{1}{2} \begin{pmatrix} 1 + \cos(r\varphi_k) & 0 \\ 0 & 1 - \cos(r\varphi_k) \end{pmatrix}.$$

They then determine φ_k by measuring the expectation value of σ_z.
"Note that in this case ensemble quantum computers have an advantage: with a single quantum computer it would be necessary to repeat the calculation several times in order to obtain a statistical estimate of $\cos(r\varphi_k)$." [11]

15.2   Review of modular arithmetic

The Chinese Remainder Theorem: Let $m_1, \ldots, m_R$ be positive integers (*moduli*) that are pairwise relatively prime:

$$\gcd(m_j, m_k) = 1 \quad \text{if } j \neq k.$$

Let $M = m_1 m_2 \cdots m_R$. For any R-tuple of integers, $(u_1, \ldots, u_R)$, there is exactly one integer u such that

$$0 \leq u < M \quad \text{and} \quad u \equiv u_j \text{ modulo } m_j \quad \text{for each } j.$$

(In particular, if u_j is restricted to the range 0 to $m_j - 1$, then $u_j = u \% m_j$, in the notation of the **C** programming language.) Addition, subtraction, and multiplication modulo M can be carried out by performing the corresponding operations on the residues modulo the respective m_j; for example,

$$(u_1, \ldots, u_R) + (v_1, \ldots, v_R) = ((u_1 + v_1) \% m_1, \ldots, (u_R + v_R) \% m_R)$$

yields the residue representation of $u + v$ modulo M.

One can think of the representation of an integer (less than M) by a string of residues as analogous to the representation of an integer by a string of decimal digits, with the important difference that under addition, subtraction, and multiplication the residues in each "place" combine only among themselves; there is no analogue of the *carry* problem of ordinary digital arithmetic.

One computational application of the Chinese remainder theorem is to exact integer (or exact rational) arithmetic [2, 8, 13, 17]. In some problems in fundamental theoretical physics or pure mathematics, it is important to do calculations exactly instead of resorting to floating-point arithmetic; this can lead to extremely "long" numbers. (For example, the last coefficient in Equation 3.65 of [16] is a fraction whose denominator has 80 decimal digits.) Since the natural word size of a computer is typically 32 bits, special programming is required when the integers exceed 2^{32}. The most obvious strategy is to represent the integers by arrays of ordinary "small" integers and to program the standard arithmetic operations on these arrays, including the necessary bookkeeping of carries and borrows, etc., among the parts. This may be good computer science, but it is rather dull mathematics. It is much more fun to approach the problem by residue arithmetic [8, 17]:

We choose R relatively prime integers, m_j, all fairly big but smaller than 2^{16}, and we do computer arithmetic with arrays of R integers, $(u_1, \ldots, u_R)$, modulo $(m_1, \ldots, m_R)$. These arrays represent all the integers from 0 up to one less than $M = m_1 \cdots m_R$, which may be much larger than 2^{32}. This representation of large integers has the advantage previously advertised, that arithmetic can be performed in parallel on the residue components, without carries and borrows and without cross terms in multiplication.

The algorithm for reconstructing the large integer in human-readable (decimal) notation is relatively complicated and should be avoided or postponed whenever possible. If the number is truly huge, even printing it out or transmitting it over a communications channel might be expensive, and the usefulness of its exact expression open to question. Instead, the advised strategy is to do all computations with the large number internally, until a useful conclusion is reached. This conclusion may be qualitative, such as that the number is exactly zero. Or, it may suffice to know the number in floating-point form, to a precision much less than the total number of digits.

15.3 A proposed "quantum" implementation

In the popular media, and sometimes among the cognoscenti, one encounters remarks like this [21]:

> [Q]uantum computers — which may be available in some 15 to 20 years — will speed drug discovery, let forecasters nail the weather with precision, and help chipmakers design circuits that are now impossibly complex. (Unfortunately, they will also allow hackers to break codes protecting secure traffic on the Internet.)

I fear that only the sentence in parentheses has been established as true. The most plausible practical applications of both the Shor factorization algorithm and the Grover search algorithms are to cryptography. (The cryptographic application of Grover is described in [3]. For doubts about other applications [database applications in the literal sense] see [22].) Beyond these two and their close relatives, what serious algorithms are there? In his public talks, Shor [19] states firmly the belief that quantum computers will always be devoted only to a few specialized tasks.

Why is quantum computation still a solution in search of problems? To be profitably implemented on a quantum computer (if we had one), an algorithm presumably must satisfy two requirements:

1. It must be parallelizable, so that identical operations are to be performed simultaneously on different terms of a superposition state vector. ("Entanglement" is often cited as the *sine qua non* of quantum computing, but the Grover algorithm demonstrates that entanglement is, in general, a secondary feature, born from the marriage of *coherent superposition* and digital architecture [1, 15].)

2. There must be a way of reading out the answer by an observation on the system — i.e., on the final state vector as a whole. Presumably this requires some kind of constructive interference of the terms in the superposition.

The second requirement is what distinguishes a quantum-computable problem from a generic parallelizable problem.

In the most intensively studied applications of quantum computing, factoring and search, most of the quantum operations are devoted to building up an eigenstate (of the computer's basic observables) that represents the result of the computation. The required Hilbert-space maneuvering gives each application an air of specialness. In Shor's words [19], "We must arrange the algorithm to make all computational paths to wrong answers destructively interfere, and those that lead to the right answer constructively interfere."

When I approached the field as a novice, I aimed to adapt some other parallel algorithm to quantum computation. It was natural to recall my previous work on modular arithmetic. The resulting preprint [9] is very short, because I considered it a waste of time (wisely, it turned out) to work out details until the basic idea was validated as sound and interesting.

The obvious strategy for the internal, parallel part of the computation was to divide the qubits of the computer into two groups (data fields), one to hold the modulus and one to hold the result of a computation modulo that modulus. (Presumably some ancillary qubits would be needed for work space, but I shall not refer to them further.) The initial state should be a sum over the moduli of states wherein the first field holds the modulus and the other is set to zero:

$$|\mathrm{i}\rangle = \sum_{j=1}^{R} |m_j\rangle \, |\mathbf{0}\rangle \, .$$

The computation should then perform on each basis state the calculation relative to the corresponding modulus, producing

$$|\mathrm{f}\rangle = \sum_{j=1}^{R} |m_j\rangle \, |u_j\rangle \, .$$

The interesting and novel aspect of this quantum algorithm was to be the output procedure. I proposed that the quantum apparatus somehow be made to emit a signal (optical or electronic) that is, for the jth pure state, a sequence of pulses with period m_j ; a pulse is to occur whenever the time (measured in some basic unit) is congruent to u_j mod m_j . The examined output signal would be the superposition of the signal from each term. Instead of arranging for constructive interference in the state of the qubits themselves and then measuring the qubits, therefore, one would look for constructive interference in the output signal. The strongest pulse would occur at time u, the number less than M that is congruent to u_j mod m_j for all j — that is, the answer! (See the figure for an example.) Thus, I announced, "The present paper proposes a

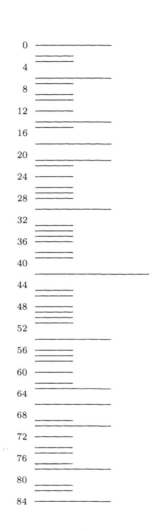

FIGURE 15.1

The pulse sequence for $R = 3$, $m_1 = 3$, $m_2 = 4$, $m_3 = 7$, $M = 84$, $u = 42$.

quantum algorithm of a different kind, in which the constructive interference that builds up the answer takes place at the level of classical waves or signals. Arguably, eventual general-purpose quantum computation is more likely to be of this type."

When writing [9] I was concerned that the precise nature of this readout procedure was still very vague; was it even physically possible? At precisely that time, however, I became aware of the paper of Jones and Mosca [11]. Immediately I concluded that the natural setting for this kind of procedure is a large ensemble of identical quantum computers, as in the NMR experiments, where each molecule is a computer. Upon measurement, each molecule independently collapses into some eigenstate. (I hoped that my pulsed signal emission could occur at this point.) If more than one eigenstate can result from the computational algorithm, then the output from the whole system is an expectation value of the basic observable, not an eigenvalue. Thus in such a computation at least some of the superposition or interference takes place at the classical level, after the quantum observations have been performed. I have already quoted the conclusion of [11] that this feature is desirable in some computations, where the sought information is a statistical property of a quantum state rather than the random outcome of a single measurement on it. "The present situation is akin to the latter," I wrote, somewhat optimistically.

The need for a pulsed signal would disappear if one could measure u on enough individual quantum computers to develop a reliable list of all the u_j. Unfortunately, bulk measurements of a liquid NMR sample would yield a useless average of the u_j, not individual values. It would be possible in principle, but absurd in practice, to code the pulses into a long string of additional qubits still at the unitary-transformation stage, so that the superposed signal with peak at the answer can indeed be obtained (as a string of expectation values) from bulk measurements. For definiteness I consider this last version in what follows. Other variations on the procedure, more complicated but more likely to be practical, have been proposed by D. M. Potts and by M. Stay (private communications).

15.4   Aftermath

After I submitted the paper [9], it occurred to me that the output of the quantum computation proposed in it would be unchanged if all the molecules decohered prematurely. That is, one could start with a mixture of equal numbers of molecules in each of the eigenstates $|m_j\rangle |\mathbf{0}\rangle$, rather than the superposition proposed above. This means that quantum superposition is not being used in any essential way (although the manipulations on the individual states presumably would use quantum gates). *This is a parallel computation by an ensemble of molecular-sized classical computers!* If the experiment is indeed performed on a superposition state, it seems unlikely that one could extract the answer by a single observation on a single molecule, since that would amount almost to "cloning" the state. (The question being asked is of the form, "Here is a state which is a superposition — with equal amplitudes and controllable phases — of eigenstates of a certain observable. *Which* eigenstates of the basis appear in the sum?" I do not know any way to answer that question experimentally; certainly measuring the observable in question will not do.) If the answer is to be obtained from bulk measurements as an expectation value, clearly the number of molecules involved must be at least equal to the number of moduli and actually much larger than that for good statistics.

Meanwhile, the referees for *Physical Review Letters* did not validate [9] as sound and interesting. Their most cogent criticism was essentially the point just made: the proposed procedure is essentially classical. One referee mentioned in passing the analogy with "liquid-state NMR which itself is not exactly quantum because entanglement has not been experimentally achieved."

I no longer believe that [9] presents a "quantum algorithm" in any serious sense of the term, nor that it has prospects of becoming a practical computational procedure. Nevertheless, it may be of some value as a gedankenexperiment in sorting out the philosophical and semantic issues surrounding those words as applied to more serious and practical procedures and in envisioning the future of general-purpose computing at the microscopic level. Let us compare and contrast the NMR implementations of Grover search with my stillborn proposal for residue arithmetic.

The Grover algorithm, whether done in liquid NMR or otherwise, is a genuinely quantum algorithm; in principle it can be performed (with high probability of success) on a single molecule in a single experimental run. In practice, however, it seems that in liquid NMR experiments one is resorting to classical parallelism — an unavoidable ensemble of *many* microscopic computers.

The "quantum" implementation of the Chinese remainder theorem seems to be *inherently* a classical parallel computation — coherent superposition is not really used. The computation could not be carried out in one run on one quantum computer (or on any number of computers smaller than the number describing the classical complexity of the computation).

The quantum counting experiment [11] is more complicated to appraise (see Appendix at the end of this Chapter). Superficially, at least, it has some features in common with the modular arithmetic procedure, because the output is an expectation value. First, since off-diagonal elements of a density matrix in the observable basis do not contribute to an expectation value, the result of the computation would be unchanged if the coherent final state were replaced by the corresponding classical mixture of the qubit eigenstates. Since unitary operations are linear, this decoherence could happen at the beginning of the computation. In modular arithmetic one would need to observe each one of these eigenstates to extract the answer; apparently the corresponding statement for the counting problem is not true. Second, many measurements must be performed. Standard probability theory [7] shows that the number of measurements needed increases quadratically with the reciprocal of the desired precision. *Note that this problem is completely separate from the $1/\epsilon$ trouble that dogs all NMR computing; it arises even if every computer in the ensemble is executing the algorithm perfectly.* Short calculations show that in both problems the standard deviations are of order unity or less, so they are not limiting factors. However, there is no requirement in the counting problem that the precision must scale with the size of the "database," whereas to discriminate the highest Chinese remainder peak from its closest competitors requires precision of order $1/R$. This is another expression of the point that every one of the moduli needs to be explicitly sampled.

Of course, there is another huge difference: papers such as [20] and [11] describe procedures that have actually been carried out in the lab (albeit on a scale too small to be of practical interest). My proposal was left so vague that an experimenter would not know how to start implementing it without extensive additional thought. In this connection it is noteworthy that another basically classical NMR computation has been fleshed out and published. According to Brüschweiler [6], "Massive parallelism can potentially be achieved by ... molecules with different spin states simultaneously perform[ing] different computations. Like DNA computing, this kind of molecular parallelism is classical in nature, but unlike DNA computing, various linear combinations of different input states are prepared and evaluated. ..." In particular, a "divide and conquer" search of a database "offers an exponential speedup over a classical search and over Grover's quantum algorithm." Purists insist that this speedup comes from the deployment of exponentially many pieces of "hardware" (but, again, in practice there are just as many molecules in NMR Grover computations).

I suggest that general-purpose ultramicroscopic computers are likely to exist a generation hence, but that for the most part they will operate in a parallel classical, not truly quantum, manner. Any speedup they achieve will thus come from their massive redundancy, not from quantum algorithms. (Nevertheless, they may exploit quantum superposition nontrivially during intermediate stages of their computations.) They will be judged by their usefulness, not the semantics of whether they fit a definition of "quantum computation." Here we should take note of the unavoidable (and usually healthy) conflict between the promoters of a particular tool and the users with a practical need. The specialists will always have a tendency to disdain solutions that do not employ their favorite fancy methods, even when those solutions are effective. Indeed, I exemplified this attitude myself earlier in this essay, by remarking that *mundane infinite-precision arithmetic is not interesting number theory*. Similarly, *microscopic classical computing is (perhaps) not interesting quantum physics*. But if the current NMR experiments eventually lead to some sizable fraction of Avogadro's number of parallel processors doing something useful in a cost-effective manner, will anyone really care whether their states are "entangled"?

Appendix

First, let A be an observable with eigenvalues λ_j and eigenvectors ψ_j. (The extension to a family of commuting observables is immediate.) Let ρ be the density matrix of the system with respect to the eigenbasis of A. Then the expectation value of a measurement of A is

$$\mathrm{Tr}\,(\rho A) = \sum_j \rho_{jj}\lambda_j = \sum_j \rho_{jj}\langle\psi_j|A|\psi_j\rangle,$$

the weighted average of the expectation values in the individual eigenstates. That is, an effective density matrix for this restricted class of measurements is

$$\rho_e = \sum_j \rho_{jj} |\psi_j\rangle\langle\psi_j|.$$

Suppppose that ρ is the outcome of a quantum computation summarized by a unitary matrix U:

$$\rho = U\rho_0 U^{-1}.$$

Then under the same time evolution ρ_e would have arisen from

$$\rho_i = \sum_j \rho_{jj} U^{-1} |\psi_j\rangle\langle\psi_j| U.$$

That is, the outcome of the computation is the same as if done on the individual states $U^{-1}|\psi_j\rangle$ (followed by the weighted average), rather than on whatever superposition and/or mixture of these initial states led to the density matrix ρ. Of course, at intermediate steps, and even at the initial step, these states need not be eigenstates of A.

Second, consider the statistics of a measurement of A in a state that is not an eigenstate. The result of each measurement is some eigenvalue λ_j. Elementary quantum mechanics teaches that the expected deviation of this result from the true expectation value $\langle A\rangle$ is

$$\sigma = \sqrt{\langle(A - \langle A\rangle)^2\rangle} = \sqrt{\langle A^2\rangle - \langle A\rangle^2}.$$

Elementary classical statistics [7] then takes over and says that the mean of a large number n of measurements will be normally distributed with mean $\langle A\rangle$ and standard deviation $\sigma/\sqrt{n}$. In other words, to assure that the measurements reproduce $\langle A\rangle$ to within an error δ, n must be of the order σ^2/δ^2.

In the NMR counting experiment [11], $A = \sigma_z$ and $\langle A\rangle = \cos r\varphi_k$. Therefore, $A^2 = 1$ and

$$\sigma = \sqrt{1 - \cos^2 r\varphi_k} = |\sin r\varphi_k| \le 1.$$

The actual goal of the measurement is not φ_k but the index k corresponding to it via

$$\sin(\varphi_k/2) = \sqrt{k/N},$$

where N is the number of items in the database and k is the unknown number of items satisfying the search criterion. Instead of measuring $\cos r\varphi_k$ to arbitrary precision, one measures it to some reasonable precision for more than one value of r and hence deduces k by a half-classical version of the well-known quantum Fourier transform (cf [4, 12]). Jones and Mosca do not discuss how many measurements are needed (in other words, how many molecules are in the smallest sample for which the computation would be possible in principle); but presumably it would be considerably smaller than N if N were large.

Turn now to the Chinese remainder theorem calculation, with the pulse sequence stored in qubits. (Such a register would be absurdly large for any computation of practical interest. We discuss this computation only because it is an unambiguous procedure, simply explained, for which the result could in principle be obtained by bulk measurements on a sample rather than observations of individual quantum computers.) The final state of the register is

$$\frac{1}{\sqrt{R}} \sum_{j=1}^{R} |j\rangle,$$

where each qubit in $|j\rangle$ contains either 0 or 1. The expectation value of the observable A_k corresponding to the kth qubit is $\frac{1}{R}$ times the number of moduli m_j for which the kth bit in $|j\rangle$ is "set." The goal is to observe for which k this expectation value equals 1 (or, more realistically, to observe enough of the pulse pattern [see Figure] to determine its position along the register; the pattern is always the same, and only the location of its centroid contains information). In this problem we have $A_k^2 = A_k$ and hence

$$\sigma = \sqrt{\langle A_k \rangle - \langle A_k \rangle^2} \leq 1,$$

the largest values occurring when $\langle A_k \rangle \approx \frac{1}{2}$, $\sigma \approx \frac{1}{2}$. For $\langle A_k \rangle = (R-1)/R$ (those pulses that tie for second place), we have

$$\sigma = \sqrt{\frac{R-1}{R^2}} \sim \sqrt{\frac{1}{R}}.$$

To distinguish the heights of the pulses, we need to measure each $\langle A_k \rangle$ to a precision $\delta \ll 1/R$ (with more slack for the shorter pulses). So to be safe we need more than $\sigma^2/\delta^2 \sim R$ measurements on each qubit, which is consistent with the intuition that we need to observe every residue u_j to determine the answer.

Acknowledgments

I thank Michael Kash, Andreas Klappenecker, Davin Potts, and Michael Stay for a variety of helpful remarks, and the Editors of this volume for inviting me to write this chapter.

References

[1] J. Ahn, T. C. Weinacht, and P. H. Bucksbaum, Information storage and retrieval through quantum phase, *Science* 287, 463–465 and 1431a (2000)

[2] I. Borosh and A. S. Fraenkel, Exact solutions of linear equations with rational coefficients by congruence techniques, *Math. Comput.* 20, 107–112 (1966)

[3] G. Brassard, Searching a quantum phone book, *Science* 275, 627–628 (1997)

[4] G. Brassard, P. Høyer, M. Mosca, and A. Tapp, Quantum amplitude amplification and estimation, `quant-ph/0005055`

[5] S. L. Braunstein, C. M. Caves, R. Jozsa, N. Linden, S. Popescu, and R. Schack, Separability of very noisy mixed states and implications for NMR quantum computing, *Phys. Rev. Lett.* 83, 1054–1057 (1999)

[6] R. Brüschweiler, Novel strategy for database searching in spin Liouville space by NMR ensemble computing, *Phys. Rev. Lett.* 85, 4815–4818 (2000)

[7] E. R. Dougherty, *Probability and Statistics for the Engineering, Computing, and Physical Sciences*, Prentice–Hall, New York, 1990, Chapter 7.

[8] S. A. Fulling, Large numbers, the Chinese remainder theorem, and the circle of fifths, `quant-ph/9911051`

[9] S. A. Fulling, A possible new quantum algorithm: arithmetic with large integers via the Chinese remainder theorem, `quant-ph/9911050`

[10] J. A. Jones, NMR quantum computation: a critical evaluation, *Fortschr. Physik* 48, 909–924 (2000)

[11] J. A. Jones and M. Mosca, Approximate quantum counting on an NMR ensemble quantum computer, *Phys. Rev. Lett.* 83, 1050–1053 (1999)

[12] A. Yu. Kitaev, Quantum measurements and the Abelian stabilizer problem, `quant-ph/9511026`, Sec. 3.

[13] D. E. Knuth, *The Art of Computer Programming*, Vol. 2, *Seminumerical Algorithms*, 2nd ed. (Addison–Wesley, Reading, MA, 1981), Sec. 4.3.2

[14] M. Linden and S. Popescu, Good dynamics versus bad kinematics: is entanglement needed for quantum computation?, *Phys. Rev. Lett.* 87, 047901 (2001)

[15] S. Lloyd, Quantum search without entanglement, *Phys. Rev. A* 61, 010301 (2000)

[16] V. B. Mandelzweig, Quasilinearization method and its verification on exactly solvable models in quantum mechanics, *J. Math. Phys.* 40, 6266–6291 (1999)

[17] D. M. Potts, The `RESIDUE_INT` software package, `http://rainbow.uchicago.edu/~dmpotts/residue_int/` (to be mirrored at `http://www.math.tamu.edu/~fulling/`)

[18] R. Schack and C. M. Caves, Classical model for bulk-ensemble NMR quantum computation, *Phys. Rev. A* 60, 4354–4362 (1999)

[19] P. Shor, various public presentations

[20] L. M. K. Vandersypen, M. Steffen, M. H. Sherwood, C. S. Yannoni, G. Breyta, and I. L. Chuang, Implementation of a three-quantumbit search algorithm, *Appl. Phys. Lett.* 76, 646–648 (2000)

[21] E. L. Wright, Quantum computers, *Business Week Online*, 30 August 1999

[22] C. Zalka, Using Grover's quantum algorithm for searching actual databases, *Phys. Rev. A* 62, 052305 (2000).

Index